魏寿昆

1936年5月在德国亚琛

1929年4月6日，魏寿昆（前排蹲者中间）于北洋大学矿冶系毕业前实习，在唐山煤矿出井后留影

1929年4月15日，魏寿昆（站立者右三）于北洋大学矿冶系毕业前实习，在本溪钢铁公司的本溪湖第三坑斜井口外留影

北洋大学发给魏寿昆的优秀毕业生斐陶斐金质奖章

魏寿昆北洋大学时期学习成绩单

1931年4月，天津，赴德留学前

1933年，德国德累斯顿工科大学实验室内

1934年1月，德国德累斯顿工科大学实验室内

1936年，自德国经意大利回国途中在维苏威火山熔岩附近

1937年10月17日，西安南秦岭的南五台山顶，魏寿昆（前排最右者）与西北联大老师们的合影

1949年春，魏寿昆（后排站立者右二）带领天津北洋大学冶金系毕业班学生参观抚顺炼油厂

1949年6月，魏寿昆（前排站立者右三）与北洋大学冶金系毕业班学生摄于北洋大学北大楼前

1949年，魏寿昆（前右站立者）指导北洋大学冶金系毕业班学生在实验室做实验

1949年6月，魏寿昆（前排站立者右三）与北洋大学冶金系毕业班学生摄于北洋大学大门口

1949年，魏寿昆为北洋大学冶金系毕业班学生讲课

1953年,与北京钢铁学院学生运动队投掷组全体合影(站立者前排右三为魏寿昆)

1954年4月,北京钢铁学院部分领导与苏联专家在一起合影(后排左二(穿白风衣者)为魏寿昆)

1962年,在北京钢铁学院成立十周年纪念大会上讲话

1979年,随冶金部赴联邦德国教育考察团在亚琛留影

1981年8月，应山西省九三学社之邀，在太原讲学时留影

1984年10月23日，应贵州省科协及贵州省九三学社之邀，赴贵阳进行冶金物化专题讲学，与贵州铝厂同志们合影（前排左二为魏寿昆）

魏寿昆（左）陪同美国麻省理工学院著名冶金物理化学教授艾洛特参观北京钢铁学院图书馆

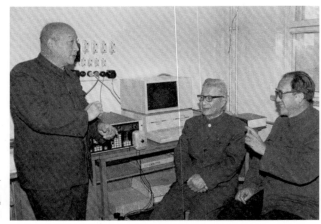

1984年，北京钢铁学院第一批科学院院士合影（中为魏寿昆，左为柯俊，右为肖纪美）

1986年3月,在家中查看资料

1986年5月,在成都科技大学参加全国第三届冶金过程动力学学术讨论会与会者留影(前排左四为魏寿昆,左五为邵象华院士)

1987年9月,80寿辰接受学院赠送的字画

1988年,考察上海宝山钢铁公司时在其高炉前合影(后排站立者左三为魏寿昆,右一为王之玺)

1988年,在北京科技大学冶金物理化学实验室与研究生讨论学术论文

1988年12月,摄于北科大家中(81岁)

1991年,摄于北京科技大学图书馆前(84岁)

1992年,85岁寿辰。在北京科技大学家中接受北科大赠送的字画。正面坐着手扶字画的魏寿昆(85岁),其旁边坐着的为夫人杨英梅

1992 年 4 月，在北科大四十周年校庆上做学术报告

1993 年，时任冶金部部长刘琪来北京科技大学视察工作合影（左四为刘琪，右三为魏寿昆）

1994 年，在北科大固体解质实验室与研究生讨论实装置

1996 年，两院院士大会魏寿昆与几位院士合影（前排左起：殷瑞钰、师昌绪、魏寿昆、徐匡迪。后排左起：王淀佐、周国治、何季麟、李依依）

1998年,在北科大家中(91岁)

2005年,在北科大家中,魏寿昆(右)与原冶金部副部长殷瑞钰院士合影

2006年,在北科大家中,魏寿昆(左)与原全国政协副主席、中国工程院院长徐匡迪院士交谈

2002年6月,在北科大西门(95岁)

本书由
北京科技大学教育发展基金会
魏寿昆科技教育基金
资助出版

北京科技大学教育发展基金会
魏寿昆科技教育基金　资助出版

魏寿昆文集
（第2卷）

姜曦　曲英　朱元凯　主编

北京
冶金工业出版社
2019

内 容 简 介

本书从数百篇魏寿昆院士的学术文章中,精选了1956~1999年期间发表的部分学术论文25篇,内容涵盖冶金溶液中组元活度研究、炉渣离子理论和选择性氧化理论三个方面。本书可供从事钢铁冶炼方面工作的科研人员、工程技术人员和高等院校相关专业师生参考阅读。

图书在版编目(CIP)数据

魏寿昆文集:第2卷/姜曦,曲英,朱元凯主编.—北京:冶金工业出版社,2019.5
ISBN 978-7-5024-8085-1

Ⅰ.①魏… Ⅱ.①姜… ②曲… ③朱… Ⅲ.①冶金工业—文集 Ⅳ.①TF-53

中国版本图书馆 CIP 数据核字(2019)第 051282 号

出版人 谭学余
地　　址　北京市东城区嵩祝院北巷39号　邮编 100009　电话 (010)64027926
网　　址　www.cnmip.com.cn　电子信箱 yjcbs@cnmip.com.cn
责任编辑　李培禄　美术编辑　彭子赫　版式设计　孙跃红
责任校对　石　静　责任印制　牛晓波

ISBN 978-7-5024-8085-1

冶金工业出版社出版发行;各地新华书店经销;三河市双峰印刷装订有限公司印刷
2019年5月第1版,2019年5月第1次印刷
169mm×239mm;21印张;6彩页;424千字;329页
100.00 元

冶金工业出版社　投稿电话　(010)64027932　投稿信箱　tougao@cnmip.com.cn
冶金工业出版社营销中心　电话　(010)64044283　传真　(010)64027893
冶金工业出版社天猫旗舰店　yjgycbs.tmall.com

(本书如有印装质量问题,本社营销中心负责退换)

《魏寿昆文集》编委会

顾　问	徐匡迪	张寿荣	殷瑞钰	罗维东
	徐金梧	张欣欣	武贵龙	杨仁树
主　任	姜　曦			
副主任	曲　英	朱元凯	张建良	
委　员	林　勤	魏文宁	张立峰	李　晶
	毛新平	李新创	张福明	朱苗勇
	吕朝伟	于成文	耿小红	吴胜利
	王广伟	焦树强	史成斌	寇明银

《魏寿昆文集》（第2卷）编委会

主　编　姜　曦　　曲　英　　朱元凯

副主编　李　晶　　包燕平　　魏文宁

编　委　张建良　　林　勤　　张立峰　　李京社

　　　　史成斌　　焦树强　　张国华　　佘雪峰

　　　　寇明银　　戴立杰　　王广伟　　王海洋

　　　　周东东　　韩宏亮　　刘　芳　　吴世磊

　　　　杨万春　　杨树峰　　吕学伟　　姜　敏

目　录

冶金溶液中组元活度研究

活度的两种标准状态与热力势 ……………………………………………………… 3
活度在钢铁冶金二相的气体—金属液化学平衡反应中的应用 …………………… 15
活度相互作用系数运算中的某些问题 ……………………………………………… 35
Thermodynamic Study of Interaction Coefficients in Multicomponent Metallic Solutions
　　by the Solubility Method ……………………………………………………… 52
Interaction Coefficients in Multicomponent Metallic Solutions at Constant Activity
　　and Constant Concentration …………………………………………………… 67

炉渣离子理论

空气顶吹过程中熔渣的气态脱硫 …………………………………………………… 81
炉渣氧化铁含量对脱硫的作用 ……………………………………………………… 99
高炉型渣脱硫的离子理论 …………………………………………………………… 107
从炉渣离子理论计算的硫分配比来看攀钢钒钛铁矿中 TiO_2 的属性 …………… 124
The Ionization Theory of Desulphurization of the Panzhihua Blast Furnace Slag …… 128
Some Advances on the Theoretical Research of Slag ………………………………… 136

选择性氧化理论

炼钢过程中铁液内磷、碳等元素氧化的热力学 …………………………………… 151
镍锍选择性氧化的热力学及动力学 ………………………………………………… 163
熔锍及熔融金属中元素选择性氧化的热力学 ……………………………………… 178
选择性氧化——理论与实践 ………………………………………………………… 191
稀土钢冶炼的物理化学问题 ………………………………………………………… 198
含钒铁水炼钢的一些物理化学问题 ………………………………………………… 210
含 Nb 及 Mn 的铁液中 Mn 对 Nb 活度系数影响的研究 ………………………… 228

A Study of the Effect of Silicon upon the Activity Coefficienf of Niobium in Liquid Iron with the Solid Electrolyte Oxygen Cell Technique ················ 237
Nb、Si 在铁液中活度相互作用的研究 ························· 243
Selective Oxidation of Elements in Metal Melt and Their Multireaction Equilibria ·· 253
Thermodynamic Study on Process in Copper Converters (The Slag-making Stage) ·· 279
纯铁液中钡-氧、钡-硫平衡的研究 ······························ 288
Fe 液中 Ba-O、Ba-S、Ba-P 平衡的研究 ························ 295
Separation of Nb from Nb-bearing Iron Ore by Selective Reduction ············ 302

附录　魏寿昆院士生平 ·· 314

后记 ·· 328

冶金溶液中组元活度研究

YEJIN RONGYE ZHONG
ZUYUAN HUODU YANJIU

活度的两种标准状态与热力势[❶]*

魏寿昆

1 活度的概念和意义

高温冶金经常存在下列四种平衡反应：（1）单相的溶液反应，表现在某一元素溶解在另一元素中，造成一溶体合金；（2）二相的气体—金属液反应；（3）二相的金属液—炉渣液反应；（4）三相的气体—炉渣液—金属液反应。在这些高温反应中，我们常应用到下列四个定律：（1）拉乌尔定律；（2）亨利定律；（3）分配定律；（4）质量作用定律。

由于高温下分子或原子间作用力的复杂关系，我们由实验中得到的反应数据，往往是不符合上列四个定律的。这样为了研究分析这些实验上的数据，我们必须采取下列任一措施：（1）在不同浓度下发现新的规律，把罗列的现象系统化起来；（2）仍用上列的四个定律，而采用"有效浓度"以代替实际的浓度；亦就是说，用一个因素把实际的浓度加以校正，使它能够符合上列四个定律。后一种方法是我们通常采用的措施。这个校正因子叫作"活度系数"，而"有效的浓度"叫作"活度"。用公式表示如下：

$$a_i = \gamma_i c_i \tag{1}$$

活度的测定必须先确定标准状态。通常我们采用两种标准状态：纯物质（溶质或溶媒）或1%的溶液。两种标准得到不同的活度值，举下面例子来解释。

我们试测定682℃时Cd-Sn合金Cd的气压[1]，数据见表1。

表1 Cd-Sn合金内Cd在682℃时的气压和活度

气压/mmHg[①]	含Cd量/%	标准状态=纯液相Cd		标准状态=1% Cd溶液	
		a_{Cd}	γ_{Cd}	a'_{Cd}	γ'_{Cd}（即f_{Cd}）
6	1	0.024	2.4	1.00	1.00
110	20	0.44	2.2	18.3	0.91
180	40	0.72	1.8	30.0	0.75
230	60	0.92	1.5	38.3	0.64

---

* 原刊于《北京钢铁学院学报》，1956（3）：103~114。

❶ 热力势是热力学函数 $\Delta G = \Delta H - T\Delta S$ 的另一种名称，现在通常称之为"吉布斯能"。

续表1

气压/mmHg[①]	含 Cd 量/%	标准状态=纯液相 Cd		标准状态=1% Cd 溶液	
		a_{Cd}	γ_{Cd}	a'_{Cd}	γ'_{Cd}（即 f_{Cd}）
245	80	0.98	1.2	40.8	0.51
250	100	1.00	1.0	41.7	0.42
P	x	$\dfrac{P}{250}$	$\dfrac{100P}{250x}$	$\dfrac{P}{6}$	$\dfrac{P}{x}$

① 1mmHg=133.322Pa。

假设理想溶液时有：

$$P_{Cd} = P^o_{Cd}(\%Cd) \tag{2}$$

那么，图 1 中虚线 a 应当代表不同溶液成分下 Cd 的气压，但实际上曲线 b 是 Cd 的气压。为了能符合方程式（2），我们采用：

$$P_{Cd} = P^o_{Cd} a_{Cd} \tag{3}$$

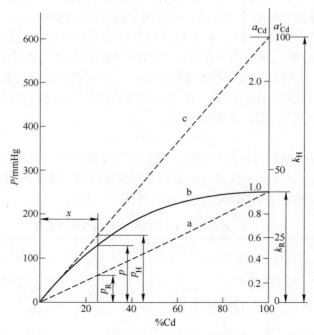

图 1　Sn 液含 Cd 量与活度的关系（682℃）
（1mmHg=133.322Pa）

计算出的活度 a_{Cd} 和活度系数 γ_{Cd} 见表 1。

从式（2）看出：

$$a_{Cd} = \gamma_{Cd}(\%Cd) \tag{4}$$

当 $\gamma_{Cd} = 1$，亦就是活度与实际浓度相等，此时即不用将实际浓度加以校正，这种状态称为标准状态。

由于：
$$a_{Cd} = P_{Cd}/P_{Cd}^o = 1$$

所以：
$$P_{Cd} = P_{Cd}^o$$

所以标准状态是纯 Cd 液体。

假若我们选定标准状态是 1%Cd 溶液，而用 a'_{Cd} 代表活度，则有：
$$P_{Cd} = P_{Cd}^{'o} a'_{Cd} \tag{5}$$

由实验数据得知，$P_{Cd}^{'o} = 6\text{mmHg}$。计算出的 a'_{Cd} 和 γ'_{Cd} 值亦附列于表 1 内。

我们可以看出，用不同标准状态计算出的活度是不同的，此点我们必须注意。

在什么时候我们采用纯物质的标准状态以计算活度呢？一般的情况是在应用拉乌尔定律的时候，即：
$$P_i = P_i^o N_i \tag{6}$$

根据同样实验数据，采用纯液体 Cd 为标准状态，利用方程式（6）计算出的 a_{Cd} 和 γ_{Cd} 见表 2。

表 2　Cd-Sn 合金内 Cd 在 682℃时的气压和活度

气压/mmHg[①]	含 Cd 量/%	N_{Cd}	a_{Cd}	γ_{Cd}
9	1	0.0106	0.024	2.39
110	20	0.12	0.44	2.09
180	40	0.42	0.72	1.71
230	60	0.61	0.92	1.505
245	80	0.81	0.98	1.205
250	100	1.00	1.00	1.0

① 1mmHg = 133.322Pa。

比较表 1 和表 2，两表内采用纯 Cd 为标准状态的活度值是一样的，而 γ 值则不同。此乃因表 1 中的 $\gamma_{Cd} = \dfrac{a_{Cd}}{x} \times 100$，而表 2 中的 $\gamma_{Cd} = \dfrac{a_{Cd}}{N_{Cd}}$。当二元素的原子量相同时，则 $N_{Cd} = \dfrac{x}{100}$，亦即两个 γ_{Cd} 值相同。

应用亨利定律时，我们经常采用 1%溶液为标准状态，此时的活度系数通常用 f_i 表示。

1%溶液的标准状态的真实意义是什么呢？我们举下列例子进一步说明。

以 H_2S/H_2 混合气体与 Fe 液作平衡实验，得下列反应：
$$[\text{FeS}] + \text{H}_2(\text{气}) \Longrightarrow \text{H}_2\text{S}(\text{气}) + \text{Fe}(\text{液})$$

或 $\qquad$ [S] + H$_2$(气) $\Longrightarrow$ H$_2$S(气) $\qquad$ (7)

[FeS] 或 [S] 代表溶在 Fe 液内的 FeS 或 S 量。

$$K' = \frac{P_{H_2S}}{P_{H_2}} \times \frac{1}{[\%S]}$$

$$\frac{P_{H_2S}}{P_{H_2}} = K'[\%S] \qquad (8)$$

式（8）可被解释相当于亨利定律；如果反应式（7）服从亨利定律，则 K' 系一常数；如不服从，则 K' 便不守常。

作 $\frac{P_{H_2S}}{P_{H_2}}$ 对 [%S] 的曲线（图 2）。表 3[2,3] 示出 1600℃ 的实验数据。我们可以看出，在 [%S]≤0.5 时，此线系一直线，亦就是说，在 [%S]≤0.5 时，平衡常数 K' 系一常数；而在 [%S]>0.5 时，平衡常数 K' 即随 [%S] 而改变。在该浓度以上质量作用定律即不能适用；如欲适用，则必须采用活度以代 [%S]。

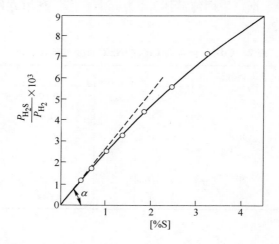

图 2　Fe 液含 S 量与 P_{H_2S}/P_{H_2} 的关系（1600℃）

亦即 $\qquad K^\circ = \frac{P_{H_2S}}{P_{H_2}} \times \frac{1}{a_S}$

$$K^\circ = \frac{P_{H_2S}}{P_{H_2}} \times \frac{1}{f'_S[\%S]} \qquad (9)$$

$$f'_S = \frac{K'}{K^\circ} \qquad (10)$$

上列两式内 K° 代表理想的真平衡常数。

表3 Fe-S 系的活度

$\dfrac{P_{H_2S}}{P_{H_2}} \times 10^3$	[%S]	K'	$f'_S = \dfrac{K'}{K^o}$	a'_S
0.65	0.25	2.6×10^{-3}	0.98	0.245
1.3	0.5	2.6×10^{-3}	0.98	0.49
2.5	1.0	2.5×10^{-3}	0.94	0.94
4.7	2.0	2.35×10^{-3}	0.88	1.76
6.6	3.0	2.2×10^{-3}	0.83	2.49
8.2	4.0	2.1×10^{-3}	0.79	3.16

我们怎样求理想的真平衡常数 K^o 呢？自图2可知：

$$K^o = \tan\alpha = 2.6 \times 10^{-3}$$

但更准确的求法是：作 K' 或 $\lg K'$ 对 [%S] 的曲线（图3）。将此曲线（事实上系一直线）外插到 %S=0，得 $\lg K' = -2.57$ 或 $K' = 2.65 \times 10^{-3}$，此即真平衡常数 K^o 的值。

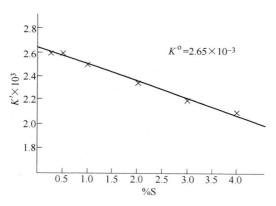

图3 Fe-S 平衡系求 K^o 的方法图

我们可以看出当 [%S] 逐渐等于零时，实际的假平衡常数逐渐接近于真平衡常数 K^o，而活度 a_S 接近于 [%S]。亦就是说：

$$\lim_{[\%S] \to 0} \frac{a_S}{[\%S]} = 1 \tag{11}$$

所以我们的标准状态是一个无限稀的理想溶液，溶质的浓度接近于零。

但由于 $a_S = f'_S [\%S]$，当标准状态时，$f'_S = 1$；同时当标准状态时，我们惯于称活度亦即 $a_S = 1$；所以必然地，标准状态时 [%S] = 1。亦就是说，通常我们称此种的标准状态是 1% 的溶液。但是我们必须指出，后者的说法是不严格的，不很科学的。更正确的更严格的说法是：标准状态是一无限稀的溶液，该时候溶质的活度等于它的浓度（通常以百分数计）。

如表1指出，同一成分的溶液采用不同标准状态的活度值不同。但在热力学中活度与热力势有数学上的关系。下面我们利用不同标准状态的活度与热力势关系解释高温的溶液生成过程。

2 热力势和活度的函数关系

钢铁冶金学者对在炼钢温度某元素溶于 Fe 液的热力势变化具有很大兴趣；特别在热力学计算上，往往我们需要知道在1600℃时 1mol 元素 A 溶于 Fe 液生成所谓的1%含［A］溶液的标准热力势变化值 $\Delta\Phi^\circ$。

$$A_{(液)} \rightarrow [A]_{(1\%)}; \quad \Delta\Phi^\circ = ?$$

依热力学计算，上列过程的热力热变化为：

$$\Delta\Phi = \overline{\Phi^\circ} + RT\ln a' - \Phi^\circ - RT\ln a \tag{12}$$

纯液体元素的活度 $a=1$，$RT\ln a = 0$，所以：

$$\Delta\Phi = \overline{\Phi^\circ} - \Phi^\circ + RT\ln a' \tag{13}$$

为了确定 $\overline{\Phi^\circ} - \Phi^\circ$ 值，我们考虑以下两种情况：

（1）该1%溶液服从拉乌尔定律：

$$a' = N_A$$

在拉乌尔定律通常以纯液体 A 为标准状态，所以：

$$\overline{\Phi^\circ} = \Phi^\circ$$

$$\Delta\Phi = RT\ln N_A$$

设 M_A 代表元素 A 的原子量，而 x 代表该溶液含 A 的百分数，则

$$N_A = \frac{\dfrac{x}{M_A}}{\dfrac{x}{M_A} + \dfrac{100-x}{55.85}}$$

当 x 很小时，分母项下的 $\dfrac{x}{M_A}$ 接近于零，可不加计算；另 $\dfrac{100-x}{55.85}$ 近似地等于 $\dfrac{100}{55.85}$。所以：

$$N_A = \frac{0.5585}{M_A}x$$

$$\Delta\Phi = RT\ln\frac{0.5585x}{M_A}$$

当 $x = 1$ 时：

$$\Delta\Phi^\circ = RT\ln\frac{0.5585}{M_A} \qquad (14)$$

Mn、Cr、Co、Ni 等元素在稀溶液内服从拉乌尔定律[4]；代入各元素的原子量，由式（14）计算出该元素 1mol 纯液相溶解到 Fe 液内生成 1%溶液的热力势变化。

$$\text{Mn}_{(液)} \rightarrow [\text{Mn}], \quad \Delta\Phi^\circ = RT\ln\frac{0.5585}{54.93} = -9.11T$$

$$\text{Cr}_{(液)} \rightarrow [\text{Cr}], \quad \Delta\Phi^\circ = RT\ln\frac{0.5585}{52.01} = -9.01T$$

$$\text{Co}_{(液)} \rightarrow [\text{Co}], \quad \Delta\Phi^\circ = RT\ln\frac{0.5585}{58.94} = -9.26T$$

$$\text{Ni}_{(液)} \rightarrow [\text{Ni}], \quad \Delta\Phi^\circ = RT\ln\frac{0.5585}{58.69} = -9.21T$$

（2）该 1%溶液服从亨利定律：

$$a' = x$$

为清楚起见，我们将式（13）的 $\overline{\Phi^\circ}$ 写作 $\overline{\Phi_H^\circ}$。$\overline{\Phi_H^\circ}$ 代表根据亨利定律标准状态亦即活度=1 时 A 在溶液内的偏热力势；换言之，亦即 A 在以 1%溶液为标准状态的偏热力势。所以有：

$$\Delta\Phi = \overline{\Phi_H^\circ} - \Phi^\circ + RT\ln x \qquad (15)$$

当 $x = 1$ 时：

$$\Delta\Phi^\circ = \overline{\Phi_H^\circ} - \Phi^\circ \qquad (16)$$

为了计算 $\overline{\Phi_H^\circ} - \Phi^\circ$ 值，我们试以上一节内的实验数据，分析含 x%Cd 的 Cd-Sn 合金液 Cd 元素的偏热力势。

根据亨利定律，1mol Cd 溶解 Sn 内造成 x%Cd 溶液时，Cd 的偏热力势为：

$$\overline{\Phi} = \overline{\Phi_H^\circ} + RT\ln a'_{Cd} \qquad (17)$$

根据拉乌尔定律，1mol Cd 溶解 Sn 内造成 x%溶液时，Cd 的偏热力势为：

$$\overline{\Phi} = \overline{\Phi_R^\circ} + RT\ln a_{Cd} \qquad (18)$$

其中 $\overline{\Phi_R^\circ}$ 为根据拉乌尔定律标准状态下，亦即活度等于 1 时 Cd 在溶液内的偏热力势；因为此标准状态为纯 Cd 液，所以 $\overline{\Phi_R^\circ}$ 实际等于纯液体 Cd 的 Φ°。因此，式（18）可改写如下：

$$\overline{\Phi} = \Phi^\circ + RT\ln a_{Cd} \qquad (19)$$

因为我们研究的是同一溶液,式(17)应等于式(16)❶,所以:

$$\overline{\Phi_H^o} + RT\ln a'_{Cd} = \Phi^o + RT\ln a_{Cd}$$

$$\overline{\Phi_H} - \Phi^o = RT\ln \frac{a_{Cd}}{a'_{Cd}} \tag{20}$$

参阅表1:

$$a_{Cd} = \frac{P}{250}$$

$$a'_{Cd} = \frac{P}{6}$$

所以:

$$\frac{a_{Cd}}{a'_{Cd}} = \frac{6}{250}$$

参阅图1:

$$\frac{P_H}{x} = 6$$

$$\frac{P_R}{x} = \frac{250}{100}$$

当 x 很小时,$\frac{x}{100}$ 可换成相当的 N_{Cd} 值。

亦即:

$$\frac{P_R}{N_{Cd}} = 250$$

所以:

$$\frac{a_{Cd}}{a'_{Cd}} = \frac{P_H}{x} \times \frac{N_{Cd}}{P_R} = \frac{P_H}{P_R} \times \frac{N_{Cd}}{x} \tag{21}$$

同时:

$$\frac{P_H}{P_R} = \frac{K_H}{K_R} \tag{22}$$

其中 $\frac{K_H}{K_R}$ 为亨利定律和拉乌尔定律二直线斜率之比。结合式(20)~式(22)可以得出:

$$\overline{\Phi_H^o} - \Phi^o = RT\ln\left(\frac{K_H}{K_R} \times \frac{N_{Cd}}{x}\right) \tag{23}$$

换用式(23)内的 N_{Cd} 值,可以得到:

$$\overline{\Phi_H^o} - \Phi^o = RT\ln\left(\frac{K_H}{K_R} \times \frac{M_{Sn}}{100M_{Cd}}\right) \tag{24}$$

结合元素 A 溶解在 Fe 液的反应,式(24)换写为:

---

❶ 从这样的推荐方法来看,热力势计算是以纯物质(液体)为最后标准的。

$$\overline{\Phi_H^\circ} - \Phi^\circ = RT\ln\left(\frac{K_H}{K_R} \times \frac{0.5585}{M_A}\right) \tag{25}$$

自式（16）得知：

$$\Delta\Phi^\circ = \overline{\Phi_H^\circ} - \Phi^\circ$$

所以：
$$\Delta\Phi^\circ = RT\ln\left(\frac{K_H}{K_R} \times \frac{0.5585}{M_A}\right) \tag{26}$$

Si、Al、Ti、Zr 及 B 等元素在稀溶液内服从亨利定律[4]。所以如果 $\frac{K_H}{K_R}$ 已知，该元素 1mol 纯液体溶解 Fe 液内造成 1%溶液的热力势变化值，可由式（26）算出。

$\frac{K_H}{K_R}$ 可用 γ° 表示。对于下列反应：

$$Si_{(液)} \longrightarrow [Si]$$

γ° 数据在 1600℃ 为 0.017[5]。根据 Chipman 最近的数据[6]，此 γ° 值可用下列间接方法求出：

$$Si_{(液)} + O_{2(气)} \longrightarrow SiO_{2(方石英)} \tag{27}$$
$$\Delta\Phi^\circ = -217700 + 47.0T$$

$$O_{2(气)} = 2[O] \tag{28}$$
$$\Delta\Phi^\circ = -55860 - 1.14T$$

式（27）- 式（28）得：

$$Si_{(液)} + 2[O] = SiO_{2(方石英)}$$
$$\Delta\Phi^\circ = -161840 + 48.14T$$
$$\Delta\Phi^\circ_{1873} = -71700 \text{cal} \text{❶} \tag{29}$$

实验数据得出，在 1600℃ 时：

$$[Si] + 2[O] = SiO_{2(方石英)}$$
$$K = [\%Si][\%O]^2 = 2.8 \times 10^5$$
$$\Delta\Phi^\circ = -RT\ln\frac{1}{K}$$
$$\Delta\Phi^\circ_{1873} = -39000 \text{cal} \tag{30}$$

式（29）- 式（30）有：

$$Si_{(液)} = [Si]$$
$$\Delta\Phi^\circ_{1873} = -32700 \text{cal} \tag{31}$$

本反应系放热反应： $\Delta H = -28500 \text{cal}$

---

❶ 热力学单位原文用卡（cal），1cal（热化学）= 4.184J。

所以反应式（31）可写成下式：

$$Si_{(液)} \rightleftharpoons [Si]$$

$$\Delta \Phi^\circ = -28500 - 2.23T \tag{32}$$

自式（26）得知：

$$\Delta \Phi^\circ = RT\ln\left(\gamma^\circ \frac{0.5585}{28.06}\right) = 4.575 T\lg\left(\gamma^\circ \frac{0.5585}{28.06}\right) \tag{33}$$

式（32）=式（33），因之可求出：

$$\lg\gamma^\circ = -\frac{6230}{T} + 1.213$$

$$\gamma^\circ_{1873} = 0.0077$$

图 4 示出 Si 溶在 Fe 液内在 1600℃ 时活度的变化情况。上左方小图为低摩尔分数时的放大图。γ° 值亦在图内表示出来。

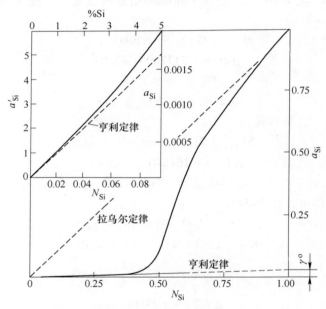

图 4 Fe 液含 Si 摩尔分数与活度的关系（1600℃）

实际上式（26）是一个一般性的公式，它代表某液体元素溶解在 Fe 液内生成含该元素 1% 的溶液的标准热力势变化 $\Delta \Phi^\circ$。它适用于亨利定律，亦可适用于拉乌尔定律；亦就是说，适用于拉乌尔定律溶解过程的标准热力势变化 $\Delta \Phi^\circ$ 可由该式导出。

$$\Delta\Phi^\circ = \overline{\Phi_R^\circ} - \Phi^\circ = RT\ln\left(\gamma^\circ \frac{0.5585}{M_A}\right) \tag{34}$$

无疑地，在适合拉乌尔定律时，$K_H = K_R$，亦即 $\gamma^\circ = 1$。同时换写 $\overline{\Phi_H^\circ}$ 为 $\overline{\Phi_R^\circ}$，我们得出：

$$\Delta\Phi^\circ = \overline{\Phi_R^\circ} - \Phi^\circ = RT\ln\frac{0.5585}{M_A} \tag{35}$$

显然地，式（35）即是式（14）；它代表在适合拉乌尔定律时 1mol 液体 A 溶解 Fe 液内成为 1% 溶液的热力势变化值。

因此，$\overline{\Phi_R^\circ}$ 可称为根据拉乌尔定律而以 1% 溶液为标准状态 A 在溶液内的偏热力势。这样在应用拉乌尔定律时，我们亦可推广地采用和亨利定律一样，以 1% 溶液为标准状态了。

3　结论

高温反应中的活度，我们经常采用两种标准状态：（1）纯物质；（2）1% 的溶液。后者标准状态实质上是一无限稀的理想溶液，溶质的浓度接近于零。

设求某元素 A 溶于 Fe 液生成含 $x\%$ A 溶液的热力势变化：

$$A_{(液)} \longrightarrow [A]_{(x\%)}$$

$$\Delta\Phi = \overline{\Phi_H^\circ} + RT\ln a' - \Phi^\circ$$

所以：

$$a' = fx$$

而：

$$\overline{\Phi_H^\circ} - \Phi^\circ = RT\ln\left(\gamma^\circ \frac{0.5585}{M_A}\right)$$

所以：

$$\Delta\Phi = RT\ln fx + RT\ln\left(\gamma^\circ \frac{0.5585}{M_A}\right)$$

$$= RT\ln\left(fx\gamma^\circ \frac{0.5585}{M_A}\right) \tag{36}$$

在生成 1% 溶液为标准状态时，即 $x=1$，则 $f=1$。所以：

$$\Delta\Phi^\circ = RT\ln\left(\gamma^\circ \frac{0.5585}{M_A}\right) \tag{37}$$

式（36）是在高温条件下计算溶液生成过程热力势变化值 $\Delta\Phi$ 最一般性的公式。式（37）用以计算在适合亨利定律时溶液生成过程的标准热力势变化值 $\Delta\Phi^\circ$。

在适合拉乌尔定律时，$\gamma^\circ = 1$，而式（38）代表该情况下溶液生成过程的标准热力势值 $\Delta\Phi^\circ$：

$$\Delta\Phi^\circ = RT\ln\frac{0.5585}{M_A} \tag{38}$$

参 考 文 献

[1] Крамров. Физко-химические Процесса Произвдства Стали. 1954: 15.
[2] Chipman J, Elliott J F. Thermodynamics in Physical Metallurgy, A. S. M., 1940: 102~143.
[3] Sherman C W, Elvander H I, Chipman J, Trans. A. I. M. E., 1950, 188: 334.
[4] Basic Open Hearth Steelmaking. 2nd ed. A. I. M. E., 1951: 635~640.
[5] Chipman J. Discussions Faraday Soc., 1948 (4): 23.
[6] Chipman J, Gokcen N A. Trans. A. I. M. E., 1953, 197: 1017.

活度在钢铁冶金二相的气体—金属液化学平衡反应中的应用[*]

魏寿昆

摘　要：在本文内作者总结了直至 1955 年文献内有关三种二相的气体—金属液化学平衡反应各国学者的研究工作（H_2S/H_2—Fe 液、H_2O/H_2—Fe 液及 CO_2/CO—Fe 液三反应），标准化了二元系及三元系平衡常数、活度及活度系数等参数的符号，并分析了 f_i^j 的物理意义。在活度系数的实际意义和应用的讨论中，作者指出 f_i^j 对脱硫和脱氧两种不同机构上的作用应采取不同的解释。利用最新的数据，作者算出 5 种元素脱氧生成物每摩尔热力势变化值 $\Delta\Phi^{\circ\prime}$，后者的次序与 e_O^M 的次序完全符合。最后作者并指出合金钢冶炼今后在此方面研究的重要性。

在钢铁冶金范围内，二相的气体—金属液化学平衡反应已被各国学者作为多年研究的对象。他们一系列的研究，多属于下列三种反应：

$$H_{2(气)} + [S] = H_2S_{(气)}$$
$$H_{2(气)} + [O] = H_2O_{(气)}$$
$$CO_{(气)} + [O] = CO_{2(气)}$$

[S] 或 [O] 代表溶在 Fe 液内的硫或氧量。实验多在高周波电炉或其他高温电炉内进行。以一定成分的 H_2S/H_2、H_2O/H_2 或 CO_2/CO 气体混合物与纯 Fe 液起化学反应，直到平衡状态，然后分析 Fe 液内含 S 或 O 量。通常分子量高低悬殊的混合气体在不同温度场内，往往表现出不均匀成分；亦即较轻的气体扩散到较热的区域，造成所谓热扩散现象，引起实验数据的不准确[1]。为了避免此现象，经常在与金属液接触之前，先把混合气体预热到实验反应的温度，或在混合气体内混入相当量的高分子量气体（通常是氩气），或两措施同时并用。研究方法扩充到 Fe 液内含有其他元素；亦就是说，研究三元系或多元系（包含 Fe 在内）金属液内其他元素相互间活度的影响。

1　H_2S/H_2—Fe 液的化学平衡反应

H_2S 和 H_2 混合气体对 Fe 液反应的平衡常数，李公达与 Chipman[2] 首先做了

---

[*] 原刊于《北京钢铁学院学报》，1956（3）：115~135。

研究。White 和 Skelly[3] 指出该研究得到的平衡常数太高。其后，Sherman、Elvander 和 Chipman[4] 又进一步地研究，得出较准确的数据。

$$H_{2(气)} + [S] \rightleftharpoons H_2S_{(气)} \tag{1}$$

$$K' = \frac{P_{H_2S}}{P_{H_2}} \times \frac{1}{[\%S]} \tag{2}$$

如表 1[4] 所示，在 [%S]>0.5 时，平衡常数 K' 即不守常。在该浓度以上质量作用定律即不能适用；如欲适用，即必须以活度代 [%S]。

亦即

$$K^o = \frac{P_{H_2S}}{P_{H_2}} \times \frac{1}{a_S} \tag{3}$$

$$K^o = \frac{P_{H_2S}}{P_{H_2}} \times \frac{1}{f'_S [\%S]} \tag{4}$$

$$f'_S = \frac{K'}{K^o} \tag{5}$$

上列两式 K^o 代表理想的真平衡常数。

表 1　Fe-S 系的活度

$\frac{P_{H_2S}}{P_{H_2}} \times 10^3$	[%S]	K'	f'_S	a'_S
0.65	0.25	2.6×10^{-3}	0.98	0.245
1.3	0.5	2.6×10^{-3}	0.98	0.49
2.5	1.0	2.5×10^{-3}	0.94	0.94
4.7	2.0	2.35×10^{-3}	0.88	1.76
6.6	3.0	2.2×10^{-3}	0.83	2.49
8.2	4.0	2.1×10^{-3}	0.79	3.16

K^o 的求法是：作 K' 或 $\log K'$ 对 [%S] 的曲线（图 1）。将此曲线（事实上是一直线）外插到 %S=0，得 $\log K' = -2.57$，因之 $K' = 2.65 \times 10^{-3}$，此即真平衡常数 K^o 的值。

活度的标准状态是无限稀溶液，[%S] 接近于零；此种标准态亦即通常所说的 1% 溶液[5]。

根据 Chipman 最近的矫正[6]，理想的真平衡常数 K^o 与温度有下列关系：

$$\log K^o = -\frac{1420}{T} - 1.83 \tag{6}$$

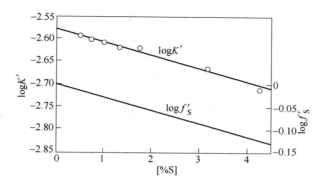

图 1　Fe 液含 S 量与实际平衡常数 K' 和活度系数 f'_S 的关系（1600℃）

而式（1）的标准热力势变化❶为：

$$\Delta \Phi^\circ = 6500 + 8.3T$$

当 $T = 1873$ K 时：

$$\log K^\circ = -2.588$$
$$K^\circ = 2.58 \times 10^{-3}$$

图 1 内 $\log f'_S$ 可用下式表示：

$$\log f'_S = -0.028 [\%S] \tag{7}$$

氩在 Fe 液内有一定的溶解度，但由于它对于活度不发生影响，所以对［S］在 Fe 内的溶液，我们通常称之为二元系溶液。今设想 Fe 液内尚有其他第三元素，亦就是说，在三元系的溶液内，该第三元素对［S］的活度是否有影响呢？在这些方面已有过好多研究，如 Fe-S-Si[7,8]、Fe-S-C[9]、Fe-S-Cu[10]、Fe-S-P[11]、Fe-S-Al[8] 和 Fe-S-Ni[6]。

如前面已指出，我们对二元系的 Fe-S 溶体，它们的平衡常数、活度及活度系数均在相应符号上加一小撇，即 K'、a'_S、f'_S 代表。今后对三元系或多元系的溶体，它们的相应参数用 K、a_S、f_S 代表。

表 2 摘自 Sherman 和 Chipman[8] 关于 Mn 加入 Fe-S 溶体内新的平衡关系的数据。我们可以看出，由于 Mn 的加入，实际的平衡常数

$$K = \frac{P_{H_2S}}{P_{H_2}} \times \frac{1}{[\%S]_{Mn}} \tag{8}$$

因之再行降低而不守常。在 Fe 内含［%S］和［%Mn］同时接近于零的三元系，实际平衡常数 K 等于理想的真平衡常数 K°：

$$\lim_{\substack{[\%S] \to 0 \\ [\%Mn] \to 0}} \frac{K}{K^\circ} = 1$$

---

❶ 本文热力势数值单位为卡（cal），1cal = 4.184J。

$$K^\circ = \frac{P_{H_2S}}{P_{H_2}} \times \frac{1}{f'_S [\%S]_{Mn}} \quad (9)$$

$$f_S = \frac{K}{K^\circ} \quad (10)$$

表 2 H₂S/H₂ 与含 [Mn] 的 Fe 液在 1600℃时的活度

实验号数	$\frac{P_{H_2S}}{P_{H_2}} \times 10^3$	[%S]	[%Mn]	log$K$	logf_S	logf'_S	logf_S^{Mn}	$K \times 10^3$	f_S	f'_S	f_S^{Mn}	a_S
M-1	4.60	2.11	0.90	-2.662	-0.085	-0.059	-0.026	2.18	0.82	0.87	0.94	1.73
M-5	2.65	1.19	1.97	-2.651	-0.074	-0.033	-0.041	0.23	0.84	0.93	0.91	1.00
M-9	1.93	0.84	1.45	-2.639	-0.062	-0.023	-0.039	2.30	0.87	0.95	0.91	0.73
M-10	1.83	0.83	2.02	-2.655	-0.078	-0.023	-0.055	2.21	0.84	0.95	0.88	0.69
M-19	1.27	0.59	3.09	-2.668	-0.091	-0.016	-0.075	2.15	0.81	0.96	0.84	0.48
M-24	0.96	0.38	0.52	-2.598	-0.021	-0.011	-0.010	2.52	0.95	0.97	0.98	0.36
M-26	0.84	0.41	3.85	-2.685	-0.108	-0.011	-0.097	2.07	0.78	0.97	0.80	0.32
M-28	0.75	0.39	5.12	-2.717	-0.140	-0.011	-0.129	1.92	0.72	0.97	0.74	0.28
M-30	0.54	0.29	6.11	-2.732	-0.155	-0.008	-0.147	1.85	0.70	0.98	0.71	0.20
M-33	0.43	0.25	6.99	-2.761	-0.184	-0.007	-0.177	1.73	0.66	0.98	0.67	0.16
M-35	0.41	0.26	7.96	-2.805	-0.228	-0.007	-0.221	1.57	0.59	0.98	0.60	0.15

下列分析可证明 f_S 包括两部分。

比较式（4）和式（9），在 $\frac{P_{H_2S}}{P_{H_2}}$ 同值的情况下：

$$f'_S [\%S] = f_S [\%S]_{Mn}$$

$$f_S = f'_S \frac{[\%S]}{[\%S]_{Mn}}$$

令

$$f_S^{Mn} = \frac{[\%S]}{[\%S]_{Mn}} \quad (11)$$

$$f_S = f'_S f_S^{Mn} \quad (12)$$

所以 f_S 由两部分组成：一部分是由于二元系内 [%S] 自行引起的活度系数 f'_S，另一部分是因为加入 Mn 后而引起对活度系数改变的影响 f_S^{Mn}。后者的物理意义代表在当两种实验有同一 $\frac{P_{H_2S}}{P_{H_2}}$ 值情况下，二元系和三元系平衡时含 S 量的比值。

由式（5）算出 f'_S（见图 1），由式（10）算出 f_S，所以 f_S^{Mn} 可以算出（见表 2）。同样的可以计算出当加入任何其他一元素如 Si、C、P、Al 或 Cu 到 Fe 液内，该第三元素对于 S 活度系数的影响。由于 Ni 在原子构造和化学性能方面很相似于 Fe，实验证明[6]加入 Ni 到 Fe 液内对 S 的活度不发生任何影响，亦即 $f_S^{Ni} = 1$。图 2 示出不同第三元素对 S 活度系数造成的影响 f_S^{III} 与该元素重量百分数的关系。

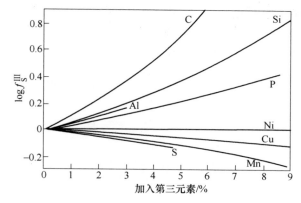

图 2　Fe 液含第三元素量与 S 的活度系数的关系（1600℃）

2　H_2O/H_2—Fe 液的化学平衡反应

H_2O 和 H_2 混合气体对 Fe 液的化学平衡反应好多学者曾做过研究。

$$H_{2(气)} + [O] \rightleftharpoons H_2O_{(气)} \tag{13}$$

$$K' = \frac{P_{H_2O}}{P_{H_2}} \times \frac{1}{[\%O]} \tag{14}$$

$$K^o = \frac{P_{H_2O}}{P_{H_2}} \times \frac{1}{f'_0[\%O]} \tag{15}$$

Fontana 和 Chipman[13] 与 Chipman 和 Samarin[14] 均证明式 (13) 的平衡常数只是温度的函数，不随 [%O] 而改变；也就是说，$f'_0 = 1$。在 1600℃ 时，$K' = K^o = 3.95$。其后 Dastur 和 Chipman[1] 综合多方实验结果得出 K^o 与温度的关系如下：

$$\log K^o = \frac{7050}{T} - 3.17 \tag{16}$$

根据式 (16)，在 1600℃ 时，$K^o = 3.93$。

好多学者根据 $f'_0 = 1$ 的结论，做了 H_2O 和 H_2 混合气体对 Fe 液含有其他第三元素，亦即三元系的平衡反应研究。但直到最近（1955 年）苏联学者 Самарин 等人[15] 根据更精确的实验，证明 $f'_0 = 1$ 的结论是不正确的。在充分研究了 H_2 气所含饱和 H_2O 汽量与通过恒温器时的气体密度和气流速度的关系后，他们做出了两组实验：第一组实验是在试样比较慢冷的情况下做的，得到的实验数据与以前各学者特别是 Dastur 和 Chipman 的数据[1] 相同；但第二组的实验是在保证试样快冷和不失掉金属中任何所含气体的条件下做的（他们精巧地设计出来一套快冷的取样方法），各试样所得的［O］量均较以往的高。同时又证明 a_0 不等于［%O］，并得出如下列经验公式所示的 f'_0 值：

$$f'_0 = 1 - (2.51 - 1.19 \times 10^{-3}T)\left(\frac{P_{H_2O}}{P_{H_2}}\right)^2 \tag{17}$$

当 $T = 1600\text{℃}$ 时：

$$f'_0 = 1 - 0.28\left(\frac{P_{H_2O}}{P_{H_2}}\right)^2 \tag{18}$$

表 3 示出不同 $\dfrac{P_{H_2O}}{P_{H_2}}$ 值下的 f'_0 值。

表 3 f'_0 值与 $\dfrac{P_{H_2O}}{P_{H_2}}$ 值的关系

$\dfrac{P_{H_2O}}{P_{H_2}}$	$0.28\left(\dfrac{P_{H_2O}}{P_{H_2}}\right)^2$	f'_0
0.1	0.0028	0.997
0.2	0.011	0.989
0.3	0.025	0.975
0.4	0.045	0.955

式（13）的理想平衡常数 K^o，Самарин 等人[15]指出与温度有下列关系：

$$\log K^o = \frac{9440}{T} - 4.536 \tag{19}$$

当 $T = 1600\text{℃}$ 时，$K^o = 3.23$。

Линчевскин[16]对此理想平衡常数 K^o 得出下列与温度的关系：

$$\log K^o = \frac{9500}{T} - 4.503 \tag{20}$$

当 $T = 1600\text{℃}$ 时，$K^o = 3.71$。

Fe 液内含有其他第三元素的三元系平衡反应，文献上已载有下列各方面的研究：Fe-O-Cr[17-20]、Fe-O-V[21-24]、Fe-O-Si[25-27]、Fe-O-Al[28,29]、Fe-O-P[30,31]、Fe-O-Ni[32]、Fe-O-Ti[33]及 Fe-O-Mn[16]。

表 4 列举第三元素引起对 O 活度系数改变的影响 $f_0^{\text{Ⅲ}}$ 的某些数据，它又给出第三元素本身的活度系数受 O 改变的影响值 $f_{\text{Ⅲ}}^0$。不可否认，由于高温下实验条件困难，各学者研究的结果存在相当的分歧。

关于 Ti 的 f_0^{Ti} 原著内未提有资料；而 Mn 的 f_0^{Mn} 值则大于 1。

表4 三元系内第三元素的 f_O^{III} 及 f_{III}^O 值（1600℃）

第三元素	f_O^{III}	f_{III}^O	来　源
Cr	$\log f_O^{Cr} = -0.034 [\%Cr]$	$\log f_{Cr}^O = -0.208 [\%O]$	Turkdogan[19]
Cr	$\log f_O^{Cr} = -0.041 [\%Cr]$	未计算	Chipman[20]
V	$\log f_O^V = -0.27 [\%V]$	$\log f_V^O = -0.86 [\%O]$	Chipman & Dastur[21]
Si	$\log f_O^{Si} = -0.02 [\%Si]$	未计算	Chipman[20]
Al	$\log f_O^{Al} = -12 [\%Al]$	$\log f_{Al}^O = -20 [\%O]$	Gokcen & Chipman[28]
P	$\log f_O^P = -0.032 [\%P]$	$\log f_P^O = -0.062 [\%P]$	Pearson & Turkdogan[30]
P(1500℃)	$\log f_O^P = -0.044 [\%P]$	$\log f_P^O = -0.085 [\%P]$	Левенец и Самарин[31]
Ni	$\log f_O^{Ni} = +0.005 [\%Ni]$	未计算	Wriedt[32]

f_{III}^O 值可由下列分析方法自 f_O^{III} 求出：

在一个三元系极稀溶液内，设 N_2 和 N_3 代表第二及第三元素的摩尔分数，当 $N_2 \to 0$ 及 $N_3 \to 0$ 时，则第二及第三元素的活度系数 f_2 和 f_3 依据 Wagner 的推论[34]，保持下列关系：

$$\frac{\partial \ln f_2}{\partial N_3} = \frac{\partial \ln f_3}{\partial N_2} \tag{21}$$

以 Fe-O-V 系为例，则：

$$\frac{\partial \ln f_O}{\partial N_V} = \frac{\partial \ln f_V}{\partial N_O}$$

由于

$$N_V = \frac{n_V}{n_V + n_O + n_{Fe}}$$

$$N_O = \frac{n_O}{n_V + n_O + n_{Fe}}$$

其中 n 代表各元素的摩尔数，而

$$n_V = \frac{[\%V]}{51}$$

$$n_O = \frac{[\%O]}{16}$$

所以有：
$$\frac{\partial \log f_O}{\partial [\%V]} = \frac{16}{51} \times \frac{\partial \log f_V}{\partial [\%O]} \tag{22}$$

试以 Chipman 和 Dastur 的经验公式推算：

$$\log f_O^V = -0.27[\%V]$$

$$\log f_O = \log f_O' + \log f_O^V$$

$$\frac{\partial \log f_O}{\partial [\%V]} = \frac{\partial \log f_O'}{\partial [\%V]} + \frac{\partial \log f_O^V}{\partial [\%V]}$$

由于 f_O' 不随 [%V] 而变化，即

$$\frac{\partial \log f_O}{\partial [\%V]} = \frac{\partial \log f_O^V}{\partial [\%V]} = -0.27$$

由式（22）转变为：

$$\partial \log f_V = \frac{51}{16}(-0.27)\partial [\%O]$$

$$\partial \log f_V = -0.86 \partial [\%O]$$

$$\log f_V = -0.86[\%O] + c$$

当 [%O] = 0、$f_V = f_V'$ 时有：

$$\log f_V = -0.86[\%O] + \log f_V'$$

$$\log f_V - \log f_V' = -0.86[\%O]$$

$$\log f_V^O = -0.86[\%O] \tag{23}$$

Fe 液内第三元素多能与 [O] 起反应，特别是好多第三元素均是强脱氧剂。下列分析方法可使我们求出氧化物的成分。以 V 为例，生成的氧化物用 VO_x 代表：

$$[V] + x[O] \Longleftrightarrow VO_{x(固)} \tag{24}$$

此外，VO_x 又可和气相 H_2O 及 H_2 的混合物发生下列平衡反应：

$$VO_{x(固)} + xH_{2(气)} \Longleftrightarrow xH_2O_{(气)} + [V] \tag{25}$$

$$K = \left(\frac{P_{H_2O}}{P_{H_2}}\right)^x a_V$$

V 与 Fe 有很相近的原子半径，故在 Fe-V 系内，我们可以估计二元素化合成一理想溶体；亦就是说，V 的活度可假定等于它的百分数浓度，即 $f_V' = 1$。在三元系内，实验数据指出，Fe 液内含 [O] 量很小，所以我们亦可能认为 [O] 对 V 的活度所施影响太小而不计；亦即 $f_V^O = 1$。这样 f_V 便等于 1，而 a_V 可用 [%V] 代替：

$$K = \left(\frac{P_{H_2O}}{P_{H_2}}\right)^x [\%V]$$

$$\log K = \log[\%V] + x\log\left(\frac{P_{H_2O}}{P_{H_2}}\right)$$

或

$$\log\left(\frac{P_{H_2O}}{P_{H_2}}\right) = -\frac{1}{x}\log[\%V] + \frac{1}{x}\log K \tag{26}$$

作 $\log\left(\frac{P_{H_2O}}{P_{H_2}}\right)$ 对 $\log[\%V]$ 的曲线图（图3），我们自图内得出一曲折的直线；同时我们得出结果：当 $[\%V]<0.2$ 时，$\frac{1}{x}=\frac{1}{2}$；当 $[\%V]=0.2\sim0.3$ 时，$\frac{1}{x}=\frac{2}{3}$；当 $[\%V]>0.3$ 时，$\frac{1}{x}=1$。因之，我们可以得到下列结论：

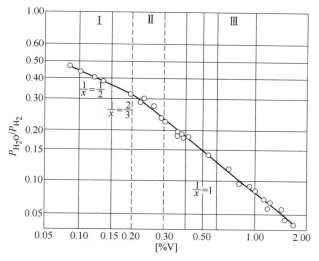

图3 Fe 液含 V 量与 P_{H_2O}/P_{H_2} 的关系（1600℃）

在 Fe 液内含 V 量很少时（0~0.2%），[O] 一部分与 Fe 化合成 FeO，一部分与 [V] 化合成 V_2O_3，而两种氧化物又结合成 FeV_2O_4，化学反应如下：

$$\frac{1}{2}FeV_2O_{4(固)} + 2H_{2(气)} \Longleftrightarrow 2H_2O_{(气)} + \frac{1}{2}Fe_{(液)} + [V] \tag{27}$$

在 Fe 液内含 V 量为 0.2%~0.3% 时，[O] 与 [V] 化合成 V_2O_3，反应如下：

$$\frac{1}{2}V_2O_{3(固)} + \frac{3}{2}H_{2(气)} \rightleftharpoons \frac{3}{2}H_2O_{(气)} + [V] \tag{28}$$

在 Fe 液内含 V 量大于 0.3% 时，[O] 与 [V] 化合成 V_2O_2，反应如下：

$$\frac{1}{2}V_2O_{2(固)} + H_{2(气)} \rightleftharpoons H_2O_{(气)} + [V] \tag{29}$$

此种分析脱氧生成物成分的方法由陈新民与 Chipman[17] 首先采用，又被以萨马林（A. M. Самарин）为首的苏联学者广泛地采用，并大加推广。利用 Gokcen 和 Chipman[25] 关于 Fe-O-Si 平衡系的数据，Морозов 和 Строганов[27] 进行同样的分析。他们指出当 1600℃ 时，在含 Si 量自 0.003% 起至 0.03% ~ 0.1% 的范围内，有 SiO_2 固体存在；而在含 Si 量为 0.1% ~ 1% 范围内，由于斜率 $1/x = 0.6$，可能有 SiO 与 SiO_2 的固溶体（见图 4）。

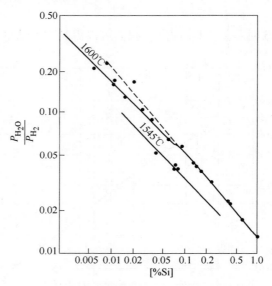

图 4　Fe 液含 Si 量与 P_{H_2O}/P_{H_2} 的关系

对于 Gokcen 和 Chipman 的 Fe-O-Al 的研究数据[28]，Морозов 和 Строганов 利用同样方法作了分析[29]，并指出该实验数据引出 [O] 与 [Al] 可生成 AlO 的结论。显然地用 Al 脱氧所得产物是 Al_2O_3，所以 Gokcen 和 Chipman 的数据不尽可靠，不可靠的原因可能是因为反应未达到平衡；或由于 [O] 与 [Al] 在 Fe 液内浓度太小，实验数据不准确。

Hilty 和 Crafts 曾用渣相—金属液反应研究 [O] 在含 Al 的 Fe 液内溶解度[35]，证明 $FeO \cdot Al_2O_{3(固)}$ 的存在。

表 5 收集了三元系内在不同成分范围下其他第三元素生成的各种氧化物，这些氧化物都是用上列分析方法推论而得，同时用非金属夹杂物分析方法，佐以金

相学和 X 射线学的分析研究加以证实。一个例外是 Hilty、Forgeng 和 Folkman[36] 的研究，他们的实验数据是用渣相—金属液的反应，研究 [O] 在 Fe-Cr 合金的溶解度而得来的。

表5　三元系在不同成分范围下生成的氧化物（1600℃）
（V、Si 及 Al 未列在内）

第三元素	成分范围	生成氧化物	来　源
Cr	0~5.5% >5.5%	$FeO \cdot Cr_2O_{3(固)}$ $Cr_2O_{3(固)}$	陈新民与 Chipman[17]
	0~6% 6%~16% >16%	$FeO \cdot Cr_2O_{3(固)}$ $Cr_2O_{3(固)}$ CrO（溶在 Fe 内）	Линчевскин и Самарин[18]
	0.06%~3% 3%~9% >9%	$FeO \cdot Cr_2O_{3(固)}$ 畸形尖晶石，可能系 $FeO \cdot Cr_2O_3$ 及 Cr_3O_4 的结合体 Cr_3O_4 及 Cr（固溶体）	Hilty、Forgeng & Folkman[36]
P	0.008%~1.2% 1.2%~3.0%	$(FeO)_3P_2O_{5(液)}$ P_2O_5（溶于 Fe 内）	Левенец и Самарин[31]
Ti	0~0.04% 0.04%~0.5%	$(FeO)_3TiO_{2(液)}$ $TiO_{2(固)}$	Ляудис и Самарин[33]
Mn	0~1.3% 1.3%~1.8% >1.8%	$xFeO \cdot yMnO_{(液)}$ $xFeO \cdot yMnO_{(液体及固体混合物)}$ $MnO_{(固)}$	Линчевскин（Самарин）[16]

3　CO_2/CO—Fe 液的化学平衡反应

Marshall 和 Chipman[37] 利用 CO_2/CO 混合气体在高温及 1~20 个大气压下研究含 O 的 Fe 液下列平衡反应：

$$CO_{(气)} + [O] = CO_{2(气)} \tag{30}$$

$$[C] + [O] = CO_{(气)} \tag{31}$$

$$[C] + CO_{2(气)} = 2CO_{(气)} \tag{32}$$

式（30）系在含 C 的 Fe 液内的平衡，所以实际的平衡常数 K_{29} 是：

$$K_{29} = \frac{P_{CO_2}}{P_{CO}} \times \frac{1}{[\%O]}$$

而理想的真平衡常数 K_{29}^o 是：

$$K_{29}^o = \frac{P_{CO_2}}{P_{CO}} \times \frac{1}{f_O[\%O]} \tag{33}$$

其中 $f_O = f_O' \times f_O^C$

如果我们采用 $f_O' = 1$

所以 $f_O^C = f_O$

作 K_{29} 对 [%C] 的曲线图,外插到 [%C] = 0,得 K_{29}^o 值。因 $f_O^C = \dfrac{K_{29}}{K_{29}^o}$,所以我们可以计算出 f_O^C 值。

此外 $K_{31} = \dfrac{P_{CO}^2}{P_{CO_2}} \cdot \dfrac{1}{[\%C]}$

而 $K_{31}^o = \dfrac{P_{CO}^2}{P_{CO_2}} \cdot \dfrac{1}{f_C[\%C]}$ (34)

$$f_C = f_C' \times f_C^O$$

由于在平衡状态下,[O] 量很小,故我们可以近似地假定 $f_C^O = 1$。

所以 $f_C' = f_C$

同样作 K_{31} 对 [%C] 的曲线图,外插到 [%C] = 0,得 K_{31}^o。因 $f_C' = \dfrac{K_{31}}{K_{31}^o}$,故 f_C' 值可算出。

Marshall 和 Chipman 算出在含 C 量为 0~0.75% 范围内,$f_C' = 1$;含 C 量大于 0.75%时,则 f_C' 随含 C 量增加而增高,f_O^C 则随含 C 量增加而降低。

Есин 和 Гельд[38] 基本同意上述结论;同时 Есин 和 Гаврилов[39] 在测定含 C 的 Fe 液的电动势时,亦证实了 f_C' 是随含 C 量增加而增高。

但在另一方面,Старк[40] 指出 Marshall 和 Chipman 所得含 C 量大于 1% 的实验数据不十分可靠,因之在含 C 量大于 1%时,有关碳及氧活度系数的结论亦不正确。

Самарин[41] 利用 Marshall 和 Chipman 的数据重新作了下列分析:式(33)可改写为:

$$\log\left(\dfrac{P_{CO_2}}{P_{CO}}\right) = \log K_{29}^o + \log f_O^C + \log[\%O]$$

式(34)可改写为:

$$\log\left(\dfrac{P_{CO}^2}{P_{CO_2}}\right) = \log K_{31}^o + \log f_C' + \log[\%C]$$

作 $\log\left(\dfrac{P_{CO_2}}{P_{CO}}\right)$ 对 $\log[\%O]$ 和 $\log\left(\dfrac{P_{CO}^2}{P_{CO_2}}\right)$ 对 $\log[\%C]$ 的曲线图,Самарин 在含 C 量小于 1%区域得出两条直线,所以他作出结论:在含 C 量小于 1% 时,f_O^C 及 f_C' 均系常数。但实际上式(31)的平衡常数并不守常:

$$K_{30} = \frac{P_{CO}}{[\%O][\%C]}$$

Самарин 解释为：在含 C 及 O 的 Fe 液内，脱氧反应可由任何下列两式进行：

$$[C] + [O] = CO_{(气)}$$
$$[C] + 2[O] = CO_{2(气)}$$

当[C]高而[O]低时，主要产物系 CO；相反，当[C]低而[O]高时，主要产物为 CO_2。K_{30} 值不守常，是由于两种反应同时发生，而并非由于碳或氧活度系数本身的变化。对于 Chipman[42] 提出 Fe 液内生成可变成分综合体来解释活度的理论，Самарин 并作了批判。

Richardson 和 Dennis[43] 在与 Marshall 和 Chipman 不同的工作条件下研究了反应式（32）：

$$[C] + CO_{2(气)} = 2CO_{(气)}$$

他们实验的条件是：含 C 量为 $0.1\% \sim 1.08\%$；$\frac{P_{CO}^2}{P_{CO_2}} = 100 \sim 800$。由于后者的值非常的大，杜绝了 Fe 液内含 O 的可能，所以他们的实验可以称为 Fe-C 二元系平衡反应。他们以纯石墨为活度的标准状态，求出 f_C' 与 [%C] 的关系。Chipman[20] 依据 1%[C] 稀溶液为标准状态重新计算了他们的数据，得出下列关系：

$$\log f_C' = \frac{358}{T}[\%C]$$

当 $T = 1600℃$ 时有： $\log f_C' = +0.19[\%C]$ (35)

同时 Chipman 再度利用他和 Marshall 关于反应式（31）的平衡常数而计算 f_O^C，得出下列的关系：

$$\log f_C' + \log f_O^C = -0.22[\%C] \quad (36)$$

式（36）- 式（35）可得：

$$\log f_O^C = -0.41[\%C] \quad (37)$$

依据式（22）

$$\log f_C^O = \frac{12}{16}(-0.41[\%C])$$

所以： $\log f_C^O = -0.31[\%C]$ (38)

根据 Turkdogan、Davis、Leake 和 Stevens[44] 最近的关于碳氧在铁液内反应的一篇研究，他们利用他们自己和 Marshall 和 Chipman 的实验数据，得出适用于 1560~1760℃ 的下列结论：

$$\log f_C = +0.22[\%C]$$
$$\log f_O = -0.49[\%C]$$

其中 f_C 及 f_O 均系式（31）反应内的三元系的活度系数。他们对二元系的活度系数及碳氧相互作用的影响并未进行分析。

Richardson 和 Dennis[45] 又用同样方法研究了 CO_2/CO 对含 Cr 及 C 的 Fe 液反应（Fe-C-Cr 系）。他们的结果 Chipman 总结如下：

$$\log f_C^{Cr} = -0.020[\%Cr] \tag{39}$$

Manley[46] 亦用同样方法研究 Fe-C-Mo 系，得出下列结论：

$$\log f_C^{Mo} = -0.002[\%Mo] \tag{40}$$

4 讨论：活度的实际意义和应用

上面我们列举了三种气体—金属液反应。通过实验数据，我们已经看出各种元素在 Fe 液内相互间作用的关系是复杂的，在二元系内，第二元素的有效浓度（活度）可能随它本身的百分数含量而改变。在三元系内，第三元素的百分数含量又对第二元素的活度施有影响。以 i 代表第二元素，j 代表第三元素，M_i 和 M_j 分别代表它们的分子量，则有：

$$\log f_i^j = \phi[\%j]$$

设令：

$$e_i^j = \frac{\partial \log f_i^j}{\partial [\%j]} \tag{41}$$

则 e_i^j 的物理意义是每加入 $1\%j$ 元素到 Fe 液内引起元素 i 活度系数改变的对数值。在互相比较中，我们更好地采用每 1mol 第三元素 j 对第二元素 i 活度系数的改变。今令：

$$\varepsilon_i^j = \frac{\partial \ln f_i^j}{\partial N_j} \tag{42}$$

则 ε 即代表每加入 $1\text{mol } j$（即元素 j 的摩尔分数 $N_j = 1$）到 Fe 液内引起元素 i 活度系数改变的自然对数值。在很稀的溶液内

$$N_i = \frac{0.5585[\%j]}{M_j} \tag{43}$$

结合式（41）~式（43）有：

$$\varepsilon_i^j = \frac{2.3}{0.5585} M_j e_i^j$$

或

$$e_i^j = 0.2425 \frac{\varepsilon_i^j}{M_j} \tag{44}$$

依据式（22）有：

$$\frac{\partial \log f_i}{\partial [\%j]} = \frac{M_i}{M_j} \frac{\partial \log f_j}{\partial [\%i]}$$

简单的数学分析证明：

$$\frac{\partial \log f_i^j}{\partial [\%j]} = \frac{M_i}{M_j} \frac{\partial \log f_j^i}{\partial [\%i]}$$

亦即

$$e_i^j = \frac{M_i}{M_j} e_j^i$$

式（43）和式（44）均对很稀的溶液有效。对图 2 中各项曲线 Chipman[20] 亦提出相应的 e_i^j 值。

表 6 和表 7 分别总结了三种气体—金属液平衡反应的 e_i^j 和 ε_i^j 值。括弧内的数值系用其他间接方法求得[20]。带有（?）号的数值表示准确度还存在有问题。

表 6　1600℃时各元素的 $e_i^j = \dfrac{\partial \log f_i^j}{\partial [\%j]}$ 值

元素 i	加入的第三元素 j										
	C	Al	Si	P	S	V	Cr	Mn	Ni	Cu	Mo
C	+0.19(?)		(+0.088)				−0.020	(−0.002)			−0.002
O	−0.41(?)	−12(?)	−0.02(?)	−0.032	(+1.00)	−0.27	−0.041	+	+0.005		
S	+0.11	0.055	0.065	+0.042	−0.028			−0.025	0.00	−0.012	

表 7　1600℃时各元素的 $\varepsilon_i^j = \dfrac{\partial \ln f_i^j}{\partial N_j}$ 值

元素 i	加入的第三元素 j										
	C	Al	Si	P	S	V	Cr	Mn	Ni	Cu	Mo
C	+9.8(?)		(+10)				−4.3	(−0.5)			−0.8
O	−21(?)	−13(?)	−2(?)	−4.1	(+1.30)	−57	−8.8	+	+1.2		
S	+6	+6.5	+7.6	+5.7	−3.8			−5.7	0.0	−3.2	

活度有什么实际意义呢？

从上列气体—金属液平衡反应首先可以推出，活度的研究进一步地解释了在钢铁冶炼过程中各种元素脱硫脱氧的作用。

表 6 或表 7 指出，当某元素的 e 或 ε 值带负号时，活度系数值小于 1，该元素的有效浓度较溶入 Fe 液内其真实浓度为小；亦就是说，第三元素的加入，使该元素的一部分在铁液内失去了化学作用力。当 e 或 ε 负数的绝对值越大时，则此作用越显著。当 e 或 ε 是正值时，表明该元素在 Fe 液内的有效浓度由于加了第三元素而增加。结合脱硫反应而论，e 或 ε 的正值对脱硫有利；亦就是说，第三元素增高了 S 的逸度。这点更可由式（11）加以说明：

$$f_S^{\text{III}} = \frac{[\%S]}{[\%S]_{\text{III}}}$$

当 $f_S^{\text{III}} > 1$ 时，$[\%S]_{\text{III}} < [\%S]$；亦即是说，在二实验有共同 $\dfrac{P_{H_2S}}{P_{H_2}}$ 值时，在 $f_S^{\text{III}} > 1$ 的情况下，加入了第三元素使铁液内在平衡时含 $[\%S]$ 量减小。在含多种元素的铁液内，不同元素活度系数引起的影响是有大致的加和性的。生铁含 C、Si 相当高，所以高炉内生铁脱硫比较容易进行。S 的活度系数用经验公式估计在生铁内达到 5.4~6.3[8]。平炉钢液内 S 的活度系数较 1 大不多，所以炼钢时脱硫反应进行比较困难。此外，Mn 在钢内有减少因 S 而生的热脆现象，但就活度影响而论，则 Mn 降低 S 的活度系数，亦即增加脱硫的困难。

此外，脱硫反应在机构上是二相的渣—金属液反应，可用下列式子表示：

$$[FeS] + (FeO) \rightleftharpoons [FeO] + (FeS)$$

或

$$[FeS] + (CaO) \rightleftharpoons [FeO] + (CaS)$$

写成离子反应式：

$$[S] + (O^{2-}) \rightleftharpoons [O] + (S^{2-}) \tag{45}$$

在大量 C、Si 存在于铁液内的时候，$[S]$ 的有效浓度由于活度系数的加大而增高，所以促使反应（45）向右方进行，亦即促进脱硫反应。同时铁液内的 $[C]$ 可降低铁液含 $[O]$ 量，使反应（45）更难向左进行，因之更有利于脱硫。

在混铁炉内，铁液内 $[Mn]$ 量有一定的脱 S 作用。这是由于当时的较低温度（1200~1300℃）促使 MnS 易于生成，通过偏析作用浮到渣面上来。高炉和平炉内的较高温度不利于 MnS 的生成，所以总的来讲，Mn 在冶炼过程中对脱硫是不利的。

谈到脱氧问题，活度的研究有更重要的意义。某元素脱氧能力的大小，可以该脱氧反应热力势的变化 $\Delta\Phi^{\circ}$ 来衡量。$\Delta\Phi^{\circ}$ 的负值越大，则脱氧能力越强，平衡常数 K 值应越大，亦即脱氧常数 m 值应越小。

$$x[M] + y[O] \rightleftharpoons M_xO_{y(\text{固或液})}$$

$$\Delta\Phi = \Phi^{\circ} - \overline{\Phi}_M^{\circ} - RT\ln a_M^x - \overline{\Phi}_O^{\circ} - RT\ln a_O^y$$

在平衡状态时 $\Delta\Phi = 0$，所以：

$$\Phi^{\circ} - \overline{\Phi}_M^{\circ} - \overline{\Phi}_O^{\circ} = RT\ln(a_M^x a_O^y)$$

$$\Delta\Phi^{\circ} = RT\ln(a_M^x a_O^y) \tag{46}$$

$$m = \frac{1}{K} = a_M^x a_O^y \tag{47}$$

亦即：

$$(f_M[\%M])^x (f_O[\%O])^y = m$$

或

$$(f_M' f_M^O[\%M])^x (f_O' f_O^M[\%O])^y = m \tag{48}$$

式（48）内［%M］及［%O］代表平衡状态下 Fe 液内含有的第三元素及氧量。我们可以看出，假若 f_M^O、f_M'、f_O' 及 f_O^M 等活度系数均可由实验测出，则不难计算出我们在实际最广泛应用的脱氧常数 m。事实上这几个活度系数往往用实验测不出来或不能准确地测出来，所以脱氧常数值经常用间接计算方法由 $\Delta\Phi^o$ 算出。

在经常计算中，f_O' 可以认为等于 1。f_O^M 可利用式（22）自 f_O^M 算出；但在绝大多数实际平衡情况下，Fe 液内所含的［O］量很少，$\log f_M^O = \phi([\%O]) = 0$，所以我们亦可认为 $f_M^O = 1$。

关于二元系各元素在 Fe 液内的活度系数 f_M' 我们已知的实验数据比较少。由于 Co、Ni、Mn、Cr 及 V 等元素的原子半径和 Fe 的相差不多，理论上可以推论这些元素互溶，因之 $f' = 1$；实验上证明这一推断是正确的。关于 Si、Al，Richardson[47] 提出在很稀的 Fe 液内：$f'_{Si} = 0.017$，$f'_{Al} = 0.025$。

由于脱氧生成物为 M_xO_y（或为与 FeO 结合成 $FeO \cdot M_xO_y$），而各氧化物未必有同一成分，由式（46）或用其他方法计算出的 $\Delta\Phi^o$ 值不能彼此比较。Pearson 及 Turkdogan[30] 曾建议以 $\overline{\Delta\Phi^o}$ 来比较，$\overline{\Delta\Phi^o}$ 代表每平均摩尔数的热力势变化，即：

$$\overline{\Delta\Phi^o} = \frac{\Delta\Phi^o}{x+y} \tag{49}$$

作者意见认为，每摩尔脱氧元素在完成脱氧反应时的热力势变化 $\Delta\Phi^{o'}$ 来比较是更适当一些，亦即：

$$\Delta\Phi^{o'} = \frac{\Delta\Phi^o}{x} \tag{50}$$

从表 6 和表 7 看出，强脱氧剂如 Al 能大大地降低氧的活度系数；而无脱氧能力的元素如 Ni 能增加氧的活度系数。e_O^M 值与 $\Delta\Phi^{o'}$ 值有何关系呢？

表 8 试作不同脱氧剂 e_O^M 与 $\Delta\Phi^{o'}$ 二值的相互比较。因为 e_O^M 值是在很稀的溶液内取得的，表 8 列举的脱氧生成物亦以在含 M 不高时生成的氧化物为限。

表 8　各脱氧元素 e_O^M 与 $\Delta\Phi^{o'}$ 的比较（1600℃）

元素 M	e_O^M	氧化物 M_xO_y 或 $FeO \cdot M_xO_y$	$\Delta\Phi^o$	$\Delta\Phi^{o'} = \dfrac{\Delta\Phi^o}{x}$	来源
Al	−12	$Al_2O_{3(固)}$	−108800	−54400	$\log m = -\dfrac{47200}{T} + 12.52$ [28]
Si	−0.02	$SiO_{2(固)}$	−39000	−39000	$\log m = -\dfrac{20270}{T} + 62.27$ [48]
C	−0.41	$CO_{(气)}$	−23100	−23100	$\log m = -\dfrac{1056}{T} - 2.131$ [44]

续表 8

元素 M	e_O^M	氧化物 M_xO_y 或 $FeO \cdot M_xO_y$	$\Delta \Phi^o$	$\Delta \Phi^{o'} = \dfrac{\Delta \Phi^o}{x}$	来　源
V	-0.27	$\dfrac{1}{2}(FeO \cdot V_2O_3)_{(固)}$	-22200	-22200	$\log m = -\dfrac{25360}{T} + 10.95$ [22]
Cr	-0.041	$\dfrac{1}{2}(FeO \cdot Cr_2O_3)_{(固)}$	-19100	-19100	$\log m = -\dfrac{53110}{T} + 26.12$ [18]

表 8 内的 $\Delta \Phi^{o'}$ 负值越大，脱氧能力越强。$\Delta \Phi^{o'}$ 的次序与 e_O^M 几乎完全一致，除去 e_O^{Si} 一项。后者的数值系 Chipman[20] 提出，它的准确度恐有问题。Pearson 和 Turkdogan[30] 指出 $e_O^{Si} = -0.87$；而依据 Kitchener[49] 的估计，则 $e_O^{Si} = -2.3$。采用上列任一值，就使 e_O^M 的次序与 $\Delta \Phi^{o'}$ 的次序完全一致。e_O^M 与 $\Delta \Phi^{o'}$ 的次序一致，值得我们注意。在这方面理论的解释尚有待进一步的研究。但我们可以肯定，由于脱硫和脱氧在机构上有所不同，活度系数应根据不同角度加以利用。

当第三元素在 Fe 液内含量较高时，脱氧作用渐渐削弱，而它的合金化作用渐形显著。同时 [O] 与第三元素能形成不同氧化物，在第三元素含量较低时，一般地生成较高级氧化物或与 FeO 共生的复杂氧化物，如 $FeO \cdot V_2O_3$、$FeO \cdot Cr_2O_3$ 等。在第三元素含量较高时，则生成较低级氧化物，如 SiO、V_2O_2、Cr_3O 等。在大多数情况下，液内含 [O] 量与第三元素含量相互间成一双曲线关系，即高 [O] 时，第三元素含量必低。但在 Cr 的脱氧反应已表现出：超过某定量 [Cr] 时，[O] 含量反而逐渐增高；如 [Cr] 再多，[O] 又逐渐降低。图 5 示出 [Cr] 与 [O] 的复杂关系[36]。同样地，在含 [P] = 1.2% 时，Fe 液内含 [O] 量亦表现一最低值[31]。这些事实对冶炼过程均有很大的意义。

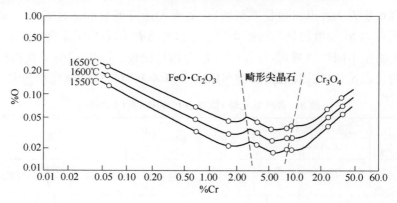

图 5　Fe 液含 Cr 量与含 O 量的关系

如以前所述，在 H_2O/H_2—Fe 液平衡反应中，利用 $\log(H_2O/H_2)$ 对 $\log[\%M]$

坐标图的分析，可由画出直线的曲折点推论不同氧化物存在的范围。简单的数学分析证明，作 $\log a_0$ 对 $\log[\%M]$ 的坐标图，亦得到同形状带曲折的直线。

从上列谈到的可以推知，多元系内各元素在 Fe 液内彼此的相互间作用关系更为复杂化。所以为了阐明合金钢热力学的性质，活度的研究有更大的必要性。

5 结束语

（1）气体—金属液平衡反应的活度研究，无论在脱硫、脱氧还是合金化热力学性质方面均有重大的意义。通过活度的研究，对钢液在高温下内在变化的实质有了进一步的了解。

（2）限于高温下反应的特殊条件，精确的实验数据的获得常有相当的困难，所以各学者所得某些有关活度的数据表现出一定程度的悬殊。改进高温实验技巧，进一步地研究各种平衡反应，尚待冶金工作者的努力。

（3）气体—金属液平衡反应上有好多未知的空白数据，特别在合金钢热力学性质方面；通过 H_2O/H_2—Fe 液反应的研究，变更 Fe 液内含合金元素的类别，如 B、Ti、Fe、Mo、W 等，将会求得更多有实际意义的宝贵数据，促使人们对合金钢生产的化学反应的规律得到进一步的掌握。

参 考 文 献

[1] Dastur, Chipman. Trans. A. I. M. E., 1949, 185: 441.
[2] Chipman, Ta Li. Trans. A. S. M., 1937, 25: 435.
[3] White, Shelly. J. Iron & Steel Inst, 1947, 155: 201.
[4] Sherman, Elvander, Chipman. Trans. A. I. M. E., 1950, 188: 334.
[5] 魏寿昆. 北京钢铁工业学院学报, 1956（3）.
[6] Cordier, Chipman. J. of Metals, 1955（8）: 905.
[7] Morris, Williams. Trans. A. S. M., 1949, 41: 1452.
[8] Sherman, Chipman. Trans. A. I. M. E., 1952, 194: 597.
[9] Morris, Buehl. Trans. A. I. M. E., 1950, 188: 317.
[10] Rocenquist, Cox. Trans. A. I. M. E., 1950, 188: 1389.
[11] Chipman, Sherman. Revue Metallurgie, 1951, 48: 613.
[12] Morris. Trans. A. I. M. E., 1952, 194: 939.
[13] Fontana, Chipman. Trans. A. S. M., 1936, 24: 313.
[14] Chipman, Samarin. Trans. A. I. M. E., 1937, 125: 881.
[15] Аверин, Поляков, Самарин. Известия АН СССР, ОТН, 1955（3）: 90.
[16] Линчевскин. 副博士论文, Самарин 教授指导, 1955.
[17] Chen, Chipman. Trans. A. S. M., 1947, 38: 70.

[18] Линчевскин, Самарин. Известия АН СССР, ОТН, 1953 (5): 691; 另见 Крамаров: Физико-химич, процессы проиэ водства стали, 1954: 136.
[19] Turkdogan. J. Iron & Steel Inst., 1954, 178: 278.
[20] Chipman. J. Iron & Steel Inst., 1955, 180: 97.
[21] Chipman, Dastur. Trans. A. I. M. E., 1951, 191: 111.
[22] Карасев, Поляков, Самарин. Известия АН СССР, ОТН, 1952 (12): 1794.
[23] Морозов, Строганов. Раскисление Мартеновской Стали, 1955: 55~57, 108~111.
[24] Самарин. Проблемы Металлургии, 1953: 76~98.
[25] Gokcen, Chipman. Trans. A. I. M. E., 1952, 194: 171.
[26] Chipman, Gokeen. Trans. A. I. M. E., 1953, 197: 1017.
[27] Морозов, Строганов. ibid: 72.
[28] Gokcen, Chipman. Trans. A. I. M. E., 1953, 197: 173.
[29] Морозов, Строганов. ibid: 91.
[30] Pearson, Turkdogan. J. Iron & Steel Inst., 1954, 176: 19.
[31] Левенец, Самарин. Доклады АН СССР, том 101, 1955 (6): 1089.
[32] Wriedt. Thesis & Massachusetts Inst. of Tech, 1954; 另见 [20].
[33] Ляудис, Самарин. Доклады АН СССР, том 101, 1955 (2): 325.
[34] Wagner. J. Chem. Physics, 1951, 19: 626; 另见 Wagner. Thermodynanics of Alloys, 1952: 45.
[35] Hilty, Crafts. Trans. A. I. M. E., 1950, 188: 414, 1342.
[36] Hilty, Forgeng, Folkman. Trans. A. I. M. E., 1955, 203: 253.
[37] Marshall, Chipman. Trans. A. S. M., 1942, 30: 695.
[38] Есин, Гельд. Физическая Химия Пкрометаллургических Процессов, 1954: 546~549.
[39] Есин, Гаврилов. Известия АН СССР, ОТН, 1950 (7): 1040.
[40] Старк. Известия АН СССР, ОТН, 1948 (5): 655.
[41] Самарин. 见 [24]; 另见: 格里古良. 北京钢铁学院学报, 1955 (1): 12.
[42] Chipman. Metal Progress, 1949, 56: 211.
[43] Richardson, Dennis. Trans, Faraday Soc., 1953, 49 (Part 1): 171.
[44] Turkdogan, Davis, Leake, Stevens. J. Iron & Steel Inst, 1955, 181: 123.
[45] Richardson, Dennis. J. Iron & Steel Inst, 1953, 175: 257.
[46] Manley. 见 Chipman [20].
[47] Richardson. J. Iron & Steel Inst, 1950, 166: 187.
[48] Морозов, Строганов. ibid: 76.
[49] Kitchener. Iron & Steel, 1954, 27: 473, 528.

活度相互作用系数运算中的某些问题[*]

魏寿昆

1 引言

在对冶金过程作热力学计算时，引用不同文献的活度相互作用系数会出现分歧。活度相互作用系数分歧的原因是：

(1) 测定原理及实验方法不同；
(2) 高温实验技术的困难及准确性；
(3) 实验数据用不同方法处理得到不同结果；
(4) 有些时候出现概念混淆不清。

在本文中主要讨论以下三个问题：

(1) 同一系列实验数据用不同方法处理，所得不同结果的表现情况；
(2) 活度相互作用系数用同浓度法和同活度法的运算方法；
(3) 二级活度相互作用系数。

2 溶解度法测定活度相互作用系数（基本为同活度法）

2.1 原理及符号

(1) 活度的标准态：纯物质，浓度用摩尔分数；重量1%溶液，浓度用重量百分数。

$$a_i = \gamma_i' N_i' = \gamma_i N_i \quad （右角角标 ' 指二元系）$$

$$\gamma_i^j = \gamma_i / \gamma_i' = N_i' / N_i \quad （右角无角标指三元系）$$

(2) 同浓度法[1]：

$$e_i^j = \left(\frac{\partial \log f_i}{\partial [\%j]} \right)_{\%i}$$

$$\varepsilon_i^j = \left(\frac{\partial \ln \gamma_i}{\partial N_j} \right)_{N_i}$$

假定由同一活度值计算出的 f_i^j 和 i 的浓度无关，在 $[i]$ 服从亨利定律时和同活度法一致。

---

[*] 在北京科技大学对冶金专业研究生和教师学术讲座的讲稿，未正式刊印。

(3) 同活度法[9,11]：

$$e_i^{*j} = \left(\frac{\partial \log f_i}{\partial [\%j]}\right)_{a_i}$$

$$\varepsilon_i^{*j} = \left(\frac{\partial \ln \gamma_i}{\partial N_j}\right)_{a_i}$$

e_i^{*j} 和 ε_i^{*j} 是 C. H. P. Lupis 所用符号；H. Schenck 所用符号则是 o_i^j 和 ω_i^j；本文采用 C. H. P. Lupis 符号。

M. Ohtani 和 N. A. Gokcen[2] 曾用 λ_i^j 表示同活度法系数，相当于 ω_i^j。

2.2 实验数据

Fe-C-j 系[1~3]、Fe-H-j 系[5]、Fe-N-j 系[6]、Fe-Ca-j 系[7]、Mn-Ca-j 系[8]、Mn-C-j 系[4]、Ni-C-j 系[4]、Co-C-j 系[4]。

2.3 计算 ε_C^{*j} 的方法

(1) J. Chipman 法：用对数数据[1]作 $\ln\gamma_C^j$ 和 N_j 对应线图。由于 $N_j = 0$ 时，$\gamma_C^j = 1$，$\ln\gamma_C^j = 0$，所以各线经过坐标原点。回归可用式 $\ln\gamma_C^j = bN_j$，所以 $\varepsilon_C^{*j} = b$。

(2) H. Schenck 法：直接采用溶解度数据[11]

$$\Delta N_C = N_C - N_C' = mN_j \tag{1}$$

或 $$\Delta[\%C] = [\%C] - [\%C]' = m'[\%j] \tag{2}$$

式中，m 和 m' 为常数。

$$\gamma_C^j = \gamma_C/\gamma_C' = N_C'/N_C$$

$$\ln\gamma_C^j = \ln[N_C'/(N_C' + mN_j)]$$

$$\left(\frac{\partial \ln\gamma_C^j}{\partial N_j}\right)_{a_C} = -\frac{m}{N_C' + mN_j}$$

所以： $$\varepsilon_C^{*j} = \left(\frac{\partial \ln\gamma_C^j}{\partial N_j}\right)_{\substack{a_C \\ N_j \to 0}} = -\frac{m}{N_C'} \tag{3}$$

同理可导出：

$$e_C^{*j} = \left(\frac{\partial \log f_C^j}{\partial [\%j]}\right)_{\substack{a_C \\ [\%j] \to 0}} = -\frac{m'}{2.30[\%C]'} \tag{4}$$

于是，用 $N_C - N_C'$ 和 N_j 对应数据作图，可以求出 m，再求出 ε_C^{*j}（应用式 (3)）。或者用 $[\%C] - [\%C]'$ 和 $[\%j]$ 数据作图，求出 m'，再求 e_C^{*j}（用式 (4)）。

式 (3) 的另一推导方法为：

$$dN_C = mdN_j$$

因为
$$\mathrm{d}\ln N_C = \frac{m}{N_C}\mathrm{d}N_j$$
$$a_C = \gamma_C N_C = \mathrm{const}$$

所以
$$\mathrm{d}\ln N_C = -\mathrm{d}\ln\gamma_C$$

$$-\left(\frac{\partial \ln\gamma_C}{\partial N_j}\right)_{a_C} = \frac{m}{N'_C + mN_j}$$

$$\left(\frac{\partial \ln\gamma_C}{\partial N_j}\right)_{\substack{a_C \\ N_j \to 0}} = -\frac{m}{N'_C}$$

由于二元系的活度系数 γ'_C 与 N_j 无关，所以可以证明：

$$\frac{\partial \ln\gamma_C}{\partial N_j} = \frac{\partial \ln\gamma'_C + \partial \ln\gamma^j_C}{\partial N_j} = \frac{\partial \ln\gamma^j_C}{\partial N_j}$$

$$\varepsilon^{*j}_C = \left(\frac{\partial \ln\gamma^j_C}{\partial N_j}\right)_{\substack{a_C \\ N_j \to 0}} = \left(\frac{\partial \ln\gamma_C}{\partial N_j}\right)_{\substack{a_C \\ N_j \to 0}} = \left(-\frac{\partial \ln N_C}{\partial N_j}\right)_{\substack{a_C \\ N_j \to 0}} = -\frac{m}{N'_C}$$

$$e^{*j}_C = \left(\frac{\partial \log f^j_C}{\partial [\%j]}\right)_{\substack{a_C \\ [\%j] \to 0}} = \left(\frac{\partial \log f_C}{\partial [\%j]}\right)_{\substack{a_C \\ [\%j] \to 0}} = \left(-\frac{\partial \log[\%C]}{\partial [\%j]}\right)_{\substack{a_C \\ [\%j] \to 0}}$$

$$= -\frac{m'}{2.30[\%C]'}$$

2.4 二级活度相互作用系数

（1）用对数数据（C. H. P. Lupis[9]推荐方法）

$$\ln\gamma^j_C = \ln\frac{N'_C}{N_C} = bN_j + cN^2_j \tag{5}$$

$$\varepsilon^{*j}_C = b$$

$$\rho^{*j}_C = c$$

（2）直接采用溶解度数据

$$N_C - N'_C = mN_j + lN^2_j \tag{6}$$

$$N_C \mathrm{d}\ln N_C = m\mathrm{d}N_j + 2lN_j\mathrm{d}N_j$$

$$\frac{\mathrm{d}\ln N_C}{\mathrm{d}N_j} = \frac{m}{N_C} + \frac{2l}{N_C}N_j \tag{7}$$

$N_j \to 0$，可消去第二项

$$\varepsilon^{*j}_C = \left(\frac{\partial \ln\gamma_C}{\partial N_j}\right)_{\substack{a_C \\ N_j \to 0}} = -\frac{m}{N'_C} \tag{8}$$

再对式（7）以 $\mathrm{d}N_j$ 求导数：

$$\frac{\mathrm{d}^2\ln N_\mathrm{C}}{\mathrm{d}N_j^2} = -\frac{m}{N_\mathrm{C}^2} \times \frac{\mathrm{d}N_\mathrm{C}}{\mathrm{d}N_j} + 2l\left(\frac{l}{N_\mathrm{C}} - \frac{N_j}{N_\mathrm{C}^2} \times \frac{\mathrm{d}N_\mathrm{C}}{\mathrm{d}N_j}\right)$$

由于

$$\mathrm{d}N_\mathrm{C}/\mathrm{d}N_j = m + 2lN_j$$

代入上式，得

$$\frac{\mathrm{d}^2\ln N_\mathrm{C}}{\mathrm{d}N_j^2} = \frac{2l}{N_\mathrm{C}} - (m + 2N_j l)\left(\frac{m}{N_\mathrm{C}^2} + \frac{2N_j l}{N_\mathrm{C}^2}\right)$$

$$\left(\frac{\partial^2 \ln\gamma_\mathrm{C}}{\partial N_j^2}\right)_{\substack{a_\mathrm{C} \\ N_j \to 0}} = -\frac{2l}{N_\mathrm{C}'} + \left(\frac{m}{N_\mathrm{C}'}\right)^2$$

按照 C. H. P. Lupis 的定义：

$$\rho^{*j}_\mathrm{C} = \frac{1}{2}\left(\frac{\partial^2 \ln\gamma_\mathrm{C}}{\partial N_j^2}\right)_{\substack{a_\mathrm{C} \\ N_j \to 0}}$$

得出：

$$\rho^{*j}_\mathrm{C} = -\frac{l}{N_\mathrm{C}'} + \frac{1}{2}\left(\frac{m}{N_\mathrm{C}'}\right)^2 \tag{9}$$

同理，对于重量 1% 浓度的标准态：

$$e^{*j}_\mathrm{C} = -\frac{m'}{2.30[\%\mathrm{C}]'} \tag{10}$$

$$r^{*j}_\mathrm{C} = -\frac{l'}{2.30[\%\mathrm{C}]'} + \frac{1}{4.60}\left(\frac{m'}{[\%\mathrm{C}]'}\right)^2 \tag{11}$$

2.5 两种标准状态的活度相互作用系数的转换

e^{*j}_i 与 ε^{*j}_i 的相互转换公式，和 e^j_i 与 ε^j_i 的相互转换公式是相同的[11,12]，即

$$\varepsilon^j_\mathrm{C} = 230\frac{M_j}{M_1}e^j_\mathrm{C} + \frac{M_1 - M_j}{M_1} \tag{12}$$

r^{*j}_i 与 ρ^{*j}_i 的相互转换公式，和 r^j_i 与 ρ^j_i 的相互转换公式是相同的[10,11]，即

$$\rho^j_\mathrm{C} = \frac{23000 M_j^2 r^j_\mathrm{C} + M_1(M_1 - M_j)\varepsilon^j_\mathrm{C} - \frac{1}{2}M_1(M_1 - M_\mathrm{C})\varepsilon^{j2}_\mathrm{C} N_\mathrm{C}' - \frac{1}{2}(M_1 - M_j)^2}{[M_1 - N_\mathrm{C}'(M_1 - M_\mathrm{C})]M_1} \tag{13}$$

式中，M_1、M_j 代表溶剂和第三元素的分子量。

2.6 计算实例

计算所用数据取自北京科技大学冶金物化实验室的研究[8]，表 1 为 Mn-Ca-Cr 系实验值，表 2 为 Mn-Ca-Al 系实验值。

表 1　Cr 对 Ca 在 Mn 液中溶解度的影响（1350℃）

%Ca	N_{Ca}	%Cr	N_{Cr}
0.15	0.002055	0	0
0.15	0.002054	0.88	0.009288
0.14	0.001915	2.59	0.027310
0.12	0.001640	4.28	0.045091
0.12	0.001635	9.48	0.099582
0.098	0.001334	11.02	0.115668
0.084	0.001143	13.40	0.140468

表 2　Al 对 Ca 在 Mn 液中溶解度的影响（1350℃）

%Ca	N_{Ca}	%Al	N_{Al}	备注
0.15	0.002055	0	0	
0.21	0.002829	1.6	0.032025	
0.27	0.003561	3.7	0.072492	
0.28	0.003656	4.7	0.091172	
0.47	0.005906	8.6	0.160536	
0.70	0.008609	10.8	0.197326	生成化合物
3.00	0.033752	20.0	0.334265	
3.29	0.036303	22.2	0.363904	

计算结果见表 3 ~ 表 5。

表 3　Mn 液中 Cr 对 Ca 活度相互作用一级系数（1350℃）

	采用对数关系	直接采用溶解度值
摩尔分数浓度	$\ln \gamma_i^j = bN_j$ $\ln \gamma_{Ca}^{Cr} = 3.6447 N_{Cr}$ $(r = 0.95)$ $\varepsilon^{*\,Cr}_{Ca} = 3.6$	$N_i - N_i^j = mN_j$ $N_{Ca} - N'_{Ca} = -0.006011 N_{Cr}$ $(r = 0.96)$ $\varepsilon^{*\,Cr}_{Ca} = 2.9$
重量百分数浓度	$\log f_i^j = b'[\%j]$ $\log f_{Ca}^{Cr} = 0.01638[\%Cr]$ $(r = 0.95)$ $e^{*\,Cr}_{Ca} = 0.016$ 由转换公式（12）计算： $\varepsilon^{*\,Cr}_{Ca} = 3.6$	$[\%i] - [\%i]' = m'[\%j]$ $[\%Ca] - [\%Ca]' = -0.004551[\%Cr]$ $(r = 0.96)$ $e^{*\,Cr}_{Ca} = -\dfrac{m'}{2.30[\%Ca]} = 0.013$ 由转换公式（12）计算： $\varepsilon^{*\,Cr}_{Ca} = 2.9$

表4　Mn 液中 Cr 对 Ca 活度相互作用二级系数（1350℃）

	采用对数关系	直接采用溶解度值
	$\ln\gamma_i^j = bN_j + cN_j^2$	$N_i - N_i' = mN_j + lN_j^2$
摩尔分数浓度	$\ln\gamma_{Ca}^{Cr} = 2.1998N_{Cr} + 12.2333N_{Cr}^2$ ($r = 0.96$) $\varepsilon^{*}{}_{Ca}^{Cr} = 2.2$ $\rho^{*}{}_{Ca}^{Cr} = 12.2$	$N_{Ca} - N_{Ca}' = -0.005349N_{Cr} - 0.005607N_{Cr}^2$ ($r = 0.96$) $\varepsilon^{*}{}_{Ca}^{Cr} = -\dfrac{m}{N_{Ca}'} = 2.6$ $\rho^{*}{}_{Ca}^{Cr} = -\dfrac{1}{N_{Ca}'} + \dfrac{1}{2}\left(\dfrac{m}{N_{Ca}'}\right)^2 = 6.1$
	$\log f_i^j = b'[\%j] + c'[\%j]^2$	$[\%i] - [\%i]' = m'[\%j] + l'[\%j]^2$
重量百分数浓度	$\log f_{Ca}^{Cr} = 0.009814[\%Cr] + 0.000582[\%Cr]^2$ ($r = 0.96$) $e^{*}{}_{Ca}^{Cr} = 0.0098$ $r^{*}{}_{Ca}^{Cr} = 0.00058$ 由转换公式（12）计算： $\varepsilon^{*}{}_{Ca}^{Cr} = 2.2$ 由转换公式（13）计算： $\rho^{*}{}_{Ca}^{Cr} = 12.1$	$[\%Ca] - [\%Ca]'$ $= -0.004016[\%Cr] - 0.000047[\%Cr]^2$ ($r = 0.99$) $e_{Ca}^{Cr} = -\dfrac{m'}{2.30[\%Ca]'} = 0.012$ $r^{*}{}_{Ca}^{Cr} = -\dfrac{l'}{2.30[\%Ca]'} + \dfrac{1}{4.60}\left(\dfrac{m'}{[\%Ca]'}\right)^2$ $= 0.00029$ 由转换公式（12）计算： $\varepsilon^{*}{}_{Ca}^{Cr} = 2.6$ 由转换公式（13）计算： $\rho^{*}{}_{Ca}^{Cr} = 6.1$

表5　Mn 液中 Al 对 Ca 活度相互作用系数（1350℃）

	一级系数	二级系数
	$\ln\gamma_i^j = bN_j$	$\ln\gamma_i^j = bN_j + cN_j^2$
采用对数关系	$\ln\gamma_{Ca}^{Al} = -7.8380N_{Al}$ ($r = 0.99$) $\varepsilon^{*}{}_{Ca}^{Al} = -7.8$	$\ln\gamma_{Ca}^{Al} = -6.2783N_{Al} - 5.1203N_{Al}^2$ ($r = 0.997$) $\varepsilon^{*}{}_{Ca}^{Al} = -6.3$ $\rho^{*}{}_{Ca}^{Al} = -5.1$
	$N_i - N_i' = mN_j$	$N_i - N_i' = mN_j + lN_j^2$
直接采用溶解度值	$N_{Ca} - N_{Ca}' = 0.07806N_{Al}$ ($r = 0.94$) $\varepsilon^{*}{}_{Ca}^{Al} = -\dfrac{m}{N_{Ca}'} = -38.0$	$N_{Ca} - N_{Ca}' = -0.02352N_{Al} + 0.33347N_{Al}^2$ ($r = 0.94$) $\varepsilon^{*}{}_{Ca}^{Al} = -\dfrac{m}{N_{Ca}'} = 11.4$ $\rho^{*}{}_{Ca}^{Al} = -\dfrac{l}{N_{Ca}'} + \dfrac{1}{2}\left(\dfrac{m}{N_{Ca}'}\right)^2 = -96.8$

从表3、表4可以看出，对于 Mn-Ca-Cr 系，无论一级系数还是二级系数，浓度采

用摩尔分数还是重量百分数计算，最后得到的 ε_i^{*j} 和 ρ_i^{*j} 值相同。但是用对数关系法（Chipman 法）得到的 ε_i^{*j}（或 ρ_i^{*j}）和用溶解度值直接法（Schenck 法）的 ε_i^{*j}（或 ρ_i^{*j}）略有不同，但还是属于同数量级。然而对于 Mn-Ca-Al 系，从表 5 可以看出，两种方法的计算结果大不相同。关于二级系数，甚至出现正负号相反的情况。

2.7 讨论

对数关系法和直接用溶解度法计算产生差异的原因如下。

2.7.1 关于一级系数

$$\ln \gamma_i^j = \ln \frac{\gamma_i}{\gamma_i'} = \ln \frac{N_i'}{N_i} = \varepsilon_i^j N_j$$

$$\varepsilon_i^j = -\frac{m}{N_i'}$$

$$\ln \frac{N_i'}{N_i} = -m \frac{N_j}{N_i'}$$

由于：
$$N_i - N_i' = mN_j$$

所以：
$$\ln \frac{N_i'}{N_i} = -\frac{N_i - N_i'}{N_j} \times \frac{N_j}{N_i'} = \frac{N_i' - N_i}{N_i'}$$

$$\ln \frac{N_i}{N_i'} = \frac{N_i - N_i'}{N_i'} = \frac{N_i}{N_i'} - 1 \tag{14}$$

利用自然对数的泰勒级数表达式：

$$\ln x = (x-1) - \frac{(x-1)^2}{2} + \frac{(x-1)^3}{3} - \cdots \quad (0 < x \leqslant 2)$$

$$\left(\frac{N_i}{N_i'} - 1\right) - \frac{1}{2}\left(\frac{N_i}{N_i'} - 1\right)^2 + \frac{1}{3}\left(\frac{N_i}{N_i'} - 1\right)^3 - \cdots = \frac{N_i}{N_i'} - 1$$

条件：
$$0 < \frac{N_{Ca}}{N_{Ca}'} \leqslant 2 \tag{15}$$

2.7.2 关于二级系数

$$N_i - N_i' = mN_j + lN_j^2$$

$$N_i = N_i'\left(1 + \frac{m}{N_i'}N_j + \frac{l}{N_i'}N_j^2\right)$$

$$\ln N_i = \ln N_i' + \ln\left(1 + \frac{m}{N_i'}N_j + \frac{l}{N_i'}N_j^2\right)$$

利用泰勒级数式：

$$\ln(1+y) = y - \frac{y^2}{2} + \cdots \quad (-1 < y \leqslant 1)$$

令：

$$y = \left(\frac{m}{N'_i}N_j + \frac{l}{N'_i}N_j^2\right)$$

则可得：

$$\ln N_i = \ln N'_i + \frac{m}{N'_i}N_j + \frac{l}{N'_i}N_j^2 - \frac{1}{2}\left(\frac{m}{N'_i}N_j + \frac{l}{N'_i}N_j^2\right)^2$$

$$\ln N_i = \ln N'_i + \frac{m}{N'_i}N_j + \left[\frac{l}{N'_i} - \frac{1}{2}\left(\frac{m}{N'_i}\right)^2\right]N_j^2$$

$$\ln \gamma_i = \ln \gamma'_i - \frac{m}{N'_i}N_j - \left[\frac{l}{N'_i} - \frac{1}{2}\left(\frac{m}{N'_i}\right)^2\right]N_j^2$$

$$\ln \gamma_i^j = -\frac{m}{N'_i}N_j - \left[\frac{l}{N'_i} - \frac{1}{2}\left(\frac{m}{N'_i}\right)^2\right]N_j^2$$

$$\ln \gamma_i^j = \varepsilon^{*j}_i N_j + \rho^{*j}_i N_j^2$$

和 $N_i - N'_i = mN_j + lN_j^2$ 对照计算 ε^{*j}_i 和 ρ^{*j}_i。条件为：

$$-1 < \left(\frac{m}{N'_{Ca}}N_j + \frac{l}{N'_{Ca}}N_j^2\right) \leqslant 1 \tag{16}$$

可见表 2 中前四行数据符合式（16）的条件。对此数据重新计算，结果见表 6。由表 6 可见，采用对数关系和直接用溶解度值所计算的结果是不同的，但属于相同的数量级。

表 6 Mn 液中 Al 对 Ca 活度相互作用系数（1350℃）（只用表 2 中前四行数据）

	一级系数	二级系数
	$\ln \gamma_i^j = bN_j$	$\ln \gamma_i^j = bN_j + cN_j^2$
采用对数关系	$\ln \gamma_{Ca}^{Al} = -703173 N_{Al}$ $(r = 0.97)$ $\varepsilon^{*Al}_{Ca} = -7.0$	$\ln \gamma_{Ca}^{Al} = -12.1027 N_{Al} + 63.1589 N_{Al}^2$ $(r = 1)$ $\varepsilon^{*Al}_{Ca} = -12.1$ $\rho^{*Al}_{Ca} = 63.2$
	$N_i - N'_i = mN_j$	$N_i - N'_i = mN_j + lN_j^2$
直接采用溶解度值	$N_{Ca} - N'_{Ca} = 0.01918 N_{Al}$ $(r = 0.98)$ $\varepsilon^{*Al}_{Ca} = -\frac{m}{N'_{Ca}} = -9.3$	$N_{Ca} - N'_{Ca} = 0.029424 N_{Al} - 0.127547 N_{Al}^2$ $(r = 0.999)$ $\varepsilon^{*Al}_{Ca} = -\frac{m}{N'_{Ca}} = -14.3$ $\rho^{*Al}_{Ca} = \frac{l}{N'_{Ca}} + \frac{1}{2}\left(\frac{m}{N'_{Ca}}\right)^2 = 40.4$
备注	数值符合 $0 < \frac{N_{Ca}}{N'_{Ca}} < 2$	数值符合 $-1 < \left(\frac{m}{N'_{Ca}}N_{Al} + \frac{l}{N'_{Ca}}N_{Al}^2\right) < 1$

2.8 小结

（1）选用两种活度标准状态的任一种，所求得的相互作用系数是一致的。

（2）直接用溶解度值计算，所得相互作用系数值不相同，但属于同一数量级。

（3）溶解度实验测定的数据，建议采用对数关系法计算活度相互作用系数。

3 化学平衡法求活度相互作用系数

化学平衡法的典型例子：

Fe-C-j 系的实验[1]

$$[C] + CO_2 = 2CO$$

Fe-S-j 系的实验[13]

$$[S] + H_2 = H_2S$$

实验方法和数据见文献 [1，13]。以 $[C] + CO_2 = 2CO$ 为例，求一级系数。

Fe-C 系：

$$K = \frac{p_{CO}^2}{p_{CO_2}} \times \frac{1}{f_C [\%C]'}$$

令 $K' = $ 平衡值，并有：

$$K' = \frac{p_{CO}^2}{p_{CO_2}} \times \frac{1}{[\%C]'}$$

作 K' 值和 $[\%C]'$ 值的对应线，外插到 $[\%C]' = 0$，该处的 $K' = K$。应用该值求出 f_C，然后按照 $\log f_C$ 和 $[\%C]'$ 的关系求 e_C^C。

Fe-C-j 系：

$$K = \left(\frac{p_{CO}^2}{p_{CO_2}}\right)' \times \frac{1}{f_C' [\%C]'} = \frac{p_{CO}^2}{p_{CO_2}} \times \frac{1}{f_C [\%C]}$$

$$f_C^j = \frac{f_C}{f_C'} = \frac{\dfrac{p_{CO}^2}{p_{CO_2}} \dfrac{1}{[\%C]}}{\left(\dfrac{p_{CO}^2}{p_{CO_2}}\right)' \dfrac{1}{[\%C]'}}$$

平衡实验有两种作法：

（1）$\dfrac{p_{CO}^2}{p_{CO_2}}$ 值变化而 $[\%C]$ 值不变，即同浓度法。$[\%C]' = [\%C]$ 时有：

$$f_C^j = \frac{\dfrac{p_{CO}^2}{p_{CO_2}}}{\left(\dfrac{p_{CO}^2}{p_{CO_2}}\right)'} = \frac{K_{(三元系)}}{K_{(二元系)}}$$

(2) $\dfrac{p_{CO}^2}{p_{CO_2}}$ 值不变而 [%C] 变化，此即同活度法。

$$f_C^j = \frac{f_C}{f_C'} = \frac{[\%C]'}{[\%C]}$$

$$f_C'[\%C]' = f_C[\%C] = a_C = \text{const}$$

作 f_C^j 和 [%j] 的对应线，以求 e_C^j 或 e^{*j}_C。

化学平衡法研究，用同浓度法较多，用同活度法须注意的要点（参见 Schenck 等对 Fe-C-Cr 系的研究[14]）：[C] + CO_2 === 2CO 的平衡常数 K，$\log K$ = 2.762，K = 578；饱和碳的活度 $a_{C(\%)}$ 按 C_{gr} = [C]%、ΔG^o = 5400 - 10.1T(cal/mol) 计算。将结果作对应线 $-\ln N_C = a + bN_{Cr}$，曲线斜率 $-b$ 为 ε^{*Cr}_C，截距 a 为 $-\ln N_C'$，该曲线即 $\ln N_C = \ln N_C' - \varepsilon^{*Cr}_C N_{Cr}$，回归计算可求得 N_C' 和 ε^{*Cr}_C。由于该研究未作二元系实验，利用文献数据，按照转换公式 $\varepsilon^{*Cr}_C = \varepsilon^{Cr}_C/(1 + \varepsilon^C_C N_C')$ 计算，可求得 ε^C_C = 11.5（转换公式见后）。

瑞典学者[15]应用的测求 e_i^j 的方法：Fe-M-X 系：

$$[M] + [X] \Longrightarrow MX_{(s)}$$

$$K = \frac{1}{f_M[\%M]f_X[\%X]}$$

$$-\log K = \log[\%M] + e_M^M[\%M] + e_M^X[\%X] + e_X^X[\%X] + e_X^M[\%M] + \log[\%X]$$

由于：

$$e_X^M \approx \frac{M_X}{M_M}e_M^M$$

$$-\log K = \log[\%M] + \log[\%X] + e_M^M[\%M] + e_X^X[\%X] + e_M^X\left([\%X] + \frac{M_X}{M_M}[\%M]\right)$$

可得到：

$$\log K - e_M^X\left([\%X] + \frac{M_X}{M_M}[\%M]\right)$$

$$= -(\log[\%M] + \log[\%X] + e_M^M[\%M] + e_X^X[\%X]) \quad (17)$$

应用式（17）作回归计算，可以求出 e_M^X 和 $\log K$。式中，M_X 和 M_M 分别代表 X 和 M 的分子量。本法实质上是同浓度法，原因是：[%M] = [%M]'，[%X] =

[%X]′。

下面再讨论二级系数的求法。

（1）对于 Fe-i 二元系：

$$\log f_i' = e_i^i [\%i] + r_i^i [\%i]^2 \tag{18}$$

按以下直线方程进行回归，可求出 e_i^i 和 r_i^i：

$$[\%i]^{-1} \log f_i' = e_i^i + r_i^i [\%i] \tag{19}$$

（2）对于 Fe-i-j 三元系：同浓度法可利用 C. Wagner 的公式：

$$\log f_i = \varphi([\%i], [\%j])$$

$$\log f_i = e_i^i [\%i] + r_i^i [\%i]^2 + e_i^j [\%j] + r_i^j [\%j]^2 + r_i^{i,j} [\%i][\%j]$$

$$\frac{\log f_i}{[\%j]} = \frac{\log f_i - e_i^i [\%i] - r_i^i [\%i]^2}{[\%j]} - r_i^{i,j} [\%i] = e_i^j + r_i^j [\%j] \tag{20}$$

M. G. Frohberg[16,17] 用迭代法，按线性公式：

$$y = a + bx$$

截距 a 为 e_i^j，斜率 b 为 r_i^j。选择一系列的 $r_i^{i,j}$ 值，求得一系列的线性方程，解出一系列的 a 和 b。再算出其误差。

斜率的误差为：

$$S_b = \frac{1}{\sum(x-\bar{x})^2} \sqrt{\frac{\sum(x-\bar{x})^2 \sum(y-\bar{y})^2 - [\sum(x-\bar{x})(y-\bar{y})]^2}{n-2}}$$

截距的误差为：

$$S_a = S_b \sqrt{\frac{\sum x^2}{n}}$$

计算 $S_y = \sqrt{S_a^2 + S_b^2}$，作 S_y 对 r_i^j 的对应曲线图，该曲线的最低值表示误差最小。于是以此相应的值代入，求得最后的 e_i^j 和 r_i^j。计算方法参见文献 [18]。

同活度法的 $\log f_i$ 函数式中没有 $r_i^{i,j}$ 项，而且 [%i] 只受 [%j] 量的限制，二元系时 [%i] 量是一常数。所以函数式 $\log f_i = f([\%i], [\%j])$ 可写为：

$$\log f_i = e_i^i [\%i] + r_i^i [\%i]^2 + e_i^{*j} [\%j] + r_i^{*j} [\%j]^2$$

$$\log f_i = \log f_i' + e_i^{*j} [\%j] + r_i^{*j} [\%j]^2$$

所以：

$$\log \frac{f_i}{f_i'} = \log \frac{[\%i]'}{[\%i]} = e_i^{*j} [\%j] + r_i^{*j} [\%j]^2$$

$$\log f_i^j = e_i^{*j} [\%j] + r_i^{*j} [\%j]^2 \tag{21}$$

对式（21）可直接回归，得出 e_i^{*j} 和 r_i^{*j}；或按照线性回归方式，求得误差小的值。

4 分配平衡法

分配平衡法应用图 1 所示的银浴进行平衡实验。银浴中 Ag、Fe 互不相溶，组元可溶于银液和铁液。但两者含 j 量不同。平衡时：

$$a_i^{Ag} = a_i^{Fe j} = a_i^{Fe}$$

式中，i 表示溶质；上角标 Ag、Fe 表示溶剂；j 表示第三元素，其浓度可变。

往 Ag 液中加入不同数量的 i 来调节 a_i^{Ag} 值：

$$a_i = \gamma_i N_i = \gamma_i' N_i'$$

$$\gamma_i^j = \frac{\gamma_i}{\gamma_i'} = \frac{N_i'}{N_i}$$

$$\ln \gamma_i^{*j} = \varepsilon_i^{*j} N_j$$

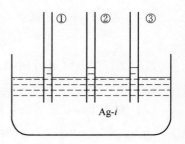

图 1　银浴示意图
(①中为 Fe-i 二元系；
②、③中为 Fe-i-j 三元系)

分配平衡法实质上是同活度法。

同样，作 $-\ln N_i$ 和 N_j 的对应值曲线图，$-\ln N_i = -\ln N_i' + \varepsilon_i^{*j} N_j$，由斜率求 ε_i^{*j}；由截距求 N_i'，即 $N_j = 0$ 时的 N_i。

S. Ban-ya[19]、D. Shroeder[20] 分别用银浴法测量 Fe-Si-C 系的分配平衡，温度 1450℃。

实验进行了 13 次，平均值：$\varepsilon_{Si}^{C} = 15.0$，1420℃，($\varepsilon_{Si}^{Si} = 13.0$)。

注意：每一个同活度值有一个 ε_{Si}^{*C}（或 ε_{Si}^{C}），实验必须作多次。

5　ε_i^{*j} 对 ε_i^j 或 e_i^{*j} 对 e_i^j 的相互转换

5.1　转换的必要性

C. Wagner 的 ε 式适用于同浓度法和稀溶液（二次项忽略）：

$$\ln \gamma_i = \ln \gamma_i^0 + \sum_{j=2}^{m} \frac{\partial \ln \gamma_i}{\partial N_j} N_j + \sum_{j=2}^{m} \frac{\partial^2 \ln \gamma_i}{\partial N_j^2} N_j^2 + \sum_{j,k=2}^{m} \frac{\partial^2 \ln \gamma_i}{\partial N_j \partial N_k} N_j N_k + \cdots \quad (22)$$

基于 Wagner 公式求出的活度和相互作用系数是同浓度法。已总结整理的相互作用系数表[21]，都是同浓度法所求的值。

由同活度法求出的活度相互作用系数应转换为同浓度法的活度相互作用系数。对于三元系 M-i-j 有：

$$\ln \gamma_i = \ln \gamma_i^0 + \varepsilon_i^i N_i + \rho_i^i N_i^2 + \varepsilon_i^j N_j + \rho_i^j N_j^2 + \rho_i^{i,j} N_i N_j + \cdots \quad (23)$$

或　$\log f_i = e_i^i [\%i] + r_i^i [\%i]^2 + e_i^j [\%j] + r_i^j [\%j]^2 + r_i^{i,j} [\%i][\%j] + \cdots$

$$(24)$$

式中，N_i 或 $[\%i]$ 用的是 三元系中 i 的浓度，似乎不合理，应当改写为 $\varepsilon_i^i N_i'$ 或

$e_i^i[\%i]'$；但由于 $N_i' = N_i$ 或 $[\%i]' = [\%i]$，所以可通用。这是 Wagner 的 ε 式是基于同浓度的原因。

5.2 转换公式

Fuwa 和 Chipman[1]首先提出相互转换的公式（对 Fe-C-j 系）：

$$\left(\frac{\partial \log f_C}{\partial c_j}\right)_{c_C} = -\left[1 + \left(\frac{\partial \log f_C}{\partial \log c_C}\right)_{c_j \to 0}\right]\left(\frac{\partial \log c_C}{\partial c_j}\right)_{a_C} \quad (25)$$

改用本文的符号：

$$\left(\frac{\partial \log f_C}{\partial [\%j]}\right)_{[\%C]} = -\left[1 + \left(\frac{\partial \log f_C}{\partial \log[\%C]}\right)_{[\%j] \to 0}\right]\left(\frac{\partial \log[\%C]}{\partial[\%j]}\right)_{a_C}$$

上式可改写为：

$$\left(\frac{\partial \log f_C}{\partial[\%j]}\right)_{[\%C]} = \left[1 + \left(\frac{\partial \log f_C}{\partial[\%C]}\right) \times 2.30[\%C]\right]\left(\frac{\partial \log f_C}{\partial[\%j]}\right)_{a_C}$$

所以：
$$e_C^j = e^{*j}_C(1 + 2.30\, e_C^C[\%C]) \quad (26)$$

这是转换公式。式（26）中的 $[\%C]$ 是 $[\%j] = 0$ 时 C 的浓度，所以最好写为：

$$e_C^j = e^{*j}_C(1 + 2.30\, e_C^C[\%C]') \quad (27)$$

Schenek 和 Frohberg 等人[22]用大体同上方法证明：

$$\ln\gamma_C = \varphi(N_C, N_j)$$

或
$$\ln\gamma_C = \varphi_1(\ln N_C, N_j)$$

$$d\ln\gamma_C = \left(\frac{\partial \ln\gamma_C}{\partial N_C}\right)_{N_j} dN_C + \left(\frac{\partial \ln\gamma_C}{\partial N_j}\right)_{N_C} dN_j$$

$$\left(\frac{\partial \ln\gamma_C}{\partial N_j}\right)_{a_C} = \left(\frac{\partial \ln\gamma_C}{\partial N_C}\right)_{N_j}\left(\frac{\partial N_C}{\partial N_j}\right)_{a_C} + \left(\frac{\partial \ln\gamma_C}{\partial N_j}\right)_{N_C}$$

$$= \left(\frac{\partial \ln\gamma_C}{\partial N_C}\right)_{N_j}\left(\frac{\partial \ln N_C}{\partial N_j}\right)_{a_C} N_C + \left(\frac{\partial \ln\gamma_C}{\partial N_j}\right)_{N_C}$$

$$\varepsilon^{*j}_C = \varepsilon_C^C(-\varepsilon^{*j}_C)N_C + \left(\frac{\partial \ln\gamma_C}{\partial N_j}\right)_{N_C}$$

$$\varepsilon^{*j}_C = \varepsilon_C^C(-\varepsilon^{*j}_C)N_C + \varepsilon_C^j$$

所以：
$$\varepsilon_C^j = \varepsilon^{*j}_C(1 + \varepsilon_C^C N_C) \quad (28)$$

Schenck 和 Frohberg[4]又用另一法推导式（28），并强调指出式中 N_C 实际上是 $N_j \to 0$ 时的 N_C'，所以式（28）应写为：

$$\varepsilon_C^j = \varepsilon^{*j}_C(1 + \varepsilon_C^C N_C') \quad (29)$$

式（29）和式（27）对应。

Lupis[9]推导了二级系数的转换式：

$$\varepsilon^{*j}_C = \frac{\varepsilon^j_C + \rho^{C,j}_C N'_C}{1 + \varepsilon^C_C N'_C + 2\rho^C_C (N'_C)^2} \tag{30}$$

式（30）中：
$$\rho^C_C = -\frac{1}{2}\varepsilon^C_C \tag{31}$$

$$\rho^{*j}_C = \frac{\rho^j_C - \varepsilon^{*j}_C \rho^{C,j}_C N'_C + \frac{1}{2}(\varepsilon^j_C)^2 [4\rho^C_C (N'_C)^2 + \varepsilon^C_C N'_C]}{1 + \varepsilon^C_C N'_C + 2\rho^C_C (N'_C)^2} \tag{32}$$

5.3 可以不用转换的情况

对于 Fe-H-j 系、Fe-N-j 系，由于 [%H]′ 或 [%N]′ 很小（或 ε^i_i 及 e^i_i 很小），式（27）中相应值可忽略不计，括号中的值近于 1，因而 $e^j_H = e^{*j}_H$，$e^j_N = e^{*j}_N$。对于 Fe-Ca-j 系、Mn-Ca-j 系，由于 Ca 在铁液或锰液中溶解度很小，也可不进行同浓度法对同活度法的转换。

5.4 关于ε^i_i和γ^0_i（自相互作用系数ε^i_i、1%浓度溶液中溶质按拉乌尔定律计算的活度系数γ^0_i）

由于 Wagner 的 ε 公式中活度相互作用系数适用同浓度法，由同活度法求出的作用系数 ε^{*j}_i 和 ρ^{*j}_i 必须转换为 ε^j_i 和 ρ^j_i 才可能估计出 γ^0_i 和 ε^i_i。

计算实例：

应用表 1 所示 Mn-Ca-Cr 系的实验数据，已计算出（见表 4）：
$$\varepsilon^{*Cr}_{Ca} = 2.2; \quad \rho^{*Cr}_{Ca} = 12.2$$

利用式（31）的关系，将有关数据代入式（30），得：
$$\varepsilon^{Cr}_{Ca} = 2.2 + 0.004512\varepsilon^{Ca}_{Ca} - 0.002055\rho^{Ca,Cr}_{Ca} \tag{33}$$

同样，将有关数据代入式（32），得：
$$\rho^{Cr}_{Ca} = 12.2 + 0.004521\rho^{Ca,Cr}_{Ca} - 0.02007\varepsilon^{Ca}_{Ca} \tag{34}$$

代入 Wagner 的 ε 公式：
$$\ln\gamma_{Ca} = -\ln N_{Ca} = \ln\gamma^0_{Ca} + \varepsilon^{Ca}_{Ca}N_{Ca} + \rho^{Ca}_{Ca}N^2_{Ca} + \varepsilon^{Cr}_{Ca}N_{Cr} + \rho^{Cr}_{Ca}N^2_{Cr} + \rho^{Ca,Cr}_{Ca}N_{Ca}N_{Cr}$$

将式（33）和式（34）代入，得：
$$-\ln N_{Ca} - 2.2N_{Cr} - 12.2N^2_{Cr} - \rho^{Ca,Cr}_{Ca}(N_{Ca}N_{Cr} - 0.002055N_{Cr} + 0.004521N^2_{Cr})$$
$$= \ln\gamma^0_{Ca} + \varepsilon^{Ca}_{Ca}(N_{Ca} - 0.5N^2_{Ca} + 0.004512N_{Cr} + 0.02007N^2_{Cr}) \tag{35}$$

式（35）也可以改写为：
$$\frac{-\ln N_{Ca} - 2.2N_{Cr} - 12.2N^2_{Cr} - \rho^{Ca,Cr}_{Ca}(N_{Ca}N_{Cr} - 0.002055N_{Cr} + 0.004521N^2_{Cr})}{N_{Ca} - 0.5N^2_{Ca} + 0.004512N_{Cr} + 0.02007N^2_{Cr}}$$
$$= \varepsilon^{Ca}_{Ca} + \frac{\ln\gamma^0_{Ca}}{N_{Ca} - 0.5N^2_{Ca} + 0.004512N_{Cr} + 0.02007N^2_{Cr}} \tag{36}$$

进行回归运算,用式(35)时的相关系数一般不理想;而用式(36)回归时,所得相关系数均接近于1。式(36)中的 $\rho_{Ca}^{Ca,Cr}$ 按 0 和 ±100 的区间进行多次计算,结果列于表7。可以看出,将 $\rho_{Ca}^{Ca,Cr}$ 认为是零所得结果是可行的。因此,计算结果为: $\varepsilon_{Ca}^{Ca} = -430$,$\ln\gamma_{Ca}^0 = 7.09$。所以:

$$\ln\gamma_{Ca} = 7.09 - 430N_{Ca} + 215N_{Ca}^2 + 0.26N_{Cr} + 3.57N_{Cr}^2$$

此式中 Ca 的自相互作用过小(负的绝对值太大)。发表在 Steel Res. 1989 年 No. 10 的文章用 $\varepsilon_{Ca}^{Ca} = -35.06$,是假定相互作用系数差别不大,将 j = Cr, Si, Ni 三相实验联合计算而得出此结果[24]。如果分别计算,则有所不同。表8列出了联合计算和分别计算的结果。

鉴于化学分析准确度不太高,$\ln\gamma_i^0$ 及 ε_i^j 以用专用实验测定为宜。

表7 $\ln\gamma_i^0$ 及 ε_i^i 的估算

项 目	式(36)					式(35)
$\rho_{Ca}^{Ca,Cr}$	0	10	100	-10	-100	0
ε_{Ca}^{Ca}	-430.2	-429.9	-428.8	-430.3	-431.6	-434.0
lnCa	7.09	7.09	7.09	7.09	7.09	7.10
r	0.997	0.997	0.997	0.997	0.997	0.83

表8 $\ln\gamma_i^0$ 及 ε_i^j 的计算(不加转换)

项 目	合并计算	单 独 计 算		
	Si, Cr, Ni	Si	Cr	Ni
ε_{Ca}^{Ca}	-35.06	-46.06	-122	-9.90
$\ln\gamma_{Ca}^0$	6.26	6.30	6.39	6.20
r	0.9996	0.999	0.999	0.999

6 结论

(1)溶解度法求活度相互作用系数,利用对数关系和直接采用溶解度值计算,所得结果不完全一致,虽然都在同一数量级。建议利用对数关系的计算法作为准绳。

(2)同活度的溶解度法计算出的相互作用系数值只有一个(因为是饱和溶液),但同活度的分配平衡法和化学平衡法,随着同活度值的不同,将有不同的相互作用系数值 ε_i^j。

(3)由同活度法得出的相互作用系数,必须转换为同浓度法的相互作用系数,才可以代入 Wagner 的 ε 关系式以计算 $\ln\gamma$。有的论文忽视这一点,未进行转换,如冀春霖等人[23,25]的文章。

(4) 银浴分配平衡法的同活度作用系数测定,必须进行多次实验,所得 ε_i^{*j} 转换成 ε_i^j 后再取平均值。不能只作几次实验就出结论。国外研究有的作 13~16 次实验。

(5) Wagner 的 ε 关系式是基于同浓度法,因此不能用下式进行计算:

$$\ln\frac{\gamma_i}{\gamma_i'} = \varepsilon_i^i(N_i - N_i') + \rho_i^i(N_i^2 - N_i'^2) + \varepsilon_i^j N_j + \rho_i^j N_j^2 + \rho_i^{i,j} N_i N_j$$

(6) 同活度法的相应 $\ln\gamma_i^j$ 式无 $\rho_i^{*i,j}$ 项,原因是 $\ln\gamma_i^{*j} = \varphi(N_j)$;而同浓度法的 $\ln\gamma_i^j$ 则等于 $\varphi(N_i, N_j)$。

(7) 除非化学分析数据能达到相对很准确,二级作用系数的可靠性不大。

(8) $\ln\gamma_i^0$ 和 ε_i^j 最好通过专用实验专门测定。由 Wagner 的 ε 关系式估算,只有在化学分析数据准确可靠时,得出的结论才是可靠的。

参 考 文 献

[1] Fuwa T, Chipman J. Trans. TMS-AIME, 1959, 215: 708.
[2] Ohtani M, Gokcen N A. Trans. TMS-AIME, 1960, 218: 533.
[3] Neumann F, Schenck H. Arch. Eisenh., 1959, 30: 477.
[4] Schenck H, Frohberg M G, Steinmetz E. Arch. Eisenh., 1963, 34: 37, 43.
[5] Weinstein M, Elliott J F. Trans. TMS-AIME, 1963, 227: 382.
[6] Pehlke R D, Elliott J F. Trans. TMS-AIME, 1960, 218: 1088.
[7] Kohler M, Engell H J, Janke D. Steel Research, 1985, 56: 419.
[8] 魏寿昆,等. Mn-Ca-j 系研究(北科大冶金物化实验室).
[9] Lupis C H P. Acta Metal, 1968, 16: 1365.
[10] Lupis C H P, Elliott J F. Acta Metal, 1966, 14: 529.
[11] Schenck H, Frohberg M G, Steinmetz E. Arch. Eisenh., 1960, 31: 671.
[12] Lupis C H P, Elliott J F. Trans. TMS-AIME, 1965, 233: 257.
[13] Ban-ya S, Chipman J. Trans. TMS-AIME, 1969, 245: 133.
[14] Schenck H, Steinmetz E, Rhee P C H. Arch. Eisenh., 1968, 39: 803.
[15] Gustafsson S, Mellberg P O. Scand. J. Met., 1980 (9): 111.
[16] Frohberg M G., Elliott J F, Hadrys H G. Arch. Eisenh., 1968, 39: 587.
[17] Hadrys H G, Frohberg M G, Elliott J F. Met. Trans., 1970 (1): 1867.
[18] Neville A D, Kennedy J B. Basic stastical methods for engineers, 1964: 178.
[19] Murakami S, Ban-ya S, Fuwa T. Tetsu-to-Hagane, 1970, 56: 32, 536.
[20] Schroeder D, Chipman J. Trans. TMS-AIME, 1964, 230: 1492.

[21] Sigworth G K, Elliott J F. Metal Science, 1974 (8): 298.
[22] Schenck H, Frohberg M G, Steinmetz E, Rutenberg B. Arch. Eisenh., 1962, 33: 223, 229.
[23] Ji Chunlin, Qi Guojun. Trans. Japan Inst. Metals, 1985, 26: 832.
[24] Wei S, Ni R, Ma Z, Cheng W. Steel Research, 1989, 60: 437.
[25] 冀春霖, 等. 钢铁, 1987, 22 (9): 16.

Thermodynamic Study of Interaction Coefficients in Multicomponent Metallic Solutions by the Solubility Method[*]

Wei Shoukun

Abstract: The calculation of activity interaction coefficients of elements in metal melt with the solubility data by logarithm formulation and the solubility equation was theoretically analyzed. Formulae for the calculation of the interaction coefficients of 2nd order for the solubility equation were derived. Calculation with the experimental data of Mn-Ca-Cr and Mn-Ca-Al systems by both ways of evaluation was made. It has been found that the values obtained from both ways of evaluation are quite different, although of the same order of magnitude. Causes for this inconsistency and the anomaly shown by the Mn-Ca-Al system were discwssed. Finally, the conversion between the interaction coefficients at constant activity and those at constant concentration in connection with the use of Wagner's formulism, was emphasized.

Activity interaction coefficients of elements in metal melt are highly indispensable in making thermodynamic analysis of the relevant metallurgical reactions. The method of studying the solubility of an element in metal melt in presence of a third element has been widely used to determine the interaction coefficients. So values of interaction coefficients for C in the systems Fe-C-j[1~7], Mn-C-j[8], Ni-C-j[8] and Co-C-j[8], for H in the system Fe-H-j[9,10], for N in the system Fe-N-j[10~18], and for Ca in the systems Fe-Ca-j[19] as well as Mn-Ca-j[20] have been reported. However, appreciable discrepancy and inconsistency between values of interaction coefficients from different sources have aroused much embarrassment among metallurgists in respect to the selection of the most reliable data for application. It is the purpose of this paper to make a study of the different ways of calculation of the interaction coefficients based on the solubility method. Causes for their discrepancy and suggestions as to eliminating or minimizing the sources of discrepancy are discussed.

---

[*] 原刊于《Steel Research》,1992,63:159~165.

1 Theoretical

1.1 1st-order interaction coefficients

The Fe-C-j system is taken as an example, in which the solubility of graphite in molten iron is measured without and with the addition of the 3rd element j (for list of symbols see Table 1). Since a_C is constant:

$$a_C = \gamma'_C N'_C = \gamma_C N_C = \text{const}$$

in which γ', γ represent the activity coefficient of C in the binary and ternary system, respectively, and N'_C, N_C represent the mole fraction of saturated C in the binary and ternary system, respectively. Accordingly:

$$\gamma^j_C = \frac{\gamma_C}{\gamma'_C} = \frac{N'_C}{N_C} \quad (1)$$

in which γ^j_C represents the effect the 3rd element j has upon the activity coefficient of C in the ternary system. The interaction coefficient of j upon C, calculated on the mole fraction basis with graphite as the standard state, is defined as:

$$\left(\frac{\partial \ln \gamma^j_C}{\partial N_j}\right)_{\substack{a_C \\ N_j \to 0}} \quad (2)$$

while the interaction coefficient, calculated on the wt.% basis with the infinitely dilute or 1 wt.% solution standard, is given by:

$$\left(\frac{\partial \lg f^j_C}{\partial \%j}\right)_{\substack{a_C(\%) \\ \%j \to 0}} \quad (3)$$

Expression (2) is designated by Ohtani and Gokcen[2] as λ^j_C, by Schenck and his co-workers[21] as ω^j_C, and by Lupis[22] as $\overset{*}{\varepsilon}^j_C$, while equation (3) is designated by Schenck as σ^j_C, and by Lupis as $\overset{*}{e}^j_C$. In this paper Lupis' designation $\overset{*}{\varepsilon}^j_C$, $\overset{*}{e}^j_C$ (or $\overset{*}{\varepsilon}^j_i$, $\overset{*}{e}^j_i$ in general) is adopted.

Evidently, in contrast to the designation at constant activity, ε^j_C and e^j_C are used to denote the interaction coefficients at constant concentration:

$$\varepsilon^j_C = \left(\frac{\partial \ln \gamma^j_C}{\partial N_j}\right)_{\substack{N_C \\ N_j \to 0}} \quad (4)$$

Table 1 List of symbols

a_i	activity of solute i
$[\%i]'$ wt.	percentage of solute i in the binary system
$[\%i]$ wt.	percentage of solute i in the ternary or multiple system
f'_i	activity coefficient of i in the binary system (infinitely dilute or 1wt. % solution standard and wt. % basis)
f_i	activity coefficient of i in the ternary or multiple system (same solution standard and concentration basis as f'_i)
f_i^j	effect of the 3rd element j upon the activity coefficient of i in the ternary system (same solution standard and concentration basis as f_i)
e_i^i	self-activity interaction coefficient of solute i (infinitely dilute or 1wt. % solution standard and wt. % basis)
e_i^j	activity interaction coefficient of j upon i in the ternary system at constant concentration (same solution standard and concentration basis as e_i^i)
$\overset{*}{e}{}_i^j$	same as e_i^j but at constant activity
r_i^i	2nd order self-activity interaction coefficient of solute i (infinitely dilute or 1 wt. % solution standard and wt. % basis).
r_i^j	2nd order interaction coefficient of j upon i in the ternary system at constant concentration (same solution standard and concentration basis as r_i^i)
r_i^{ij}	2nd order cross interaction coefficient of i and j upon i at constant concentration (same solution standard and concentration basis as r_i^i)
$\overset{*}{r}{}_i^j$	same as r_i^j but at constant activity
N'_i	mole fraction of solute i in the binary system
N_i	mole fraction of solute i in the ternary or multiple system
γ'_i	activity coefficient of i in the binary system (pure substance standard and mole fraction basis)
γ_i	activity coefficient of i in the ternary or multiple system (pure substance standard and mole fraction basis)
γ_i^j	effect of the 3rd element j upon the activity coefficient of i in the ternary system (pure substance standard and mole fraction basis)
γ_i^o	Raoultian activity coefficient of solute i at infinitely dilute or 1 wt. % solution standard
ε_i^i	self-activity interaction coefficient of solute i (pure substance standard and mole fraction basis)
ε_i^j	activity interaction coefficient of j upon i in the ternary system at constant concentration (pure substance standard and mole fraction basis)
$\overset{*}{\varepsilon}{}_i^j$	same as ε_i^j but at constant activity
ρ_i^i	2nd order self-activity interaction coefficient of solute i (pure substance standard and mole fraction basis)
ρ_i^j	2nd order interaction coefficient of j upon i in the trenary system at constant concentration (pure substance standard and mole fraction basis)
$\overset{*}{\rho}{}_i^j$	same as ρ_i^j but at constant activity
M_l	molecular weight of the solvent
M_i	molecular weight of solute i
M_j	molecular weight of 3rd element j

$$e_i^j = \left(\frac{\partial \lg f_C^j}{\partial \%j}\right)_{\substack{\%C \\ \%j \to 0}} \tag{5}$$

Since γ_C' is independent of N_j, it can be proved that:

$$\frac{\partial \ln \gamma_C^j}{\partial N_j} = \frac{\partial \ln \gamma_C - \partial \ln \gamma_C'}{\partial N_j} = \frac{\partial \ln \gamma_C}{\partial N_j} \tag{6}$$

Since $a_C = \gamma_C N_C = $ constant, $d\ln\gamma_C = -d\ln N_C$. Accordingly:

$$\overset{*j}{\varepsilon}_C = \left(\frac{\partial \ln N_C'/N_C}{\partial N_j}\right)_{\substack{a_C \\ N_j \to 0}} = \left(\frac{\partial \ln N_C}{\partial N_j}\right)_{\substack{a_C \\ N_j \to 0}} \tag{7}$$

By plotting $\ln\gamma_C^j$ (or $\ln\gamma_C$) against N_j for lower concentration of j, a linear relationship is usually obtained. The slope determined graphically or mathematically by regression analysis is the value $\overset{*j}{\varepsilon}_C$:

$$\ln\gamma_C^j = \ln N_C'/N_C = \overset{*j}{\varepsilon}_C N_j \tag{8}$$

Similarly,

$$\lg f_C^j = \lg[\%C]' - \lg[\%C] = e_C^{*j}[\%j] \tag{9}$$

This method of evaluation is the usual aproach followed by Fuwa and Chipman[1] as well as Ohtani and Gokcen[2]. Various authors such as Sanbongi and Ohtani[23], Turkdogan[24] and Schenck with his co-workers[4] have established the fact that for the system Fe-C-j at graphite saturation the following equations hold true for lower concentration of j:

$$\Delta N_C = N_C - N_C' = mN_j \tag{10}$$
$$\Delta[\%C] = [\%C] - [\%C]' = m'[\%j] \tag{11}$$

in which m, m' represent the proportional constants for the mole fraction concentration and the wt. % concentration, respectively. From equation (10) or (11) interaction coefficient of j upon C can be derived. With combination of equations (1) and (10):

$$\gamma_C^j = \frac{N_C'}{N_C' + mN_j}$$

$$\frac{\partial \ln \gamma_C^j}{\partial N_j} = \frac{-\partial \ln\left(\frac{N_C' + mN_j}{N_C'}\right)}{\partial N_j} = -\frac{m}{N_C' + mN_j}$$

When N_j approaches zero:

$$\overset{*j}{\varepsilon}_C = \left(\frac{\partial \ln\gamma_C^j}{\partial N_j}\right)_{\substack{a_C \\ N_j \to 0}} = -\frac{m}{N_C'} \tag{12}$$

Equation (12) is originally derived by Schenck, Frohberg, Steinmetz and Rutenberg[5], but it can also be derived in another way as given in[21] as follows:

$$N_C = N_C' + mN_j$$

$$dN_C = m dN_j$$

$$d\ln N_C = \frac{m}{N_C} dN_j$$

$$d\ln N_C = \left(\frac{m}{N'_C + mN_j}\right) dN_j$$

$$\left(\frac{\partial \ln \gamma_C}{\partial N_j}\right)_{\substack{a_C \\ N_j \to 0}} = -\frac{m}{N'_C}$$

which is the same as equation (12).

Similar derivation for the interaction coefficients on wt. % basis is given as equation (13):

$$\overset{*}{e}{}_C^j = -\frac{m'}{2.30[\%C]'} \tag{13}$$

Schenck and his co-workers[3~6] have experimentally found values of m and m' for a number of elements, and reported values of $\overset{*}{\varepsilon}{}_C^j$ and $\overset{*}{e}{}_C^j$ (designated as ω_C^j and o_C^j in the original literature) calculated by equations (12) and (13). Systematic relations between $\overset{*}{\varepsilon}{}_C^j$ and the periodic table for the systems Fe-C-j, Mn-C-j, Ni-C-j, and Co-C-j have also been discussed[7].

1.2 Interaction coefficients of the 2nd order

As mentioned above, the straight line relationship of ΔN_C vs N_j holds true only for a limited range of concentration of the 3rd element j. For example, values of m' (for wt. % concentration) for a higher range of concentration, which are quite different from those for lower range of concentration, have been given for j = Si, P, Ni, Cr, Co, W, Ti, S, V, Cu, Al, Mo and Mn[25]. A parabolic concentration relationship has been suggested by Wang[26], so the following equation might be written:

$$\Delta N_C = N_C - N'_C = m N_j + l N_j^2 \tag{14}$$

in which m, l denote proportional constants.

$$\frac{dN_C}{dN_j} = m + 2l N_j \tag{15}$$

$$\frac{d\ln N_C}{dN_j} = \frac{m}{N_C} + \frac{2l N_j}{N_C} \tag{16}$$

$$\overset{*}{\varepsilon}{}_C^j = \left(\frac{\partial \ln \gamma_C}{\partial N_j}\right)_{\substack{a_C \\ N_j \to 0}} = -\frac{m}{N'_C} \tag{17}$$

By comparing equations (17) and (12), it is obvious that the same formula can be used for calculating $\overset{*}{\varepsilon}{}_C^j$ both of the parabolic concentration and of the linear concentration.

By differentiating equation (16) again with dN_j and simplifying with equation (15), the following equation is obtained:

$$\frac{d^2\ln N_C}{dN_j^2} = \frac{2l}{N_C} - (m + 2N_j l)\left(\frac{2N_j l}{N_C^2} + \frac{m}{N_C^2}\right)$$

$$\left(\frac{\partial^2 \ln \gamma_C}{\partial N_j^2}\right)_{\substack{a_C \\ N_j \to 0}} = -\frac{2}{N_C'}l + \left(\frac{m}{N_C'}\right)^2 \tag{18}$$

According to Lupis[22]:

$$\overset{*j}{\rho}_C = \frac{1}{2}\left(\frac{\partial^2 \ln \gamma_C}{\partial N_j^2}\right)_{\substack{a_C \\ N_j \to 0}}$$

$$\overset{*j}{\rho}_C = -\frac{l}{N_C'} + \frac{1}{2}\left(\frac{m}{N_C'}\right)^2 \tag{19}$$

Equations (17) and (19) have been derived by yet another approach by Fu and Zhai[27], but the present way of derivation is more direct.

Similarly, for the wt. % concentration:

$$\Delta[\%C] = [\%C] - [\%C]' = m'[\%j] + l'[\%j]^2$$

$$\overset{*j}{e}_C = -\frac{m}{2.30[\%C]'} \tag{20}$$

$$\overset{*j}{r}_C = -\frac{l'}{2.30[\%C]'} + \frac{1}{4.60}\left(\frac{m'}{[\%C]'}\right)^2 \tag{21}$$

Another way of evaluation of interaction coefficients of the 2nd order is to use the logarithm formulation by extending equations (8) or (9) with a 2nd order term as suggested by Lupis[22]:

$$\ln \gamma_C' = \ln \frac{N_C'}{N_C} = \overset{*j}{\varepsilon}_C N_j + \overset{*j}{\rho}_C N_j^2 \tag{22}$$

$$\lg f_C^j = \lg \frac{[\%C]'}{[\%C]} = \overset{*j}{e}_C [\%j] + \overset{*j}{r}_C [\%j]^2 \tag{23}$$

The conversion between $\overset{*j}{\varepsilon}_C$ and $\overset{*j}{e}_C$ is exactly the same as that between ε_C^j and e_C^j with the following formula[21,22,28]:

$$\overset{*j}{\varepsilon}_C = 230\frac{M_j}{M_l}\overset{*j}{e}_C + \frac{M_l - M_j}{M_l} \tag{24}$$

The conversion formula between $\overset{*j}{\rho}_C$ and $\overset{*j}{r}_C$ is given according to Lupis[22] as below:

$$\overset{*j}{\rho}_C = \frac{23000 M_j^2 \overset{*j}{r}_C + M_l(M_l - M_j)\overset{*j}{\varepsilon}_C}{[M_l - N_C'(M_l - M_C)]M_l} \times$$

$$\frac{-\frac{1}{2}M_l(M_l - M_C)(\overset{*j}{\varepsilon}_C)^2 N'_C - \frac{1}{2}(M_l - M_j)^2}{[M_l - N'_C(M_l - M_C)]M_l} \quad (25)$$

in which M_l, M_C, M_j represent the molecular weights of the solvent, the solute C and the 3rd element j, respectively.

1.3 Numerical calculation with the experimental data

Experimental data for the systems Mn-Ca-Cr and Mn-Ca-Al are taken from our laboratory[29], details of experimental procedure being given elsewhere[20]. Tables 2 and 3 give the effect of Cr and Al upon the solubility of Ca in liquid Mn at 1350℃, respectively. Table 4 gives the results of evaluation of $\overset{*}{\varepsilon}_{Ca}^{Cr}$ and $\overset{*}{e}_{Ca}^{Cr}$ by both the logarithm formulation (equations (8) and (9)) and the solubility equation (equations (10) and (11)). Table 5 gives the results of evalution $\overset{*}{\varepsilon}_{Ca}^{Cr}$ and $\overset{*}{\rho}_{Ca}^{Cr}$ as well as $\overset{*}{e}_{Ca}^{Cr}$ and $\overset{*}{r}_{Ca}^{Cr}$ by the same two ways of evaluation. It is evident that results of calculation with the mole fraction basis of concentration are identical with those with the wt.% basis of concentration. But it is to be emphasized that the result obtained by the logarithm formulation does differ from that obtained by the solubility equation, inspite of the fact that the two values are of the same order of magnitude. In view of correlation and minimizing the discrepancy between interaction coefficients from different sources, it might be highly preferable to adopt a standard way of evaluation from the same data of measurements.

Table 2 Effect of Cr upon the solubility of Ca in Mn at 1350℃

%Ca	N_{Ca}	%Cr	N_{Cr}
0.15	0.002055	0	0
0.15	0.002054	0.88	0.009288
0.14	0.001915	2.59	0.027310
0.12	0.001640	4.28	0.045091
0.12	0.001635	9.48	0.099582
0.098	0.001334	11.02	0.115668
0.084	0.001143	13.40	0.140468

Table 3 Effect of Al upon the solubility of Ca in Mn at 1350℃

%Ca	N_{Ca}	%Al	N_{Al}
0.15	0.002055	0	0
0.21	0.002829	1.6	0.03205
0.27	0.003561	3.7	0.072492
0.28	0.003656	4.7	0.091172

Continued Table 3

%Ca	N_{Ca}	%Al	N_{Al}
0.47	0.005906	8.6	0.160536
0.70	0.008609	10.8	0.197326
3.00	0.033752	20.0	0.334265
3.29	0.036303	22.2	0.363904

Table 4 1st order interaction coefficients of Cr upon Ca in Mn at constant activity at 1350℃

Logarithm formulation	Solubility equation
(a) mole fraction basis	
$\ln\gamma_i^j = bN_j$	$N_i - N_i' = mN_i$
$\ln\gamma_{Ca}^{Cr} = 3.6447 N_{Cr}$	$N_{Ca} - N_{Ca}' = -0.006011 N_{Cr}$
$(r = 0.95)$	$(r = 0.96)$
$\overset{*}{\varepsilon}_{Ca}^{Cr} = 3.6$	$\overset{*}{\varepsilon}_{Ca}^{Cr} = -\dfrac{m}{N_{Ca}'} = 2.9$
(b) wt. % basis	
$\lg f_i^j = b'[\%j]$	$[\%i] - [\%i]' = m'[\%j]$
$\lg f_{Ca}^{Cr} = 0.01638[\%Cr]$	$[\%Ca] - [\%Ca]' = -0.004551[\%Cr]$
$(r = 0.95)$	$(r = 0.96)$
$\overset{*}{\varepsilon}_{Ca}^{Cr} = 0.016$	$\overset{*}{\varepsilon}_{Ca}^{Cr} = -\dfrac{m'}{2.30[\%Ca]'} = 0.013$
by conversion with equation (24),	by conversion with equation (24),
$\overset{*}{\varepsilon}_{Ca}^{Cr} = 3.6$	$\overset{*}{\varepsilon}_{Ca}^{Cr} = 2.9$

Table 5 2nd order interaction coefficients of Cr upon Ca in Mn at constant activity at 1350℃

Logarithm formulation	Solubility equation
(a) mole fraction basis	
$\ln\gamma_i^j = bN_j + cN_j^2$	$N_i - N_i' = mN_i + lN_j^2$
$\ln\gamma_{Ca}^{Cr} = 2.1998 N_{Cr} + 12.2333 N_{Cr}^2$	$N_{Ca} - N_{Ca}' = -0.005349 N_{Cr} - 0.005607 N_{Cr}^2$
$(r = 0.96)$	$(r = 0.96)$
$\overset{*}{\varepsilon}_{Ca}^{Cr} = 2.2$	$\overset{*}{\varepsilon}_{Ca}^{Cr} = -\dfrac{m}{N_{Ca}'} = 2.6$
$\overset{*}{\rho}_{Ca}^{Cr} = 12.2$	$\overset{*}{\rho}_{Ca}^{Cr} = -\dfrac{l}{N_{Ca}'} + \dfrac{1}{2}\left(\dfrac{m}{N_{Ca}'}\right)^2 = 6.1$

Continued Table 5

Logarithm formulation	Solubility equation
(b) wt. % basis	
$\lg f_i^j = b'[\%j] + c'[\%j]^2$	$[\%i] - [\%i]' = m'[\%j] + l'[\%j]^2$
$\lg f_{Ca}^{Cr} = 0.09814[\%Cr] + 0.000582[\%Cr]^2$	$[\%Ca] - [\%Ca]' = -0.004016[\%Cr] - 0.000047[\%Cr]^2$
$(r = 0.96)$	$(r = 0.99)$
$e_{Ca}^{*Cr} = 0.0098$	$\varepsilon_{Ca}^{*Cr} = -\dfrac{m'}{2.30[\%Ca]'} = 0.012$
$r_{Ca}^{*Cr} = 0.00058$	$r_{Ca}^{*Cr} = -\dfrac{l'}{2.30[\%Ca]'} + \dfrac{1}{4.60}\left(\dfrac{m'}{[\%Ca]'}\right)^2 = 0.00029$
by conversion with equation (24) and (25),	by conversion with equation (24),
$\varepsilon_{Ca}^{*Cr} = 2.2$	$\varepsilon_{Ca}^{*Cr} = 2.6$
$\rho_{Ca}^{*Cr} = 12.1$	$\rho_{Ca}^{*Cr} = 6.1$

In comparison with Table 4 (a), regression with the equation $\ln\gamma_i = a + bN_j$ or $-\ln N_i = a + bN_j$ gives:

$$-\ln N_{Ca} = 6.1698 + 3.8132 N_{Cr}$$

Obviously, $\varepsilon_{Ca}^{*Cr} = 3.8$, and the intercept equal to $-\ln N'_{Ca}$, which means $N'_{Ca} = 0.002092$ or $[\%Ca]' = 0.1525$, which agrees with the experimental value of 0.15. Consequently, it can be concluded that calculation of ε_{Ca}^{*Cr} either with $\ln\gamma_{Ca}^{Cr}$ or with $\ln\gamma_{Ca}$ would give the same results within the limits of the experimental error.

From Table 6, which gives the interaction coefficients of Al upon Ca in Mn at 1350℃, it can be shown that for the 1st order interaction coefficients, great deviation exists between the value calculated by the logarithm formulation and that by the solubility equation, while for the 2nd order interaction coefficients the values calculated by the two different ways of evaluation differ not only in sign, but also in the order of magnitude appreciably.

Causes for this inconsistency might be explained by the following discussion.

Table 6 Interaction coefficients of Al upon Ca in Mn at constant activity at 1350℃

Logarithm formulation	Solubility equation
1st order coefficients	
$\ln \gamma_i^j = bN_j$	$N_i - N'_i = mN_j$
$\ln \gamma_{Ca}^{Al} = -7.8380 N_{Al}$	$N_{Ca} - N'_{Ca} = 0.07806 N_{Cr}$
$(r = 0.99)$	$(r = 0.94)$
$\varepsilon_{Ca}^{*Al} = -7.8$	$\varepsilon_{Ca}^{*Al} = -\dfrac{m}{N'_{Ca}} = -38.0$

Continued Table 6

Logarithm formulation	Solubility equation
2nd order coefficients	
$\ln r_i^j = bN_j + cN_j^2$	$N_i - N_i' = mN_j + lN_j^2$
$\ln \gamma_{Ca}^{Al} = -6.2783 N_{Al} - 5.1203 N_{Al}^2$	$N_{Ca} - N_{Ca}' = -0.02352 N_{Al} + 0.33347 N_{Al}^2$
$(r = 0.997)$	$(r = 0.94)$
$\overset{*}{\varepsilon}{}_{Ca}^{Al} = -6.3$	$\overset{*}{\varepsilon}{}_{Ca}^{Al} = -\dfrac{m}{N_{Ca}'} = 11.4$
$\overset{*}{\rho}{}_{Ca}^{Al} = -5.1$	$\overset{*}{\rho}{}_{Ca}^{Al} = -\dfrac{l}{N_{Ca}'} + \dfrac{1}{2}\left(\dfrac{m}{N_{Ca}'}\right)^2 = -96.8$

2 Discussion

To study the relation between the two ways of evaluation of the 1st order interaction coefficients, combine equation (8) with (12):

$$\ln \frac{N_i'}{N_i} = -\frac{m}{N_i'} N_j$$

Incorporate equation (10) into the above equation and rearrange:

$$\ln \frac{N_i}{N_i'} = \frac{N_j}{N_i'} - 1 \qquad (26)$$

In accordance with the series:

$$\ln x = (x - 1) - \frac{1}{2}(x - 1)^2 + \frac{1}{3}(x - 1)^3 - \cdots \qquad (0 < x \leq 2)$$

Equation (26) after expansion becomes:

$$\left(\frac{N_i}{N_i'} - 1\right) - \frac{1}{2}\left(\frac{N_i}{N_i'} - 1\right)^2 + \frac{1}{3}\left(\frac{N_i}{N_i'} - 1\right)^3 - \cdots = \frac{N_i}{N_i'} - 1 \qquad (27)$$

Both sides of equation (27) are identical on condition that (1) terms of higher orders are neglected, and (2) restriction for the series, namely for the present case equation (28), is fulfilled:

$$0 < \frac{N_{Ca}}{N_{Ca}'} \leq 2 \qquad (28)$$

Similarly, as regards the interaction coefficients of the 2nd order, rearrange equation (14):

$$N_i = N_i'\left(1 + \frac{m}{N_i'} N_j + \frac{l}{N_i'} N_j^2\right)$$

$$\ln N_i = \ln N'_i + \ln\left(1 + \frac{m}{N'_i}N_j + \frac{l}{N'_i}N_j^2\right) \tag{29}$$

In accordance with the series:

$$\ln(1+y) = y - \frac{1}{2}y^2 + \cdots \qquad (-1 < y \leq 1)$$

expand equation (29) and rearrange:

$$\ln N_i = \ln N'_i + \frac{m}{N'_i}N_j + \left[\frac{l}{N'_i} - \frac{1}{2}\left(\frac{m}{N'_i}\right)^2\right]N_j^2 + \cdots$$

or

$$\ln\gamma_i = \ln\gamma'_i - \frac{m}{N'_i}N_j + \left[-\frac{l}{N'_i} + \frac{1}{2}\left(\frac{m}{N'_i}\right)^2\right]N_j^2 + \cdots \tag{30}$$

By comparison with equations (17) and (19), equation (30) becomes:

$$\ln\gamma_i = \ln\gamma'_i + \overset{*}{\varepsilon}^j_i N_j + \overset{*}{\rho}^j_i N_j^2 + \cdots$$

or

$$\ln\gamma^j_i = \overset{*}{\varepsilon}^j_i N_j + \overset{*}{\rho}^j_i N_j^2 + \cdots$$

which is the same as equation (22) when terms of higher order are omitted. Hence, it is evident that equation (14) can be transformed into equation (22), or in other words, the two ways of evaluation of $\overset{*}{\varepsilon}^j_i$ and $\overset{*}{\rho}^j_i$ are identical, on condition that (1) terms of higher order are neglected, and (2) restriction for the series, namely in the present case equation (31) is fulfilled:

$$-1 < \left[\frac{m}{N'_{Ca}}N_j + \frac{l}{N'_{Ca}}N_j^2\right] \leq 1 \tag{31}$$

By re-examining Table 3, it can be shown that only the first four values fulfil the conditions given by equations (28) and (31). Recalculation with these four values as shown in Table 7 yields results of the same order of magnitude with no contradiction as to sign. The discrepancy between the two ways of evaluation might be ascribed to the omission of terms of higher order in the mathematical treatment.

The anomaly shown by the behaviour of higher values of Al in Mn might be caused by the formation of intermetallic compounds between Ca and Al. It can be inferred that elements which decrease the solubility of Ca in Mn and are manifested with a positive value of $\overset{*}{\varepsilon}^j_{Ca}$. Would not be able to form intermetallic compounds with Ca.

Table 7 Interaction coefficients of Al upon Ca in Mn at constant activity at 1350℃
(calculated with only the first four values shown in Table 3)

Logarithm formulation	solubility equation
1st order coefficients	
$\ln\gamma_i^j = bN_j$	$N_i - N_i' = mN_j$
$\ln\gamma_{Ca}^{Al} = -7.0317 N_{Al}$	$N_{Ca} - N_{Ca}' = 0.001918 N_{Cr}$
$(r = 0.97)$	$(r = 0.98)$
$\overset{*}{\varepsilon}_{Ca}^{Al} = -7.0$	$\overset{*}{\varepsilon}_{Ca}^{Al} = -\dfrac{m}{N_{Ca}'} = -9.3$
2nd order coefficients	
$\ln r_i^j = bN_j + cN_j^2$	$N_i - N_i' = mN_j + lN_j^2$
$\ln\gamma_{Ca}^{Al} = -12.1027 N_{Al} - 63.1589 N_{Al}^2$	$N_{Ca} - N_{Ca}' = 0.02942 N_{Al} - 0.12755 N_{Al}^2$
$(r = 1.0)$	$(r = 0.999)$
$\overset{*}{\varepsilon}_{Ca}^{Al} = -12.1$	$\overset{*}{\varepsilon}_{Ca}^{Al} = -\dfrac{m}{N_{Ca}'} = -14.3$
$\overset{*}{\rho}_{Ca}^{Al} = 63.2$	$\overset{*}{\rho}_{Ca}^{Al} = -\dfrac{l}{N_{Ca}'} + \dfrac{1}{2}\left(\dfrac{m}{N_{Ca}'}\right)^2 = 164.6$

3 Conversion of interaction coefficients at constant activity to those at constant concentration

In the Mn-Ca-j system, the equation:
$$\ln\gamma_{Ca} = \ln\gamma_{Ca}' + \overset{*}{\varepsilon}_{Ca}^{j} N_j + \overset{*}{\rho}_{Ca}^{j} N_j^2$$
holds true for saturated solutions of Ca in Mn at constant a_{Ca}. For unsaturated solutions with N_{Ca} from O to $N_{Ca(sat.)}$. Wagner's formulism should be used to express in γ_{Ca}.

In general, for a multiple system, Wagner's formulism[30] is written as follows:
$$\ln\gamma_i = \ln\gamma_i^o + \varepsilon_i^i N_i + \rho_i^i N_i^2 + \varepsilon_i^j N_j + \rho_i^j N_j^2 + \rho_i^{i,j} N_i N_j + \varepsilon_i^k N_k + \rho_i^k N_k^2 +$$
$$\rho_i^{i,k} N_i N_k + \rho_i^{j,k} N_j N_k + \cdots \qquad (32)$$

In this formulism, N_i is multiplied by the binary coefficients ε_i^i, ρ_i^i by the ternary cross coefficients ρ_i^{ij} and ρ_i^{ik} and so on, but actually in calculation only the value N_i from the multiple system is to be used for substitution. This can only be comprehensible when the calculations for the evaluation of ε_i^i, ρ_i^i, ε_i^j, ρ_i^j, ρ_i^{ij} etc. are all conducted at constant N_i. Hence, it is to be emphasized that Wagner's formulism is really applicable to dilute, unsaturated solution of multiple system only at constant concentration. Any interaction coefficients which are evaluated at constant activity should be converted into those at constant concentration before they can be introduced into the Wagner's formulism.

The conversion between the two kinds of interaction coefficient was first investigated by Fuwa and Chipman[1], then by Schenck and his co-workers[5], and lately by Lupis[22], who gave the following formulae:

$$\overset{*}{\varepsilon}{}_i^j = \frac{\varepsilon_i^j + \rho_i^{i,j} N_i'}{1 + \varepsilon_i^i N_i' + 2\rho_i^i N_i'^2} \tag{33}$$

$$\overset{*}{\rho}{}_i^j = \frac{\rho_i^i - \overset{*}{\varepsilon}{}_i^j \rho_i^{i,j} N_i' + \frac{1}{2}\overset{*}{\varepsilon}{}_i^{j2}(4\rho_i^i N_i'^2 + \varepsilon_i^i N_i')}{1 + \varepsilon_i^i N_i' + 2\rho_i^i N_i'^2} \tag{34}$$

In the above equations N_i' represents the mole fraction concentration of i at $N_j = 0$, i. e. of the binary system at constant activity. With neglect of terms of the 2nd order, equation (33) becomes:

$$\varepsilon_i^j = \overset{*}{\varepsilon}{}_i^j(1 + \varepsilon_i^i N_i')$$

which had been derived by Schenck and his associates[5].

ρ_i^i is connected with ε_i^i by the relation $\rho_i^i = -\frac{1}{2}\varepsilon_i^{i[22]}$. Hence, for elements with $\varepsilon_i^i = 0$, $\rho_i^i = 0$.

In the Fe-Ca-j system, ε_{Ca}^{Ca} is reported as being equal to $0^{[10]}$. Since as a metal Mn behaves chemically with a great similarity to Fe, it might be appropriate to predict ε_{Ca}^{Ca} in Mn being equal to 0. Hence, for the Mn-Ca-j system:

from equation (33) $\quad \varepsilon_{Ca}^j = \overset{*}{\varepsilon}{}_{Ca}^j - \rho_{Ca}^{Ca,j} N_{Ca}'$ \hfill (35)

from equation (34) $\quad \rho_{Ca}^j = \overset{*}{\rho}{}_{Ca}^j + \rho_{Ca}^{Ca,j} \overset{*}{\varepsilon}{}_{Ca}^j N_{Ca}'$ \hfill (36)

The Wagner's formulism for the Mn-Ca-j system is:

$$\ln\gamma_i = \ln\gamma_{Ca}^\circ + \varepsilon_{Ca}^{Ca} N_{Ca} + \rho_{Ca}^{Ca} N_{Ca}^2 + \varepsilon_{Ca}^j N_j + \rho_{Ca}^j N_j^2 + \rho_{Ca}^{Ca,j} N_{Ca} N_j$$

Substitute equations (35) and (36) into the formulism, note that the 2nd and 3rd terms on the left side of the equation are equal to 0, and rearrange:

$$\ln\gamma_{Ca} = \ln\gamma_{Ca}^\circ + \overset{*}{\varepsilon}{}_{Ca}^j N_j + \overset{*}{\rho}{}_{Ca}^j N_j^2 + \rho_{Ca}^{Ca,j} N_j(N_{Ca} - N_{Ca}') + N_{Ca}' N_j^2 \overset{*}{\varepsilon}{}_{Ca}^j \rho_{Ca}^{Ca,j}$$

In the above equation, $N_j(N_{Ca} - N_{Ca}')$ and $N_{Ca}' N_j^2$ are extremely small, so the last two terms on the left side of the equation could be neglected. The Wagner's formulism then takes the form:

$$\ln\gamma_{Ca} = \ln\gamma_{Ca}^\circ + \overset{*}{\varepsilon}{}_{Ca}^j N_j + \overset{*}{\rho}{}_{Ca}^j N_j^2 \tag{37}$$

which holds true for dilute, unsaturated solutions with concentration N_{Ca} ranging from 0 to saturation. It is to be mentioned that for the case in which ε_i^i is not equal to 0, a more complicated mathematical treatment would be needed for the conversion between the two kinds of interaction coefficients.

By comparison of equations (37) with (22), it can be seen that

$$\ln\gamma_{Ca}^\circ = -\ln N_{Ca}' = 6.187 \quad \text{at } 1350^\circ C$$

With the assumption that the solution of Ca in Mn is a regular solution:

$$Ca_{(1)} = [Ca]_{\%(Mn)}$$

$$\Delta G^\circ = RT \ln \gamma_{Ca}^\circ + RT \ln \frac{0.5494}{40.08}$$

$$\Delta G^\circ = 83500 - 35.66T (\text{J/mol})$$

This value holds true in the neighbourhood of 1350℃, and agrees very well with the value previously reported else where[20].

4 Concluding remarks

Two ways of evaluation of interaction coefficients of elements in metal melt with the solubility measurements are prevalent in literature, namely the logarithm formulation and the solubility equation. It has been shown by this investigation that with the same experimental data the two ways of evaluation yield values that are quite different, although of the same order of magnitude. To avoid unnecessary inconsistency between values from different sources, it would be recommended to adopt the logarithm formulation as the standard method for calculation, just because it follows directly from the definition of the interaction coefficients.

It is to be emphasized that Wagner's formulism is justified for application to dilute, unsaturated solutions at constant concentration. All the interaction coefficients found at constant activity should be converted into those at constant concentration before they can be introduced into the Wagner's formulism.

References

[1] Fuwa T, Chipman J. Trans. TMS-AIME, 1959, 215:708.
[2] Ohtani M; Gokcen N A. Trans. TMS-AIME, 1960, 218:533.
[3] Neumann F, Schenck H. Arch. Eisenhlittenwes, 1959, 30:477.
[4] Neumann F, Schenck H, Patterson W. Giesserei Tech-wiss. Beihefte Nr, 1959, 23:1217.
[5] Schenck H, Frohberg M G, Steinmetz E, Rutenberg B. Arch. Eisenhlittenwes, 1962, 33:223, 229.
[6] Schenck H, Gloz M, Steinmetz E. Arch. Eisenhlittenwes, 1970, 41:1.
[7] Schenck H, Frohberg M G, Steelmaking: The Chipman Conference, The M.I.T. Press; Cambridge, Mass., U.S.A., 1965:95.
[8] Schenck H, Frohberg M G, Steinmetz E. Arch. Eisenhlittenwes, 1963, 34:37, 43.
[9] Weinstein M Elliott J F. Trans. TMS-AIME, 1963, 227:382.
[10] Sigworth G K, Elliott J F. Metal Scie. 1974, 8:298.
[11] Pehlke R D, Elliott J F. Trans. TMS-AIME, 1960, 218:1088.
[12] Evans D B, Pehlke R D. Trans. TMS-AIME, 1964, 230:1651, 1657.

[13] Evans D B,Pehlke R D. Trans. TMS-AIME,1965,233:1620.
[14] Wada H,Pehlke R D. Met. Trans,1977(8B):443,675.
[15] Wada H,Pehlke R D. Met. Trans,1978(9B):441.
[16] Wada H,Pehlke R D. Met. Trans,1979(10B):409.
[17] Wada H,Pehlke R D. Met. Trans, 1980(11B):51.
[18] Wada H,Pehlke R D. Met. Trans, 1981(12B):333.
[19] Kohler M,Engeli H J,Janke D. Steel Res. 1985,56:419.
[20] Wei Shoukun, Ni Ruiming; Ma Zhongting, Cheng Wu. Steel res. 1989,60:437.
[21] Schenck H,Frohberg M G,Steinmetz E. Arch. Eisenhlittenwes,1960,31:671.
[22] Lupis C H P. Acta Metal, 1968,16:1365.
[23] Sanbongi K. Ohtani M. Res. Inst. Mineral Dressing and Met. , Tohoku Univ. 1955(11):2.
[24] Turkdogan E T,Leake L E. J. Iron Steel Inst. (London), 1955,179:39.
[25] Neumann F,Schenck H,Patterson W. Giesserei,1960,47:25.
[26] Wang Zhichang. Acta Metal. Sinica, 1980,16:195.
[27] Fu Chongyue,Zhai Yuchun. Erzmetall,1989,42:522.
[28] Lupis C H P,Elliott J F. Trans. TMS-AIME,1965,233:257.
[29] Cheng Wu. Thermodynamics of Related Elements in Mn-alloys and their Dephosphorization under Reducing Atmosphere, Dr. Eng. thesis. University of Science and Technology Beijing 1991.
[30] Wagner C. Thermodynamics of Alloys, Addison-Wesley Press. Reading, Mass. U.S.A, 1952:53.

Interaction Coefficients in Multicomponent Metallic Solutions at Constant Activity and Constant Concentration[*]

Wei Shoukun

Abstract: Determination of interaction coefficients of solutes in metal melt with the solubility measurements by two ways of evaluation, namely, the logarithm formulation and the solubility equation, was studied. The interaction coefficients of Ca in the Mn-Ca-Cr system were calculated as an example. It has been found that with the same experimental data the two ways of evaluation give quite different results, although the values are of the same order of magnitude. Determination of interaction coefficients in metallic melt by four kinds of equilibrium methods, namely, the gas-metal equilibrium, the electrochemical equilibrium, the gas-liquid dissolution equilibrium and the distribution equilibrium, was discussed. It has been ascertained that through change in condition of treatment of experimental data, interaction coefficients either at constant concentration or at constant activity can be evaluated, a general relationship between the two ways of evaluation being derived. Correct use of Wagner's formalism as regards the conversion of interaction coefficient at constant activity to that at constant concentration was emphasized.

Key words: interaction coefficient, multicomponent metallic solution

1 Introduction

Interaction coefficients of elements in multicomponent metallic solutions are widely used in making thermodynamic analysis of the relevant metallurgical reactions. Various experimental methods of determination of the interaction coefficients have been derived. But discrepancy and scatter of results for the same system from different sources have often been met in the literature. Different ways of approach in the elaboration of the same experimental data for the same system, arisen either from different ways of calculation or from conceptional confusion or misunderstanding might be a main cause for this inconsistency. It is the purpose of this paper to make a study on this aspect, the two ways of

---

[*] 原发表于《第六届中日双边钢铁科技会议论文集》,日本千叶,1992:1~10.

evaluation by the solubility method at constant activity, as well as the evaluation by the equilibrium methods at constant concentration and at constant activity being investigated.

2 The solubility method at constant activity

The solubility method has been used for determination of interaction coefficients in the systems Fe-C-j [1~5], Mn-C-j [6], Ni-C-j [6], Co-C-j [6], Fe-H-j [7,8], Fe-N-j [9,10], Fe-Ca-j [11] and Mn-Ca-j [12]. From the effect of the 3rd element j upon the change in the solubility of the solute i in the metal melt Me, two ways of evaluation of the interaction coefficients at constant activity, namely, the logrithm formulation and the solubility equation, are prevalent in the literature.

2.1 The logarithm formulation

In the system Me-i-j, the solubility of i in the metal melt Me is measured in and without the presence of the 3rd element j. In these saturated solutions, a_i = constant.

$$a_i = \gamma'_i N'_i = \gamma_i N_i$$

in which γ'_i, γ_i represent the activity coefficient of i in the binary and ternary system respectively, with pure substance as the standard. And N'_i, N_i represent the solubility of i expressed in mol fraction for the binary and ternary system respectively.

The effect brought by the presence of j upon i is designated as

$$\gamma^j_i = \frac{\gamma_i}{\gamma'_i} = \frac{N'_i}{N_i}$$

By plotting $\ln\gamma^j_i$ or $\ln(N'_i/N_i)$ against N_j, the slope at $N_j \to 0$ is the interaction coefficient $\overset{*}{\varepsilon}{}^j_i$, the asterisk being uesd to designate the value at constant activity.

$$\overset{*}{\varepsilon}{}^j_i = \left[\frac{\partial \ln\gamma^j_i}{\partial N_j}\right]_{a_i; N_j \to 0} = \left[\frac{\partial \ln \frac{N'_i}{N_i}}{\partial N_j}\right]_{a_i; N_j \to 0}$$

Usually, regression analysis is carried out with abandonment of those points deviating much from the linear relationship. The regression equation takes the form of

$$\ln\gamma^j_i = \ln \frac{N'_i}{N_i} = \overset{*}{\varepsilon}{}^j_i N_j \tag{1}$$

Equ. (1) might be called as the logarithm formulation for the evaluation of $\overset{*}{\varepsilon}{}^j_i$. The corresponding equation for the wt. % concentration is:

$$\lg f^j_i = \lg \frac{[\%i]'}{[\%j]} = \overset{*}{e}{}^j_i [\%j] \tag{2}$$

2.2 The solubility equation

It has been well known [13,14] that the change in solubility due to the presence of the 3rd element j is directly proportional to the amount of j:

$$\Delta N_i = N_i - N'_i = mN_j \tag{3}$$

in which m is the proportional constant. $\overset{*}{\varepsilon}{}_i^j$ is related with m by:

$$\overset{*}{\varepsilon}{}_i^j = -\frac{m}{N'_i} \tag{4}$$

Equ. (3) might be named as the solubility equation and has been used by Schenck and his associates [3~6] for calculating the interaction coefficients of the Me-C-j systems. The corresponding equations for the weight concentration and the 1 wt. % solution standard are:

$$[\%i] - [\%i]' = m'[\%j] \tag{5}$$

$$\overset{*}{\varepsilon}{}_i^j = \frac{m'}{2.30[\%i]'} \tag{6}$$

2.3 The 2nd order interaction coefficients

According to Lupis [15] the logarithm formulation is written as:

$$\ln\gamma_i^j = \overset{*}{\varepsilon}{}_i^j N_j + \overset{*}{\rho}{}_i^j N_j^2 \tag{7}$$

The coefficients $\overset{*}{\varepsilon}{}_i^j$ and $\overset{*}{\rho}{}_i^j$ can be found by the regression analysis.

The solubility equation could be written as:

$$N_i - N'_i = mN_j + lN_j^2 \tag{8}$$

$$\overset{*}{\varepsilon}{}_i^j = -\frac{m}{N'_i} \tag{9}$$

$$\overset{*}{\rho}{}_i^j = -\frac{l}{N'_i} + \frac{1}{2}\left(\frac{m}{N'_i}\right)^2 \tag{10}$$

Similarly, for the 1 wt. % solution standard, the logarithm formulation is:

$$\lg f_i^j = \overset{*}{e}{}_i^j[\%j] + \overset{*}{r}{}_i^j[\%j]^2 \tag{11}$$

For the solubility equation,

$$[\%i] - [\%i]' = m'[\%j] + l'[\%j]^2 \tag{12}$$

$$\overset{*}{e}{}_i^j = -\frac{m'}{2.30[\%i]'} \tag{13}$$

$$\overset{*}{r}{}_i^j = -\frac{l'}{2.30[\%i]'} + \frac{1}{4.60}\left(\frac{m'}{[\%i]'}\right)^2 \tag{14}$$

Derivation of Equ. (9), (10), (13) and (14) have been given elsewhere [16].

2.4 Conversion fomulae

The conversion between $\overset{*}{\varepsilon}{}_i^j$ and $\overset{*}{e}{}_i^j$ is exactly the same as ε_i^j and e_i^j (at constant concentration) with the following formulae:

$$\overset{*}{\varepsilon}{}_i^j = 2.30 \frac{M_j}{M_l} \overset{*}{e}{}_i^j + \frac{M_l - M_j}{M_l} \qquad (15)$$

The conversion formula between $\overset{*}{\rho}{}_i^j$ and $\overset{*}{r}{}_i^j$ is given by Lupis[15]:

$$\overset{*}{\rho}{}_i^j = \frac{23000 M_j^2 \overset{*}{r}{}_i^j + M_l(M_l - M_j) \overset{*}{\varepsilon}{}_i^j - \frac{1}{2} M_l(M_l - M_i)(\overset{*}{\varepsilon}{}_i^j)^2 N_i^2 - \frac{1}{2}(M_l - M_j)^2}{[M_l - N_i'(M_l - M_i)] M_l} \qquad (16)$$

In the above formulae, M_l, M_i and M_j represent the molecular weight of the solvent, solute and the 3rd element respectively.

The conversion from the interaction coefficient at constant activity to that at constant concentration has been studied by Fuwa and Chipman[1], Mori et al[17], Schenck etal[4] and Lupis[15], who gave the following formulae:

$$\overset{*}{\varepsilon}{}_i^j = \frac{\varepsilon_i^j + \rho_i^{i,j} N_i'}{1 + \varepsilon_i^i N_i' + 2\rho_i^i (N_l')^2} \qquad (17)$$

$$\overset{*}{\rho}{}_i^j = \frac{\rho_i^j - \overset{*}{\varepsilon}{}_i^j \rho_i^{i,j} N_i' + \frac{1}{2}(\overset{*}{\varepsilon}{}_i^j)^2 [4\rho_i^i (N_i')^2] + \varepsilon_i^i N_i']}{1 + \varepsilon_i^i N_i' + 2\rho_i^i (N_1')^2} \qquad (18)$$

With neglect of terms of higher orders, Equ. (17) becomes:

$$\varepsilon_i^j = \overset{*}{\varepsilon}{}_i^j (1 + \varepsilon_i^i N_i') \qquad (19)$$

Which is the same formula already derived by Mori[17], Schenck and their associates[4]. The corresponding formula on the wt.% basis is accordingly:

$$e_i^j = \overset{*}{e}{}_i^j (1 + 2.30 e_i^j [\%i]') \qquad (20)$$

Obviously, in case that the self interaction coefficient ε_i^i or $e_i^i = 0$, or the concentration for the binary system is extremely small, then $\varepsilon_i^j = \overset{*}{\varepsilon}{}_i^j$ or $e_i^j = \overset{*}{e}{}_i^j$.

2.5 Numerical calculation

The experimental data for the system Mn-Ca-Cr (Table 1[12]) has been used as an example for calculation. The first order interaction coefficients $\overset{*}{\varepsilon}{}_{Ca}^{Cr}$, $\overset{*}{e}{}_{Ca}^{Cr}$ are given in Table 2[16], and the 2nd order interaction coefficients $\overset{*}{\rho}{}_{Ca}^{Cr}$, $\overset{*}{r}{}_{Ca}^{Cr}$ in Table 3[16].

Table 1 Effect of Cr upon the solubility of Ca in Mn at 1350℃

%Ca	N_{Ca}	%Cr	N_{Cr}
0.15	0.002055	0	0
0.15	0.002054	0.88	0.009288
0.14	0.001915	2.59	0.027310
0.12	0.001640	4.28	0.045091
0.12	0.001635	9.48	0.099582
0.098	0.001334	11.02	0.115668
0.084	0.001143	13.40	0.140468

Table 2 1st order interaction coefficients of Cr upon Ca in Mn at constant activity at 1350℃

(a) Mole fraction basis	
Logarithm formulation	Solubility equation
$\ln \gamma_i^j = bN_j$	$N_j - N_i' = mN_j$
$\ln \gamma_{Ca}^{Cr} = 3.6447 N_{Cr}\ (r=0.95)$	$N_{Ca} - N_{Ca}' = -0.006011 N_{Cr}\ (r=0.96)$
$\overset{*}{\varepsilon}_{Ca}^{Cr} = 3.6$	$\overset{*}{\varepsilon}_{Ca}^{Cr} = -\dfrac{m}{N_{Ca}'} = 2.9$

(b) wt.% basis	
Logarithm formulation	Solubility equation
$\lg f_i^j = b'[\%j]$	$[\%i] - [\%i]' = m'[\%j]$
$\lg f_{Ca}^{Cr} = 0.01638[\%Cr]\ (r=0.95)$	$[\%Ca] - [\%Ca]' = -0.004551[\%Cr]\ (r=0.96)$
$\overset{*}{e}_{Ca}^{Cr} = 0.016$	$\overset{*}{e}_{Ca}^{Cr} = -\dfrac{m'}{2.30[\%Ca]'} = 0.013$
By conversion with Equ. (15),	By conversion with Equ. (15),
$\overset{*}{\varepsilon}_{Ca}^{Cr} = 3.6$	$\overset{*}{\varepsilon}_{Ca}^{Cr} = 2.9$

Table 3 2nd order interaction coefficients of Cr upon Ca in Mn at constant activity at 1350℃

(a) Mole fraction basis	
Logarithm formulation	Solubility equation
$\ln \gamma_i^j = bN_j + cN_j^2$	$N_i - N_i' = mN_j + lN_j^2$
$\ln \gamma_{Ca}^{Cr} = 2.1998 N_{Cr} + 12.2333 N_{Cr}^2\ (r=0.96)$	$N_{Ca} - N_{Ca}' = -0.005349 N_{Cr} - 0.005607 N_{Cr}^2\ (r=0.96)$
$\overset{*}{\varepsilon}_{Ca}^{Cr} = 2.2$	$\overset{*}{\varepsilon}_{Ca}^{Cr} = -\dfrac{m}{N_{Ca}'} = 2.6$
$\overset{*}{\rho}_{Ca}^{Cr} = 12.2$	$\overset{*}{\rho}_{Ca}^{Cr} = -\dfrac{l}{N_{Ca}'} + \dfrac{1}{2}\left(\dfrac{m}{N_{Ca}'}\right)^2 = 6.1$

Continued Table 3

(b) wt. % basis

Logarithm formulation	Solubility equation
$\lg f^j_i = b'[\%j] + c'[\%j]^2$	$[\%i] - [\%i]' = m'[\%j] + l'[\%j]^2$
$\lg f^{Cr}_{Ca} = 0.009814[\%Cr] +$ $0.000582[\%Cr]^2 (r=0.96)$	$[\%Ca] - [\%Ca]' = -0.004016[\%Cr] -$ $0.000047[\%Cr]^2 (r=0.99)$
$\overset{*}{e}{}^{Cr}_{Ca} = 0.0098$	$\overset{*}{e}{}^{Cr}_{Ca} = -\dfrac{m'}{2.30[\%Ca]'} = 0.012$
$\overset{*}{r}{}^{Cr}_{Ca} = 0.00058$	$\overset{*}{r}{}^{Cr}_{Ca} = -\dfrac{l'}{2.30[\%Ca]'} + \dfrac{1}{4.60}\left(\dfrac{m'}{\%[Ca]'}\right)^2 = 0.00029$
By conversion with Equ. (15) and (16),	By conversion with Equ. (15) and (16),
$\overset{*}{\varepsilon}{}^{Cr}_{Ca} = 2.2$	$\overset{*}{\varepsilon}{}^{Cr}_{Ca} = 2.6$
$\overset{*}{\rho}{}^{Cr}_{Ca} = 12.1$	$\overset{*}{\rho}{}^{Cr}_{Ca} = 6.1$

It has been shown that from the same experimental data of the solubility measurements, the two ways of evaluation, namely, the logarithm formulation and the solubility equation for calculating the interaction coefficients give quite different results, although the values are of the same order of magnitude. To avoid the discrepancy arising from different ways of evaluation, it would be preferable to choose a standard way of evaluation. It might be recommended to choose the logarithm formulation as the standard way of evaluation because of its direct relationship with the definition of the interaction coefficient.

3 The equilibrium methods

3.1 The gas—metal chemical equilibrium

The Fe-S-j system is taken as an example. A mixture of H_2S and H_2 gas was equilibrated with [S] in liquid iron[18~20].

$$H_2 + [S] \rightleftharpoons H_2S$$

$$K = \left(\frac{p_{H_2S}}{p_{H_2}}\right)' \frac{1}{f'_S [\%S]'} = \left(\frac{p_{H_2S}}{p_{H_2}}\right) \frac{1}{f_S [\%S]} \quad (21)$$

The prime refers to the binary system.

Let
$$K_{bin} = \left(\frac{p_{H_2S}}{p_{H_2}}\right)' \frac{1}{[\%S]'} \quad \text{and} \quad K_{tern} = \left(\frac{p_{H_2S}}{p_{H_2}}\right) \frac{1}{[\%S]} \quad (22)$$

K is found by extrapolating the curve of K_{bin} vs. [%S] to zero concentration.

$$f_S^j = \frac{f_S}{f_S'} = \frac{K_{tern}}{K_{bin}} \qquad (23)$$

For a certain value of [%S], K_{tern} is calculated from the experimental data, while K_{bin} for the same value of S can be read from the curve of K_{bin} plotted against [%S]. The value f_S^j thus found is at constant concentration of S. By plotting $\lg f_S^j$ against [%j], the slope at [%j] = 0 equals the interaction coefficient e_S^j, Usually regression analysis through those points pertaining to the linear relationship is made:

$$\lg f_S^j = b[\%j]$$

and $e_S^j = b$. The regression line passes naturally through the origin. Recommended values of $e_i^j(\varepsilon_i^j)$ as given in the literature[21~24] are supposed to be the conventional interaction cofficients at constant concentration.

Should the evaluation of f_i^j (an asterisk being added for distinction) is required, then $a_S' = a_S$

$$\overset{*}{f}_S^j = \frac{f_S}{f_S'} = \frac{[\%S]'}{[\%S]} \qquad (24)$$

Since f_S' and f_S can be calculated from the experimental data, curves of a_S against [%S] for both the binary and ternary systems can be drawn. For a certain value of ternary [%S], the corresponding value of binary [%S]' at constant activity can be read from the activity curves. Thus different values of $\overset{*}{f}_S^j$ are calculated, and $\overset{*}{e}_S^j$ will be evaluated as usually.

From Equ. (21), it can be shown that for constant activity:

$$\left(\frac{p_{H_2S}}{p_{H_2}}\right)' = \left(\frac{p_{H_2S}}{p_{H_2}}\right)$$

So the evaluation of interaction coefficient at constant activity is really based on the performance of the equilibrium experiment at constant gas pressure ratio. Furthermore, from Equ. (21):

$$\left(\frac{p_{H_2S}}{p_{H_2}}\right)' \bigg/ \left(\frac{p_{H_2S}}{p_{H_2}}\right) = \frac{a_S'}{a_S}$$

By comparson with Equ. (23):

$$f_S^j = \frac{a_S}{a_S'} \cdot \frac{[\%S]'}{[\%S]} \qquad (25)$$

Equ. (25) is the general equation for calculating f_S^j.

At constant concentration:

$$f_S^j = \frac{K_{tern}}{K_{bin}} = \frac{a_S}{a_S'} \qquad (26)$$

At constant activity:

$$f_S^j = \frac{[\%S]'}{[\%S]} \tag{27}$$

Therefore with the experimental data from both the binary and the ternary systems, by change in condition of treatment of data, either the interaction coefficient at constant concentration or at constant activity can be evaluated.

3.2 The electrochemical equilibrium

For the determination of the interaction coefficient of Nb in the Fe-Nb-Mn system with the solid electrolyte cell technique, the following cell assembly is used:

$$\text{Mo} | \text{Mo, MoO}_2 \parallel \text{ZiO}_2(\text{MgO}) \parallel [\text{Nb}], \text{NbO}_2 | \text{Mo, Mo-cermet}$$

The overall reaction is:

$$[\text{Nb}] + 2[\text{O}] \rightleftharpoons \text{NbO}_{2(s)}$$

$$K = \frac{1}{a'_{Nb} a_O'^2} = \frac{1}{a_{Nb} a_O^2}$$

K is found by extrapolation at zero concentration of Nb. Due to the small content of [O] in the melt, the effect of f_{Nb}^O is neglected.

$$f_{Nb}^{Mn} = \frac{a_O'^2 [\%Nb]'}{a_O^2 [\%Nb]} = \frac{a_{Nb}}{a'_{Nb}} \times \frac{[\%Nb]'}{[\%Nb]} \tag{28}$$

Equ. (28) is similar to Equ. (25), and evaluation of e_{Nb}^{Mn} and $\overset{*}{e}_{Nb}^{Mn}$ [25,26] as well as e_{Nb}^{Si} and $\overset{*}{e}_{Nb}^{Si}$ [27] have been reported elsewhere.

3.3 The gas-liquid dissolution equilibrium

For dissolution of a monoatomic gas, Henry's law holds true.

$$M(g) \longrightarrow [M]$$

$$K = \frac{f_M' [\%M]'}{p_M'} = \frac{f_M [\%M]}{p_M}$$

K is really the Henry's constant and can be found by extrapolation as usually. In the same way as the chemical equilibrium,

$$f_M^j = \frac{p_M [\%M]'}{p_M' [\%M]} = \frac{a_M [\%M]'}{a_M' [\%M]} \tag{29}$$

For dissolution of a diatomic gas, Sievert's law holds true, and derivation leads to the same Equ. (29). The latter is similar to Equ. (25) for calculation of e_M^j and $\overset{*}{e}_M^j$. The isopiestic methods [28, 29] belongs to this category.

3.4 The distribution equilibrium

Since Fe and Ag are insoluble in the molten state, distribution equilibrium of a certain element i which is both soluble in Fe and Ag is often used to study the activity of i. If a third element j which is soluble in Fe but not in Ag, the distribution equilibrium can be used to evaluate the interaction coefficient of j upon i. Usually, this method is used to evaluate the interaction coefficient at constant activity, but that at constant concentration could also be calculated as well.

For the Fe-Si-C[30, 31],

$$[Si]_{Ag} \rightleftharpoons [Si]_{Fe}$$

$$K = \frac{f'_{Si(Fe)} [\%Si]'_{Fe}}{f'_{Si(Ag)} [\%Si]'_{Ag}} = \frac{f_{Si(Fe)} [\%Si]_{Fe}}{f_{Si(Ag)} [\%Si]_{Ag}}$$

Since the solubility of Si in Ag is rather small, it might be assumed that $f'_{Si(Ag)} = f_{Si(Ag)} = 1$. The true equilibrium constant K is found by plotting $[\%Si]'_{Fe}/[\%Si]'_{Ag}$ against $[\%Si]'_{Fe}$ and extrapolating to zero concentration of Si in Fe.

$$f^j_{Si} = \frac{a_{Si(Ag)} [\%Si]'_{Fe}}{a'_{Si(Ag)} [\%Si]_{Fe}} = \frac{a_{Si(Fe)} [\%Si]'_{Fe}}{a'_{Si(Fe)} [\%Si]_{Fe}} \qquad (30)$$

Equ. (30) is similar to Equ. (25). Hence, it might be concluded that for the different equilibrium methods, be they physical or chemical, both the interaction coefficients at constant concentration and at constant activity could be evaluated. As the Wagner's formalism[32] is used conventionally at constant concentration, there would be no necessity for calculating the interaction coefficients at constant activity by the equilibrium methods.

4 Restriction of the constant-activity evaluation

4.1 The gas—metal chemical equilibrium

For the reaction $[C] + CO_2 = 2CO$, should the equilibrium be operated under constant p^2_{CO}/p_{CO_2}, then the condition for evaluating the interaction coefficient at constant activity is warranted. Table 4 is quoted from Schenck, Steinmetz and Rhee[33] with some additional calculations made by the present author. It is to be noted that:

(1) The experiments were conducted at constant $p_{CO_2}(p_{CO} + p_{CO_2} = 1\text{atm.})$, hence p^2_{CO}/p_{CO_2} being constant.

(2) No experiments of the binary system were conducted. For conversion with Equ. (19), $\varepsilon^C_C = 11.5$ being taken from the literature.

(3) In the calculation of $\overset{*}{\varepsilon}{}_C^{Cr}$, no γ_C^{Cr} was used, only the experimental data of the ternary system being used. A regression line of
$$-\ln N_C = a + bN_{Cr}$$
was calculated. The slope b, which equals $-\partial \ln N_C / \partial N_{Cr}$, is the $\overset{*}{\varepsilon}{}_C^{Cr}$ and the intercept equals $-\ln N'_C$. This is due to the fact that since $\partial \ln \gamma'_C / \partial N_{Cr} = 0$, therefore $\partial \ln \gamma_C / \partial N_{Cr} = \partial \ln \gamma_C^{Cr} / \partial N_{Cr}$. Because of constant activity, $\partial \ln \gamma_C = - \partial \ln N_C$.

(4) For every isoactivity line of $-\ln N_C = a + bN_{Cr}$, there is a definite value of $\overset{*}{\varepsilon}{}_C^{Cr}$.

(5) This constant-activity evaluation is subjected to the restriction that a great number of experiments should be performed, each of the different $\overset{*}{\varepsilon}{}_C^{Cr}$ values should be converted into ε_C^{Cr} separately, and these ε_C^{Cr} values should be finally averaged.

Table 4 ε_C^{Cr} for the Fe-C-Cr system at constant activity

$t/℃$	p_{CO_2}/atm	$a_{(\%)}$	Regression line $-\ln N_C = a + bN_{Cr}$	$\overset{*}{\varepsilon}{}_C^{Cr}$	N'_C ($N_{Cr}=0$)	ε_C^{Cr}
1600	4.6×10^{-5}	37.8	Saturated solution	-1.36	0.2100	-4.64
1600	0.0022	0.78	$-\ln N_C = 3.6266 - 3.10 N_{Cr}$	-3.10	0.0266	-4.05
1600	0.0043	0.40	$-\ln N_C = 4.1792 - 4.26 N_{Cr}$	-4.26	0.0153	-5.00
1600	0.0086	0.20	$-\ln N_C = 4.8469 - 7.36 N_{Cr}$	-7.36	0.0078	-8.02

4.2 The distribution equilibrium

In the literature the distribution method is often used for evaluation of interaction coefficient at constant activity with only the experimental data of the ternary system. For the Fe-Si-j system:

$$K = \frac{\gamma_{Si(Fe\text{-}j)} N_{Si(Fe\text{-}j)}}{\gamma_{Si(Ag)} N_{Si(Ag)}}$$

Since the concentration of the Ag bath is kept constant, $a_{Si(Ag)} = a_{Si(Fe\text{-}j)}$ = constant. The concentration $N_{Si(Fe\text{-}j)}$ can be changed by adding different amount of j in the Fe bath, and regression analysis of the isoactivity line $-\ln N_{Si} = a + bN_j$ was made, its slope being equal to $\overset{*}{\varepsilon}{}_{Si}^{j}$, and its intercept equal to $-\ln N'_{Si}$. The $\overset{*}{\varepsilon}{}_{Si}^{j}$ should be converted into ε_{Si}^{j} with N'_{Si} for the binary system and ε_{Si}^{Si} from the literature. To insure the reliability of the averaged ε_{Si}^{j}, a great number of experiments have to be done. So for the Fe-Si-C system, 13 experiments by Schroeder and Chipman [30] and 19 experiments by Murakami, Ban-ya and Fuwa [31] have been reported.

5 Discussion

Since the interaction coefficients at constant concentration and at constant activity for the same system are quite different in magnitude, it would not be justifiable to list these values from different sources together for comparison, as occurred sometimes in the literature. As Wagner's formalism is originally justified for application at constant concentration, no interaction coefficients at constant activity should be introduced into the Wanger's formalism before being converted into those at constant concentration.

As already mentioned for the distribution equilibrium with the Ag bath, a great number of experiments giving more isoactivity lines, and the conversion from $\overset{*}{\varepsilon}{}_i^j$ into ε_i^j, should be accomplished. But in some recent papers using the distribution equilibrium to evaluate the interaction coefficients at constant activity for the systems Fe-Si-C[34], Fe-Cr-C[35], Fe-Ti-C[36] and Fe-Ti-C-i[37], two important points have escaped the notice of the original authors:

(1) The distribution equilibrium in the Ag bath was carried out with only 2 or at most 4 experiments. With these few experiments no reliability of the results could be credited.

(2) The interaction coefficients at constant activity were directly introduced into the Wanger's formalism for further elaboration without conversion. This procedure is theoretically questionable.

Because of these, their results are doubtful and might not be accepted.

6 Concluding remarks

Attention should be paid to avoid conceptional confusion and misunderstanding between interaction coefficients at constant concentration and at constant activity.

Excepting the solubility method for determining the interaction coefficient at constant activity, it would be always preferable to determine the interaction coefficient at constant concentration, in order to save the trouble of conversion.

For the solubility method of saturated solution, it might be suggested to adopt the logarithm formulation for calculating $\overset{*}{\varepsilon}{}_i^j$ to minimizing the discrepancy.

For unsaturated solution, should any method at constant activity, for instance, the distribution equilibrium, be used for determining the interaction coefficient, care should be taken to perform a great number of experiments in order to insure the reliability of the results.

References

[1] Fuwa T, Chipman J. Trans. TMS-AIME, 1959, 215: 708.

[2] Ohtani M,Gokcen N A.: TMS-AIME 1960,218:533.
[3] Neumann F,Schenck H. Arch. Eisenh. ,1959,30:477.
[4] Schenk H,Frohberg M G Steinmetz E Rutenberg B. Arch. Eisenh. ,1962,33:223,229.
[5] Schenck H,Gloz M Steinberg E. Arch. Eisenh. ,1970,41:1.
[6] Schenck H,Froberg M G Steinberg E Arch. Eisenh. 1963,34:37,43.
[7] Weinstein M, Elliott J F. Trans. TMS-AIME,1963,227:382.
[8] Sigworth G K Elliott J F. Metal Science 1974,8:298.
[9] Pehlke R D Elliott J F. Trans. TMS-AIME,1960,218:1088.
[10] Wada H Pehlke R D. Met. Trans. ,1981,12B:333.
[11] Koehlor M Engell H J Janke D. steel research,1985,56:419.
[12] Shoukun Wei, Ruiming Ni, Zhongting Ma, Wu Cheng. Steel Research,1989,60:437.
[13] Sangbongi K, Ohtani M. Res. Inst. Mineral Dressing and Met. , Tohoku Univ. 1955(11):2.
[14] Turkdogan E T,Leake L E. J. Iron Steel Inst. (London),1955,179:39.
[15] Lupis C H P. Acta Metal,1968,16:1365.
[16] Shoukun Wei. steel research,1992,63(4).
[17] Mori T, Aketa K, Ono H,Sugita H,Tetsu-to-Hagane 1959,45:929; ibid, 1960,46:1429.
[18] Sherman C W, Elvander H I, Chipman J. Trans. AIME,1950,188:334.
[19] Sherman C W,Chipman J. Trans. TMS-AIME,1952,194:597.
[20] Ban-ya S,; Chipman J. Trans. TMS-AIME,1969,245:133.
[21] Elliott J F, Gleiser M, Ramakrishna V. Thermochemistry for Steelmaking, Addison-Wesley Publ. Co. , Reading, Mass. , U. S. A. , 1963, II :564.
[22] Schenck H, Steinmetz E. Stahl E. Sonderbericht,1968(7).
[23] Sigworth G K,Elliott J F. see ref. 8; Elliott J F: E. F. Proceedings,1974,32:62.
[24] Matsushita Y. Sakao H. Steelmaking Data Source Book, Gor den and Bleach, New York,1988.
[25] Wei Shoukun,Zhang Shengbi, Tong Ting, Tan Zanlin. Iron Steel (CSM),1984,19(7):1.
[26] Zhang Shengbi, Tong Ting, Wang Jifang, Wei Shounkun. Acta Met. Sinica,1984,20:348.
[27] Tong Ting,Wei Shoukun, Zhang Shengbi, Hu Mingfu. Acta Met. Sinica,1987,23:B47.
[28] Meysson M, Rist A. Revue Met,1965:1127.
[29] Speer M C, Parlee A D. AFS Cast Metals Res. J. ,1972:122.
[30] Schroeder D, Chipman J. Trans. TMS-AIME,1960,230:1492.
[31] Murakami S,Ban-ya S,Fuwa T. Tetsu-to-Hajgane,1970,56:536.
[32] Wagner C. Thermodynamics of Alloys, Addision-Wesley Publ. Co. , Reading, Mass. U. S. A. , 1952:53.
[33] Schenck H,Steinmetz E,Rhee P C H: Arch. Eisenh,1968,39:803.
[34] Ji Chunlin; Qi Guojun. Trans. Japan Inst. Metals 1985,26:832.
[35] Ji Chunlin, Yu Rongxiang, Zhang Guofan, Liu Sulan: Iron and Steel (CSM),1987,22(9):16.
[36] Guo Yuanchang, Wang Changzhen, Yu Hualong. Met. Trans. ,1990,B, 2(B):537.
[37] Guo Yuanchang, Wang Changzhen. Met. Trans. ,1990, B, 21B:543.

炉渣离子理论
LUZHA LIZI LILUN

空气顶吹过程中熔渣的气态脱硫[*]

魏寿昆　王国忱

摘　要：利用空气顶吹研究熔渣的气态脱硫动力学，着重研究了熔渣ΣFeO含量、j值 $[j = N_{Fe_2O_3}/(N_{Fe_2O_3} + N_{FeO})]$ 以及碱度对气态脱硫速度的影响。对气态脱硫反应的机理及反应速度的限制性环节作了分析和讨论。

气态脱硫的反应如下：

$$(S^{2-}) + \frac{3}{2}O_2 = (O^{2-}) + SO_2$$

渣中的铁离子是氧的传递媒介而非脱硫组分：

$$6(Fe^{3+}) + (S^{2-}) + 2(O^{2-}) = 6(Fe^{2+}) + SO_2$$

$$6(Fe^{2+}) + \frac{3}{2}O_2 = 6(Fe^{3+}) + 3(O^{2-})$$

最后，对金属直接的气态脱硫、炉渣的气态脱硫和炉渣脱硫的矛盾，以及气态脱硫在转炉炼钢工艺上的应用问题进行了简要的讨论。

炼钢过程中的气态脱硫，文献上早有报道。在搅炼炉炼制熟铁时，脱硫高达90%～95%，其中大部分以气态挥发[1]。Herty[2]在研究平炉炼钢自炉气吸硫问题时指出，对有一定含硫量的废钢，炉气燃料有一定的含硫值，超过此值时则废钢自炉气吸硫，而低于此值时，则废钢的硫挥发；Herty、Belyea和Burkart[3]又给出炉渣含硫量与炉气含SO_2量的平衡关系，根据后者可以估计炉渣是否自炉气吸硫，抑有气态脱硫。表1列举近年来不同炼钢方法气态脱硫的数值，这些数值都是用炉料平衡计算出来的，显然不够准确。炉气内含有来自炉料的硫（通常以SO_2存在），Ступарь[11]利用S^{35}在平炉内进行试验予以证实；对侧吹碱性转炉炉气含有SO_2也曾进行过直接测定[8]。此外，Щимон、Абросимов和Трубин[12]进行过氧气顶吹钢水的实验研究，本文作者之一和其他合作者[9]在感应炉内进行过压缩空气顶吹生铁水的研究，均肯定了气态脱硫的存在。

---

[*] 原刊于《金属学报》，1965，8：419～434。

表 1　不同炼钢方法的气态脱硫

炼钢方法	占生铁水或熔池总硫量/%		来源
	气态脱硫	炉渣脱硫	
酸性侧吹转炉	9.6	22.9	[4]
平炉，炉头喷氧，低硫混合煤气，含硫 2~3g/m³（标态）	40.0 或以上	20.0	[5]
平炉，炉头喷氧，高硫混合煤气，含硫 20g/m³（标态）	22.0	40.0 或以上	[5]
氧气顶吹转炉（5t）	39.4	26.6	[6]
氧气顶吹转炉（50t）	11.7	37.4	[6]
回转炉（Rotor）	15.0	51.0	[7]
碱性侧吹转炉	45.5~50.0	27.9~33.8	[8]
碱性侧吹转炉	25.0~35.0	35.0	[9]
电炉，熔化期及氧化期吹氧，还原期薄渣吹氧	41.0	16.0	[10]

关于含硫气体与不同类型炉渣的高温平衡实验，已有不少报道，例如与硅酸盐及铝酸盐[13]、含 FeO 的硅酸盐[14]、平炉渣[15]、FeO-FeS 渣[16]以及铁酸钙[17]等。这些研究都证实 Richardson 和 Fincham[13]的结论，即随着炉气氧分压的不同，炉渣内的硫能以不同形态而存在。

当 $p_{O_2}<10^{-5}\sim 10^{-6}$ 大气压时：

$$\frac{1}{2}S_{2(气)} + (O^{2-}) = \frac{1}{2}O_{2(气)} + (S^{2-})$$

当 $p_{O_2}<10^{-3}\sim 10^{-4}$ 大气压时：

$$\frac{1}{2}S_{2(气)} + \frac{3}{2}O_{2(气)} + (O^{2-}) = (SO_4^{2-})$$

也即：

$$(S^{2-}) + 2O_2 = (SO_4^{2-})$$

在更高的 p_{O_2}（>0.1 大气压）时，渣内也可生成 $(S_2O_7^{2-})$ [17]。

关于气态脱硫动力学研究，文献是比较少的。潘鸿芳和邹元爔[17]对转炉渣，Чучмарев、Есин 和 Добрыдень[19]对氧气顶吹 CaO-Al₂O₃-SiO₂ 渣，以及 Neuhaus、Langhammer 和 Kosmider 等人[20]对氧气顶吹平炉渣等方面的气态脱硫有关动力学方面进行过研究。

由于本研究的初期工作证明了吹炼铁水时气态脱硫主要通过熔渣进行，本文着重研究利用空气顶吹熔渣的气态脱硫动力学，特别研究熔渣成分——氧化铁含量及碱度，对气态脱硫速度的影响。

1 实验设备及方法

实验在密封的立式感应炉内进行（图1）。熔化炉渣时采用石墨套筒间接加热。

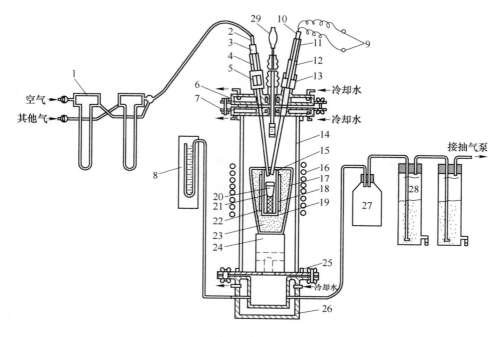

图1 实验的设备系统

1—流量计；2—石英吹气管；3，11—橡皮塞；4，12—玻璃升降管；5，13—橡皮管；
6，7—炉盖；8—炉内压力计；9—热电偶；10—石英套管；14—石英粗管；
15，17—镁砂坩埚；16—感应圈；18—石墨套筒；19，23—镁砂；20—炉渣；
21—氧化铝坩埚；22—氧化铝座垫；24—坩埚座垫；25—橡皮垫圈；
26—水冷炉底；27—除尘器；28—气体吸收器；29—取样器

空气先由压缩空气机送入储气瓶，瓶内维持4个大气压，经一系列干燥瓶后，以2.3L/min的流量通过1.5mm内径的石英管进入炉内，排出的气体系统连以抽气泵，使炉内经常维持10mmH_2O（98.1Pa）的正压。空气吹入在两种情况下进行：（1）直接吹到渣面（或铁液面），此时石英管口距液面约5mm（个别情况20mm），空气对液面有明显的搅动；（2）不直接吹到渣面，此时石英管口距液面约300mm。

对金属-熔渣-气体三相系实验采用氧化镁坩埚，金属重250g，炉渣25g。对熔渣-气体二相系实验采用氧化铝坩埚，炉渣重70g。

炉渣由化学纯药品配制，在空气或氮气气氛下熔化，渣中硫由自制的硫化钙

（以硫化氢通入氢氧化钙制成）引入。碱度（%CaO/%SiO$_2$）采用 1.1~1.2、2.3~2.8、4.6~5.8 以及 6~7 等四种。对高碱度炉渣，为了提高流动性适当增加 Al$_2$O$_3$ 含量，多的达 36%~43%。炉渣 ΣFeO 含量由 0.8%以下至 24%。

排出的气体经除尘器后由 3% H$_2$O$_2$ 溶液吸收，以标定的氢氧化钠溶液滴定。为了避免溶液内所含 CO$_2$ 的影响（在吹炼金属时），用甲基红及次甲基蓝为指示剂。

由于炉渣含硫可能以硫化物及硫酸物存在，采用 CO$_2$ 气燃烧法测定全硫，并与重量法相比较。经进一步的研究，以 CO$_2$ 气燃烧测定全硫量比以 O$_2$ 气燃烧测定量更较为准确（在炉渣含硫 0.6%~0.7%时，前者较后者约高 10%），这符合 Fincham 和 Richardsont[21] 的研究结果。采用发生法以测定同炉渣样以硫化物存在的硫，全硫与此值之差即代表以硫酸物存在的硫量。

通过炉料平衡计算发现，由炉气吸收的硫量和炉渣失去的硫量相差 2%~11%。这可能由于下列原因造成：（1）气相中可能有微量 SO 或 S$_2$ 未能被吸收成 H$_2$SO$_4$；（2）气道系统可能有微量 S 凝集；（3）气道系统有些漏气；（4）取样器及玻璃升降管凸出炉外部分存有死角，该地炉气未被吸收。我们只在三相系实验采用排气系统以吸硫，其他实验只根据炉渣不同时间的含硫量作计算。

温度用铂铑热电偶测定，熔渣顶吹过程中温度波动控制在±10℃之内（个别情况达±15℃）。

硫含量分析的绝对误差为±0.006%~0.015%。

顶吹过程中，由于氧化铝坩埚的侵蚀，实验终止时熔渣的 Al$_2$O$_3$ 含量有些增加，这样引起硫含量的冲淡，造成-1.5%~3.0%的相对误差。

2 实验结果及分析

2.1 金属-熔渣-气体三相系实验

此实验分两部分平行进行：（1）生铁熔化 10min 后引入空气直吹液面；（2）同样成分的生铁熔化后加入不含硫的炉渣，待 10min 后再引入空气直吹渣面。生铁成分（%）为：

C	Si	Mn	S
2.8	1.4	0.02	0.42

炉渣成分（%）为：

CaO	SiO$_2$	Al$_2$O$_3$	MgO	ΣFeO
31.82	26.88	13.36	8.06	13.56

空气流量 2.3L/min。吹炼温度 1530℃。从图 2 得出，不加炉渣的生铁全

部脱硫只 31.0%。其中气态脱硫占 10.1%，而加入炉渣的生铁，在 10min 内即有 38.0%的硫进入炉渣，而一经吹入空气，气态脱硫即猛烈进行，同时促进金属液的硫进入炉渣从而再进一步地气化，全部脱硫达 74.0%，而气态脱硫占 39.3%，几乎为不加炉渣的金属气态去硫的 4 倍。这样可以看出，吹炼过程中金属的气态脱硫主要通过炉渣进行。图 2 也指出气态脱硫的速度在脱碳末期大为增加。

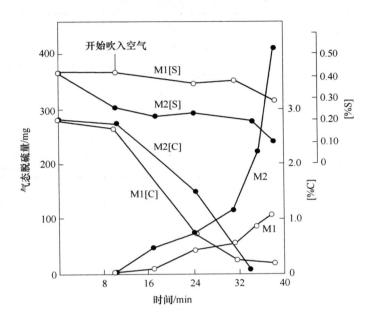

图 2　吹炼过程中金属含［S］、［C］量及气态脱硫量与吹炼时间的关系

M1—金属熔化 10min 后引入空气吹到液面；

M2—金属熔化后如入炉渣，10min 后再引入空气吹到渣面

2.2　对不同氧化铁含量的熔渣空气直吹渣面的实验

表 2 给出一个吹炼记录。炉渣成分（%）为：

CaO	SiO$_2$	Al$_2$O$_3$	MgO	ΣFeO	碱度（%CaO/%SiO$_2$）
27~40	25~35	14~16	4~9	<0.8~24	1.1~1.2

可以看出，在顶吹过程中不仅渣内的硫气化，同时渣内 Fe$_2$O$_3$ 量逐渐增加，而 FeO 量逐渐减小，ΣFeO 量由于氧化铝坩埚的熔蚀（使 Al$_2$O$_3$ 量稍为增加）稍为减小，但可认为基本上不变。Fe^{3+} 及 Fe^{2+} 的变化可用 $j = \dfrac{N_{Fe_2O_3}}{N_{Fe_2O_3} + N_{FeO}}$ 表示。由

（%S）对时间 τ 的曲线可求出 v_S，也即每分钟的去硫速度 $-\dfrac{d(\%S)}{d\tau}$。图 3 给出 v_S 对（%S）的关系。很明显，渣内含硫量越高，气态脱硫速度越大；而对同含硫量的熔渣，ΣFeO 越高，气态脱硫速度越大。Ⅳ 组实验曲线是一直线，说明对含氧化铁很少或无氧化铁的熔渣，气态脱硫是一级反应。

表 2　顶吹期间熔渣成分的变化[①]

取样时间 /min	熔渣成分/%							
	CaO	SiO_2	Al_2O_3	MgO	ΣFeO	FeO	Fe_2O_3	S
0	27.5	24.82	15.19	4.42	22.88	17.46	6.02	0.231
1	—	—	—	—	—	—	—	0.160
2	—	—	—	—	—	—	—	0.113
7	—	—	—	—	—	—	—	0.043
14	—	—	—	—	22.18	7.33	16.50	0.019
21	—	—	—	—	21.48	4.75	18.60	0.005

① 炉号 126。

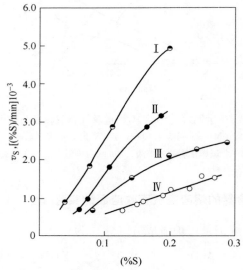

图 3　气态脱硫速度与熔渣含硫量的关系
（空气直吹渣面；流量：2.3L/min；温度：1430~1460℃）
Ⅰ—ΣFeO：21%~24%，j：0.14~0.63，碱度 1.1；
Ⅱ—ΣFeO：13%~15%，j：0.22~0.69，碱度 1.2；
Ⅲ—ΣFeO：8%~9%，j：0.29~0.61，碱度 1.2；
Ⅳ—ΣFeO：<0.8%，j 的变化忽略不计，碱度 1.1

2.3 温度对不含氧化铁炉渣气态脱硫的影响

对于无氧化铁（指很少氧化铁）的炉渣，由于图3指出：

$$v_S = -\frac{d(\%S)}{d\tau} = k_1(\%S), \min^{-1} \tag{1}$$

因之，可以导出：

$$\log(\%S) = -k_2\tau + \log(\%S^0) \tag{2}$$

式中 τ——顶吹时间，min；
$(\%S^0)$——顶吹开始时熔渣的含硫量；
k_1，k_2——两个常数，$\min^{-1}$，其相互关系是：

$$k_1 = 2.30\, k_2 \tag{3}$$

图 4 给出不同温度下熔渣含硫量的对数值与顶吹时间的关系，证实二者有直线关系。所用炉渣成分（%）为：

CaO	SiO₂	Al₂O₃	MgO	ΣFeO	S
31.08	25.38	11.16	9.06	0.78	0.22~0.36

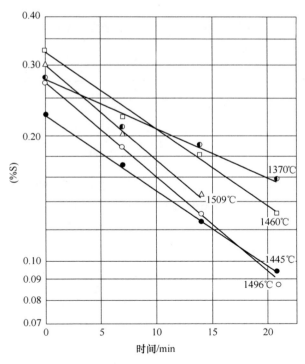

图 4 不同温度下熔渣含硫量与顶吹时间的关系
（空气直吹渣面；流量：2.3L/min；碱度：1.2；ΣFeO：0.78%）

吹入空气的石英管距渣面 20mm。

由于反应是在气-渣界面上发生，界面面积应予考虑，又根据炼钢物理化学工作者的惯例，v_S 通常以 g/min 表示，也即：

$$-\frac{d(\%S)}{d\tau}\frac{W}{100} = Ak_3(\%S), \text{g/min}$$

或

$$-\frac{d(\%S)}{d\tau} = \frac{100Ak_3}{W}(\%S), \text{min}^{-1} \quad (4)$$

式中　A——气-渣界面面积，cm^2；

　　　W——渣的重量，g；

　　　k_3——比反应速度常数，$g/(cm^2 \cdot min)$。

比较式（1）、式（3）及式（4）可得：

$$k_3 = \frac{2.30W}{100A}k_2 \quad (5)$$

对我们的实验，$W = 70g$，$A = 13.85 cm^2$。由式（2）的斜率 k_2 根据式（5）计算出不同温度下的 k_3 见表 3。可以看出，提高温度对气态脱硫有利。

表 3　不同温度下的比反应速度常数 k_3

温度/℃	1370	1445	1460	1496	1509
$k_3/g \cdot (cm^2 \cdot min)^{-1}$	0.00135	0.00177	0.00238	0.00294	0.00327

作 $\log k_3$ 对 $1/T$ 图，经用最小自乘法处理，得一直线关系，相关系数 $r = 0.97$，见图 5。

$$\log k_3 = -\frac{8190}{T} + 2.09$$

根据 Arrhenius 方程式求出气态脱硫反应的活化能 $E = 37500 cal/mol$，与 Чучмарев 等人[19]的研究结果，$E = 30000 cal/mol$ 很接近。

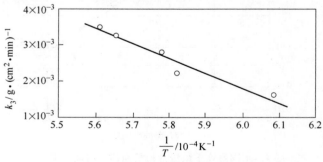

图 5　比速度常数与 $1/T$ 的关系

2.4 三氧化铁在气态脱硫中的作用

图 6 表示一个对基本上同类型的炉渣（$\Sigma FeO\ 0.8\%$，碱度 1.1）不加 Fe_2O_3 粉和加入 $5g\ Fe_2O_3$ 粉（占渣量的 7.2%）的不同顶吹实验。可以看出，加入 Fe_2O_3 粉在 3~4min 内即能使硫降低到很低，证实了 Neuhaus 等人[20]类似研究的结果。加入的 Fe_2O_3 粉在熔渣上迅速地熔化而进行分解，自炉子的窥视孔可以明显地看到渣面的沸腾现象。

为了肯定在中性气氛下，渣内 Fe_2O_3 有无直接与硫起反应的可能，又进行用氩气顶吹熔渣的实验，石英吹管口距渣面 20mm。表 4 给出硫量降低的结果，同时指出 Fe_2O_3 值

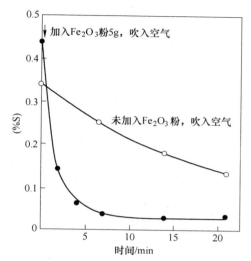

图 6 加入 Fe_2O_3 粉与不加 Fe_2O_3 粉顶吹熔渣的实验
（空气直吹渣面；流量：2.3L/min；温度：1445~1455℃）

随氩气吹入时间加长而减少，相反地 FeO 增加，说明在熔渣内 Fe_2O_3 对硫有氧化作用。St. Pierre 等人[14]曾用高 Fe_2O_3 高硫的铁酸钙在 SO_2 气氛下停留一定的时间，所得的结果和我们的相符合。

表 4 氩气顶吹实验（1485℃）①

取样时间/min	熔渣含量/%					
	CaO	SiO_2	Al_2O_3	FeO	Fe_2O_3	S
0	43.76	6.13	42.85	3.11	3.61	0.35
7	—	—	—	6.10	0.21	0.21
21	—	—	—	5.96	0.05	0.19

① 炉号 140。

为了避免 Fe_2O_3 的变化引起气态脱硫的影响，进行了另一组实验：空气不吹到而远离渣面，相距 300mm，流量仍与以前相同。炉渣含 ΣFeO 为 14%~15%，但 j 有不同值，碱度（% CaO/% SiO_2）为 6.0~6.2。为了保证炉渣有良好的流动性，Al_2O_3 含量提高到 36%~37%。吹炼结果证明 Fe_2O_3 量基本上不变，Δj 自 0 ~ 0.08，而 v_S 对（%S）均系一直线（图 7），说明在 ΣFeO 及 j 值基本上不变的条件下，气态脱硫是一级反应。

对图 7 的 Ⅰ~Ⅳ 四条直线的 j 取平均值，作 v_S 对此平均 j 值的曲线图，得出，v_S 与 j 值正比的关系（图 8），其中（S）= 0.4%线采用一部分的外插值。

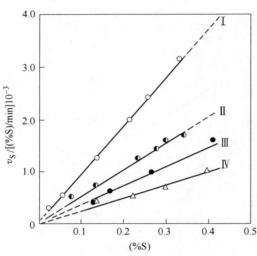

图 7 气态脱硫速度与熔渣含硫量的关系
（空气不吹到渣面；流量：2.3L/min；温度：
1460~1490℃；ΣFeO：14%~15%；碱度：6.0~6.2）
Ⅰ—j=0.61；Ⅱ—j=0.27~0.33；
Ⅲ—j=0.10~0.17；Ⅳ—j=0.05~0.08

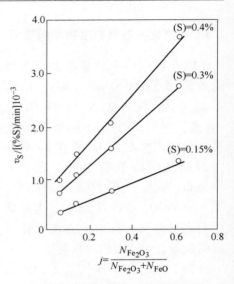

图 8 气态脱硫速度与 j 值的关系
（空气不吹到渣面；流量：2.3L/min；
温度：1460~1490℃；ΣFeO：
14%~15%；碱度：6.0~6.2）

令 v_S 的表示法改换为 $N_{CaS}/\min$（其因次 = $\min^{-1}$），作 $\log v_S$ 对 $\log(N_{CaS})(N_{Fe_2O_3})$ 的曲线（图9），经用最小自乘法处理，得出下列直线关系（相关系数 r = 0.92）：

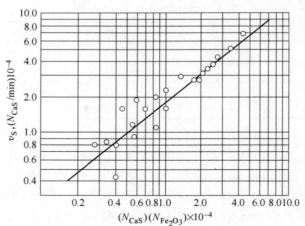

图 9 气态脱硫速度与熔渣 Fe_2O_3 及含硫量的关系
（空气不吹到渣面；流量：2.3L/min；温度：1460~1490℃；
ΣFeO：14%~15%；碱度：6.0~6.2）

$$\log v_S = 0.8053\log(N_{CaS})(N_{Fe_2O_3}) - 0.4965$$

或

$$v_S = 0.32\,(N_{CaS}N_{Fe_2O_3})^{0.81}, \min^{-1} \tag{6}$$

根据图 7

$$v_S = k'(\%S), \min^{-1} \tag{6a}$$

式中，k' 为常数。在 ΣFeO 及 j 基本上守常的条件下，式（6）可改写为：

$$v_S = k'' N_{CaS}^{0.81}, \min^{-1} \tag{6b}$$

式中，k'' 为常数。式（6b）中变量 N_{CaS} 的指数不再是 1，可能由于该组实验中 ΣFeO 及 j 实际上在一小范围内波动，（%S）换成 N_{CaS} 引起小的离差，以及实验上存有误差所致。

2.5 炉渣碱度对气态脱硫的影响

采用三种碱度、$\Sigma FeO \approx 8\%$ 而开始的 $j \approx 0.54$ 的熔渣进行空气吹到渣面的实验。吹炼的平均温度 1500℃。图 10 指出（%S）与顶吹时间的关系（每个点代表三个试验的平均值），说明碱度的增加对气态脱硫不利。同样的结论也可从比较图 3 及图 7 中两条曲线 Ⅱ 得出；两条曲线 Ⅱ 的 ΣFeO 基本上相同，虽由于空气顶吹距液面高度不同，但图 3 曲线 Ⅱ 的 v_S 较图 7 曲线 Ⅱ 的 v_S 大两倍以上，主要是由于前者的碱度是 1.2，而后者的碱度是 6。Larsen 和 Sordahl[22] 的氧气顶吹实验也得到相同的结果。

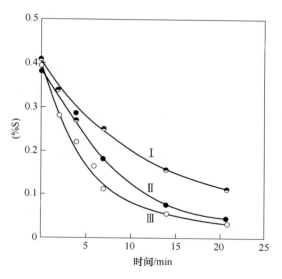

图 10 不同碱度的熔渣含硫量与顶吹时间的关系
Ⅰ—碱度 5.2；Ⅱ—碱度 2.5；Ⅲ—碱度 1.2

3 讨论

3.1 炉渣含硫以硫化物抑以硫酸物存在的问题

炉渣含硫同时以硫化物及硫酸物存在，已有较多的报道。冷凝的平炉渣[23~26]及电炉渣（包括氧化渣及还原性白渣）[25]含有硫酸物已多次被证实。本文作者之一及其他合作者[19]证明了侧吹碱性转炉的炉渣含的硫酸物占全部含硫量的 10%~60%。Чучмарев 等[19]指出，他们采用的 CaO-Al_2O_3-SiO_2 炉渣，无论在还原气氛熔化后的试样，抑在不同期间以氧气吹炼的试样，都含有 0.14%~0.15%以硫酸物存在的硫。我们的研究证明，在熔化后的渣样及吹炼过程中的渣样冷却后均含有不同量的硫酸物（表 5）。这些硫酸物是熔渣在吹炼期间抑或在冷凝过程中生成，涉及熔渣气态脱硫的机理。在本工作实验条件下，p_{O_2} = 0.2 大气压，根据 Richardson 等人[13]的研究和其他人的证实[14,17]，在平衡条件下熔渣内能生成硫酸物，问题是：在气流经常地流动和排出的条件下，硫酸物如果生成，在高温下是否稳定？Neuhaus 等人[20]对纯硫酸钙在 1500~1600℃加入不同氧化物，如 Fe_2O_3、FeO、SiO_2、Al_2O_3 及 CaO 等进行过研究。他们得到的结论是：凡有可与 CaO 结合的氧化物存在时，例如 SiO_2、Fe_2O_3 等，$CaSO_4$ 极易分解成 SO_3，后者再分解成 SO_2 及 O_2；单纯的 $CaSO_4$ 也可分解，但较困难，而加入 CaO 的 $CaSO_4$ 分解最难。为了明确在本工作实验条件下熔渣内 $CaSO_4$ 是否分解，特制备了加入 $CaSO_4$ 的炉渣，用它进行空气顶吹，结果是，随着炉渣碱度和最初含硫量的高低，$CaSO_4$ 有不同的分解速度，分解量可达 95% 以上。因之，可以认为，在高温和适当氧化气氛下，硫酸物如果生成，也是不稳定而立即分解的。冷凝下的炉渣（特别是还原气氛冶炼的炉渣）所含的硫酸物，极可能是在冷却过程中受空气氧化而生成。Archibald、Leonard 和 Mincer[27]在研究高炉渣冷凝过程时，对硫酸物生成的机理作出了解释。值得注意的是，分析炉渣含硫量不能用发生法，而应采用分析全硫的重量法或更较敏捷的 CO_2 气燃烧法[21]。这点在本文实验部分也已提到。

表 5 某些高硫炉渣硫酸物含量

炉 号	以硫酸物加入的总含硫量 /%	以硫酸物存在的流量 /%	备 注
94-1	1.10	0.00	还原气氛下熔化
94-3	0.16	0.02	空气顶吹 14min
94-4	0.12	0.035	空气顶吹 21min
96-4	0.12	0.076	空气顶吹 21min
1-a	1.15	0.10	还原气氛下熔化
4-a	1.39	0.13	还原气氛下熔化
17-a	0.74	0.04	还原气氛下熔化

3.2 熔渣气态脱硫的机理及限制性环节

根据上面推论，含硫熔渣在氧化气氛下虽有机会生成硫酸物，但由于它在高温下不稳定，只可能成为气态脱硫的媒介。所以，对不含氧化铁的熔渣，空气吹炼的反应可能是：

$$(CaS) + 2O_2 \longrightarrow (CaSO_4)$$

$$(CaSO_4) \longrightarrow (CaO) + SO_2 + \frac{1}{2}O_2$$

$$(CaS) + \frac{3}{2}O_2 \longrightarrow (CaO) + SO_2 \tag{7}$$

或

$$(S^{2-}) + 2O_2 \longrightarrow (SO_4^{2-})$$

$$(SO_4^{2-}) \longrightarrow SO_2 + (O^{2-}) + \frac{1}{2}O_2$$

$$(S^{2-}) + \frac{3}{2}O_2 \longrightarrow SO_2 + (O^{2-}) \tag{8}$$

式（8）的反应实际上是下列步骤的总和：

$$2O_2 =\!=\!= 2O_2^* \tag{8a}$$

$$(S^{2-}) =\!=\!= (S^{2-*}) \tag{8b}$$

$$(S^{2-*}) + 2O_2^* \longrightarrow (SO_4^{2-}) \longrightarrow SO_2 + (O^{2-}) + \frac{1}{2}O_2 \tag{8c}$$

$$(S^{2-}) + \frac{3}{2}O_2 =\!=\!= SO_2 + (O^{2-}) \tag{8}$$

式中 (S^{2-*})——气-渣界面的硫离子；

O_2^*——气-渣界面的氧分子。

如果硫的气化是通过硫酸物的分解，那么就没有必要像邹元爔等人[18]那样假定有 SO 生成。但鉴于 SO_4^{2-} 的生成需要两个分子的 O_2^*，因而还不能排斥有其他中间产物分解成 SO_2 的可能。因此，要确定熔渣气态脱硫的机理，尚有待于进一步的研究。

设假定硫离子扩散是反应式（8）最慢的限制性环节，则根据菲克定律：

$$v_S = -\frac{d(\%S)}{d\tau} = \frac{D}{e\delta}[(\%S) - (\%S^*)], s^{-1} \tag{9}$$

式中 D——硫离子的扩散系数，cm^2/s；

δ——边界层厚度，cm；

e——熔渣厚度，cm。

在通常情况下，熔渣内部硫离子浓度远远大于界面的硫离子浓度，也即$(\%S) \gg (\%S^*)$。所以：

$$v_S = \frac{D}{e\delta}(\%S), s^{-1} \tag{10}$$

式（10）指出气态脱硫是一级反应。采用 $D = 2.5 \times 10^{-6} cm^2/s^{[28]}$，本文的 $e = 3cm$，并采用1500℃时 k_3 的数据，将式（10）及式（4）的单位和因次换成相同而使之相等，计算出的 $\delta = 0.8 \times 10^{-3} cm$。这个数值和文献一般给出的数量级尚属符合。

所以根据上面的计算，可以认为，由实验证实的、不含氧化铁（或很少氧化铁）的熔渣的气态脱硫反应可用式（8）表示，而硫离子扩散的步骤是这个反应的限制性环节。上面计算出的较小活化能值（37500cal/mol）也支持这个论点。

对于含氧化铁的熔渣，氩气顶吹实验证明 Fe_2O_3 能参加脱硫反应，可能有下列步骤：

$$4(Fe_2O_3) + (CaS) = (CaSO_4) + 8(FeO)$$

$$(CaSO_4) = (CaO) + SO_2 + \frac{1}{2}O_2$$

$$2(FeO) + \frac{1}{2}O_2 = (Fe_2O_3)$$

$$3(Fe_2O_3) + (CaS) = (CaO) + SO_2 + 6(FeO) \tag{11}$$

或

$$6(Fe^{3+}) + (S^{2-}) + (O^{2-}) = 6(Fe^{2+}) + SO_2 \tag{12}$$

由于氩气冲入渣内，使得反应式（11）或式（12）能在渣内部进行，形成所谓的"容积反应"，但因熔渣内 Fe^{3+} 浓度随时间逐渐减小，气态脱硫也逐渐减弱。

对含氧化铁的熔渣而空气不吹到渣面时，实验证明 j 值基本上不变，气态脱硫也是一级反应（见图7）。由于气流不冲入渣内，渣内部缺乏气相核心，反应式（11）或式（12）只能在气-渣界面也即熔渣表面进行。同时空气可再氧化 Fe^{2+} 成 Fe^{3+}，使反应循环进行，实际上 Fe^{2+} 作为氧的传递剂：

$$6(Fe^{2+}) + \frac{3}{2}O_2 \longrightarrow 6(Fe^{3+}) + 3(O^{2-}) \tag{13}$$

合并式（12）及式（13）：

$$(S^{2-}) + \frac{3}{2}O_2 \longrightarrow SO_2 + (O^{2-})$$

实质上在气-渣界面进行的气态脱硫反应仍是式（8），熔渣的 j 值自然没有什么变化。因之，对含有一定量 ΣFeO 的熔渣，当空气不吹到渣面时，此一级反应的

限制性环节仍是 S^{2-} 的扩散。

当空气吹到渣面，气流冲入渣内，空气与熔渣接触面增大，反应式（8）、式（12）及式（13）均能在渣内部进行，脱硫速度大大加速，而且 j 值随吹炼时间而增加，这样 S^{2-} 的扩散已不再是限制性环节，而气态脱硫反应已不可能是一级反应了（见图3）。

图7指出，原含 Fe^{3+} 较多，也即 j 值较大的熔渣的 v_S 较大，可能由于 Fe^{3+} 能增加 S^{2-} 的扩散系数所致。

从式（6）也可看出，增加熔渣 Fe_2O_3 含量能提高气态脱硫的速度。

最后对熔渣内有无 Fe^{3+} 作一说明。我们认为，作为一个有强氧化势的氧化剂，Fe^{3+} 是有可能在熔渣内存在的，特别在气-渣界面上。对高碱度的熔渣，Fe^{3+} 很可能被结合为 $Fe_2O_4^{2-}$ 或 $Fe_2O_5^{4-}$，降低了氧化势；同时这些复离子与 Fe^{3+} 存有动平衡关系：

$$Fe_2O_4^{2-} \rightleftharpoons 2Fe^{3+} + 4O^{2-}$$

$$Fe_2O_5^{4-} \rightleftharpoons 2Fe^{3+} + 5O^{2-}$$

高碱度熔渣内这些复离子的较稳定性，可能是气态脱硫速度下降的一个原因。

3.3 气态脱硫在转炉炼钢工艺上的应用问题

首先讨论两个问题：（1）金属直接气态脱硫的可能性与条件；（2）炉渣脱硫与通过炉渣气态脱硫存在的矛盾。

采用含 [C] 3.0%，[S] 0.1% 的铁液和较新的热力学数据[29]，考虑到 [S] 的表面活性大100倍[30]以及有关的活度系数 f_S 及 f_C，对金属气态脱硫有关的反应的 ΔF 进行计算。从表6可以看出，代表间接氧化的反应式（D），纵然大量炉气使 SO_2 含量冲淡而降低到体积百分数 0.1%（p_{SO_2} = 0.001 大气压）而铁液含氧量又接近饱和，从热力学角度来看，仍是不可能进行的。比较代表直接氧化的反应式（C）和式（F），可以看出，硫可以直接自铁液气化，但应保证最有利的条件：（1）SO_2 降低到 0.001 大气压之下，方能使式（C）的 ΔF 接近于代表铁氧化的式（F）的 ΔF；（2）铁液内不含有较多量的 Si、Mn 或 C，因为这些元素都能立即消耗引入的氧，而使硫氧化不可能 [比较反应式(H)]。同时由于金属液内铁原子远较硫原子为多，引入的氧大部与 Fe 化合而被吸收为 [O]，使得硫自金属直接气化的机会大为减小。因之，金属的气态脱硫只好通过炉渣进行，这在实验部分已充分地予以证实。

高碱度对炉渣脱硫有利。对转炉炼钢而言，氧化铁含量对炉渣脱硫影响不大。但高碱度对气态脱硫则不利，而高 $N_{Fe^{3+}}$ 对气态脱硫则有利。因之，通过炉

表6 金属直接气态脱硫有关反应的 ΔF 值

编号	反 应 式	备 注
(A)	$[S] + O_{2(气)} = SO_{2(气)}$; $\Delta F^\circ = -54610 + 12.00T$	标准状态
(B)	$[S] + 2[O] = SO_{2(气)}$; $\Delta F^\circ = 1390 + 13.38T$	标准状态
(C)	$[S] + O_{2(空气)} = SO_{2(气)}$; $\Delta F = -54610 - 4.90T$	$p_{O_2} = 0.2$ 大气压,$p_{SO_2} = 0.001$ 大气压,[S] 表面活性大 100 倍
(D)	$[S] + 2[O] = SO_{2(气)}$; $\Delta F = 1390 - 0.30T$	$p_{SO_2} = 0.001$ 大气压,[S] 表面活性大 100 倍,[O] = 0.20%(接近饱和)
(E)	$O_{(气)} = 2[O]$; $\Delta F^\circ = -56000 - 1.38T$	标准状态
(F)	$O_{2(空气)} = 2[O]$; $\Delta F = -56000 - 4.58T$	$p_{O_2} = 0.2$ 大气压,[O] = 0.2%(接近饱和)
(G)	$2[C] + O_{2(气)} = 2CO_{(气)}$; $\Delta F^\circ = -66700 - 20.34T$	标准状态
(H)	$2[C] + O_{2(空气)} = 2CO_{(气)}$; $\Delta F = -66700 - 29.47T$	$p_{O_2} = 0.2$ 大气压,$p_{CO} = \frac{1}{3}$ 大气压

注:ΔF 值单位为 cal,1cal = 4.184J。

渣气态脱硫和炉渣脱硫本身存有一定的矛盾。一般来讲,炉渣氧化铁含量太高意味着炼钢的消耗太大,因之,不应过度地使 $N_{Fe^{3+}}$ 太高。此外,气态脱硫如欲在渣内部进行,必须在渣内先有气相核,方能使反应迅速进行。沸腾时,CO 穿过熔渣能促进含有 Fe^{3+} 的熔渣按式(12)气态脱硫,但另一方面,它又能使 Fe^{3+} 还原,使气态脱硫不利。

从上面分析可以看到,正确运用气态脱硫与炉渣脱硫,以促进脱硫总效率的提高,是一项较复杂的问题。对转炉炼钢工作者来说,一旦硫自金属进入炉渣,即已完成脱硫任务。问题在于如何通过气态脱硫使炉渣含硫量降低,从而促使硫自金属更多、更快地进入炉渣,使脱硫总效率提高。可以看出,充分发挥气态脱硫的作用和转炉炼钢的工艺操作有密切关系,例如氧化气流的吹入方式(顶吹、侧吹或底吹),吹入熔渣抑吹入金属,氧化气流的浓度和压力,喷嘴距液面的距离,喷嘴的形状以及造渣制度等等。转炉炼钢工作者应根据现用的吹炼方法及具体情况,适当地调整操作程序,制定适当的工艺规程,以促进气态脱硫,从而提高脱硫总效率。

4 结论

(1) 空气顶吹过程中熔渣气态脱硫按下列方程式进行:

$$(S^{2-}) + \frac{3}{2}O_2 = SO_2 + (O^{2-})$$

熔渣的铁离子作为氧的传递媒介：

$$6(Fe^{3+}) + (S^{2-}) + 2(O^{2-}) = 6(Fe^{2+}) + SO_2$$

$$6(Fe^{2+}) + \frac{3}{2}O_2 = 6(Fe^{3+}) + 3(O^{2-})$$

对不含氧化铁或含一定量的 ΣFeO 而在空气顶吹过程中 $j(=\frac{N_{Fe_2O_3}}{N_{Fe_2O_3}+N_{FeO}})$ 值保持不变的炉渣（当空气不吹到渣面时），气态脱硫是一级反应，反应速度很可能受硫离子扩散所控制。对含氧化铁熔渣而空气吹到渣面时，由于空气的搅动作用，反应式（8）、式（12）及式（13）不只在熔渣表面进行，而且能在渣内部进行，同时反应式（13）进行的比重较反应式（12）为大，渣内 j 值随吹炼时间增加，气态脱硫速度随 j 值的增加而加大。

（2）在空气顶吹的氧化气氛下，S^{2-} 有可能氧化成 SO_4^{2-}，但后者在实验高温下不稳定而分解。冷凝后的炉渣所含的 SO_4^{2-} 极大可能是在冷却过程中受空气氧化而生成。采用燃烧法分析炉渣所含的全硫，以 CO_2 气燃烧较以纯 O_2 燃烧给出更较准确的结果。

（3）提高温度，提高炉渣 ΣFeO 及 Fe^{3+} 含量，均对熔渣的气态脱硫有利。提高炉渣碱度则相反地对气态脱硫不利。

（4）实验结果证明金属的气态脱硫主要通过炉渣进行。热力学计算也指出，只有含 Si、Mn 及 C 较少的金属液在气流强烈流动及排出条件下才能有直接的气态脱硫。

（5）炉渣气态脱硫和炉渣脱硫本身存有一定的矛盾。转炉炼钢工作者应根据现用的吹炼方法及具体情况，调整操作工艺以促进气态脱硫而提高脱硫总效率。

本工作承北京钢铁学院炼钢实验室有关人员大力协助，该院中心实验室进行了大量的试样分析工作，特此致谢。

参 考 文 献

[1] Osann B. Lehrbuch der Eisenhüttenkunde，(Verlag von w. Engelmann, Leipzig, 1926)，Bd. 2, S：46~47.
   另见 Карнаухов М М. Металлургия стали，(ОНТИ, 1934)，Часть 2，стр. 126.
[2] Herty Jr. C H. Trans. AIME, 1929, 84：260.
[3] Herty Jr. C H, Belyea A R, Burkatt E H. Trans. AIME, 1925, 71：512.

另见 Christopher C F, Freeman H, Sanderson J F. Deoxidation of Steel, Memorial Volume to Hetty Jr. C H, (AIME, 1957): 9.

[4] Крянин Н Р. 小型贝氏炉俄式炼钢法. 杨永宜译. 北京: 北京科技出版社, 1952: 4.

[5] Молонов Г Д. *Изв. высш. учебн. заведений. Черн. металлургия*, 1959, (12): 31.

[6] 前原繁, 森田重明, 広瀬丰. 铁と钢, 1958, 44: 1062.
前原繁, 若林一男, 成田进. 1960, 46: 1187.

[7] Graef R. *Stahl Eisen*, 1957: 1.

[8] 上海市冶金工业局中心试验室, 中国科学院冶金陶瓷研究所. 全国转炉会议资料汇编. 上海: 上海科技出版社, 1959, 第一辑: 79~98.

[9] 魏寿昆, 王鑑, 王光雍, 万天骥. 北京钢铁学院 1952~1962 年论文集 (炼钢部分): 1~20.

[10] 北京钢铁学院电冶金教研组资料. 另见文献 [9].

[11] Ступарь С Н. *Сталь*, 1957 (8): 707.

[12] Шимон Ш, Абросимов Е В, Трубин К Г. Применение радиоактивных изотопов в металлургии. Металлургиздат, 1955: 146~177.

[13] Richardson F D, Fincham C J B. *J. Iron Steel Inst.* (London), 1954, 178: 4.

[14] St. Pierre G R, Chipman J. *Trans. AIME*, 1956, 206: 1474.

[15] Молонов Г Д. *Изв. высш. учебн. заведений. Черн. металлургия*, 1958 (8): 53.

[16] Dewing E W, Richardson F D. *J. Iron Steel Inst.* (London), 1960, 194: 446.

[17] Turkdogan E T, Darken L S. *Trans. AIME*, 1961. 221: 464.

[18] 潘鸿芳, 邹元爔. 科学通报, 1963 (7): 58.

[19] Чучмарев С К Есин О А Добрыдень А А *Изв. высш. учебн. заведений. Черн. металлургия*, 1962 (7): 12.

[20] Neuhaus H, Langhammer H J, Kosmider H, et al. *Arch. Eisenhüttenw*, 1962, 33: 505.
另见 *Stahl Eisen*, 1962: 1279.

[21] Fincham C J B, Richardson F D. *J. Iron Steel Inst.* (London), 1952, 172: 53.

[22] Larsen B M, Sordahl L O. Physical Chemistry of Process Metallurgy, ed. G. R. St. Pierre, Interscience, 1961, pt. 2: 1141~1179.

[23] Schenck H. Physikalische Chemie der Eisenhüttenprozesse, (Verlag von J. Springer, Berlin. 1934). Bd. 2, S: 177~178.

[24] Darken L S, Larsen B M. *Trans. AIME*, 1942. 150: 87.

[25] Speith K G, vom Ende H, Mahn G. *Stahl Eisen*, 1958: 27.

[26] Speith K G, vom Ende H, Bardenheuer F, Mahn G. *Stahl Eisen*, 1959: 926.

[27] Archibald W A, Leonard L A, Mincer A M A. *J. Iron Steel Inst.* (London), 1962, 200: 113.

[28] Воронцов Е С, Есин О А. *Изв. АН СССР, Отд. техн. н.*, 1958 (2): 152.

[29] AIME. Electric Furnace Steelmaking, Interscience, 1963, vol. 2: 134~136.

[30] Явойский В И, Вишкарев А Ф. *Изв. высш. учебн. заведении. Черн. металлургия*, 1960 (5): 39.

炉渣氧化铁含量对脱硫的作用*

魏寿昆

摘 要：根据由炉渣完全离子化理论导出的硫分配比公式：

$$\frac{(\%S)}{[\%S]} = \frac{2.56 f_S}{\gamma_{S^{2-}}} \times \frac{\sum n_+ \sum n_-}{n_{FeO}}$$

在充分引用活度系数的条件下，研究炉渣氧化铁含量对脱硫的作用。文内采用三种碱度的炉渣进行了计算，结果证明文献中前人对 FeO 作用四种分歧的评价不是相互矛盾的，而相反地是有内在联系的；它们说明 FeO 在不同条件下显示不同的作用，而这些不同的作用都可用上列硫分配比公式统一地表示出来。最后，对该公式能否适用于碱性化铁炉及侧吹碱性转炉炉渣，以及对该公式存在的某些缺点和问题进行了讨论。

炉渣氧化铁含量对脱硫的作用，文献上存在分歧的结论。早在 1934 年 Bardenheuer 和 Geller[1] 即指出纯氧化铁有脱硫的作用，1600℃ 时硫的分配比 (%S)/[%S] 为 3.6。Grant 和 Chipman[2] 研究平炉渣的脱硫作用，指出渣中氧化铁含量在 3%～70% 范围内变化对硫分配比没有影响。但高炉渣[3] 及电炉渣[4] 脱硫的研究，均指出氧化铁含量越低，硫分配比越高。利用平炉渣数据，Bishop、Lander、Grant 和 Chipman[5] 又进行处理分析，确定在碱度 $(N_{CaO} + N_{MgO} + N_{MnO})/(N_{SiO_2} + N_{Al_2O_3} + N_{PO_{2.5}})$ 小于 2.4 范围内，炉渣氧化铁含量增高，则硫分配比加大。Rocca、Grant 和 Chipman[4] 曾根据分子理论讨论了氧化铁含量对硫分配比的影响，但他们未能成功地用一种脱硫反应或者一个硫分配比公式统一地解释上述氧化铁的四种分歧的作用。Самарин、Шварцман 和 Темкин[6] 利用炉渣完全离子化理论，通过计算证实纯氧化铁渣的硫分配比为 3.6。Борнацкий[7] 对碱度为 2 的炉渣进行同样的计算，证明当氧化铁含量由 20% 降到 10% 时，硫分配比无大变化；但当氧化铁含量低于 7% 时，则硫分配比有显著的增大；而当氧化铁含量为 1% 时，硫分配比达 25.5。他的计算未联系到纯氧化铁渣的脱硫作用，同时更未考虑到在某些一定情况下，氧化铁含量越高，脱硫作用越好。本文旨在通过理

---

* 本文曾在 1962 年 12 月全国第一次冶金过程物理化学学术报告会上宣读。原刊于《金属学报》，1964，7：157~164。

论计算，证明由炉渣完全离子化理论导出的硫分配比公式，在引用活度系数的条件下，可以统一地解释上列的四种分歧的结论，并对该公式在冶炼工艺实践应用上作进一步的讨论。

众所周知[6~8]，铁液内[S]按分配定律进入炉渣：

$$[FeS] = (Fe^{2+}) + (S^{2-}) \tag{1}$$

而炉渣内 Fe^{2+} 结合 O^{2-} 又按分配定律进入铁液：

$$(Fe^{2+}) + (O^{2-}) = [FeO] \tag{2}$$

所以脱硫反应可写为：

$$[FeS] + (O^{2-}) = (S^{2-}) + [FeO]$$

或

$$[S] + (O^{2-}) = (S^{2-}) + [O] \tag{3}$$

假定炉渣内酸根离子有 SiO_4^{4-}、AlO_3^{3-} 和 PO_4^{3-}，则无论根据式（1）或式（3）均可导出：

$$\frac{(\%S)}{[\%S]} = \frac{32 L_S f_S}{\gamma_{S^{2-}}} \times \frac{\sum n_+ \sum n_-}{n_{FeO}} \tag{4}$$

$$\sum n_+ = n_{CaO} + n_{MgO} + n_{MnO} + n_{FeO}$$

$$\sum n_- = n_{CaO} + n_{MgO} + n_{MnO} + n_{FeO} - n_{SiO_2} - n_{Al_2O_3} - n_{P_2O_5} + n_S$$

式中，各摩尔数 n 均指100g炉渣内各组分的摩尔数值；L_S 为硫的分配常数，用式（5）表示：

$$L_S = \frac{\gamma_{Fe^{2+}} N_{Fe^{2+}} \gamma_{S^{2-}} N_{S^{2-}}}{[\%S]} \tag{5}$$

f_S 为铁液内硫的活度系数；$\gamma_{S^{2-}}$ 为渣液内硫离子的活度系数，而

$$\lg \gamma_{S^{2-}} = 1.53 \sum N_{SiO_4^{4-}} - 0.17^{[8]} \tag{6}$$

$$\sum N_{SiO_4^{4-}} = \frac{n_{SiO_2} + 2n_{P_2O_5} + 2n_{Al_2O_3}}{\sum n_-} \tag{7}$$

各式中的摩尔分数 N 均按焦姆金（Темкин）[9]法计算。

根据文献[6]，$L_S = 0.052 \sim 0.095$。如按 $L_S = 0.08$ 计算，则式（4）变为：

$$\frac{(\%S)}{[\%S]} = \frac{2.56 f_S}{\gamma_{S^{2-}}} \times \frac{\sum n_+ \sum n_-}{n_{FeO}} \tag{8}$$

作者选定三种不同碱度的，符合碱性化铁炉、侧吹碱性转炉及平炉典型操作的炉渣，其成分见表1，并按式（8）计算硫分配比。在计算时，炉渣的氧化铁含量作适当的变更，而其他氧化物的比例，包括碱度，则基本上维持不变。计算的实例另行发表，计算的综合结果见表2。为了相互比较起见，f_S 都按1计算。图1及图2分别给出 $\sum n_+ \sum n_- / n_{FeO}$ 和 $(\%S)/[\%S]$ 对 $(\%FeO)$ 的关系。

表 1 计算用的炉渣成分

炉渣	炉渣成分/%								%(CaO+MgO)/%(SiO$_2$+P$_2$O$_5$)
	SiO$_2$	P$_2$O$_5$	Al$_2$O$_3$	CaO	MgO	MnO	FeO	S	
碱性化转炉	23.9	—	5.95	49.75	16.90	1.0	1.0	1.5	2.79
侧吹碱性转炉	5.3	3.5	6.3	43.6	10.8	—	30.0	0.5	6.18
平炉	14.8	1.4	1.6	41.3	4.1	6.6	30.0	0.2	2.80
平炉	26.2	—	3.5	35.1	5.0	—	30.0	0.2	1.53

表 2 不同碱度时 $\sum n_+ \sum n_-/n_{FeO}$ 和硫分配比对 %FeO 的关系

项目	$\sum n_+ \sum n_-/n_{FeO}$			(%S)/[%S]		
碱度 %FeO	1.53	2.79	6.18	1.53	2.79	6.18
0.5	62.6	176.0	—	—	92.1	—
1.0	—	88.3	—	—	47.1	—
1.5	—	59.2	—	—	32.2	—
2.0	16.5	44.6	—	—	24.8	—
3.0	—	29.7	—	—	16.5	—
5.0	5.78	17.9	26.9	—	10.2	39.5
10	4.02	—	—	—	—	—
20	2.46	4.91	6.80	0.3	5.6	11.1
30	1.97	3.50	4.57	0.6	5.1	8.6
50	—	—	2.74	—	—	6.3
60	1.50	1.98	2.30	2.0	4.5	5.9
80	1.42	—	—	3.5	—	—
90	1.41	—	—	3.6	—	—
100	1.39	1.39	1.39	3.6	3.6	3.6

由理论计算得到以下结论：

(1) 氧化铁对 $\sum n_+ \sum n_-/n_{FeO}$ 起着双方面作用，但由于 n_{FeO} 对分母所引起的作用较大，总的来讲，随着 n_{FeO} 的增加，$\sum n_+ \sum n_-/n_{FeO}$ 值逐渐变小。这说明增加炉渣的氧化铁含量对脱硫在绝大多数情况下（低碱度炉渣除外）是不利的。当 n_{FeO}

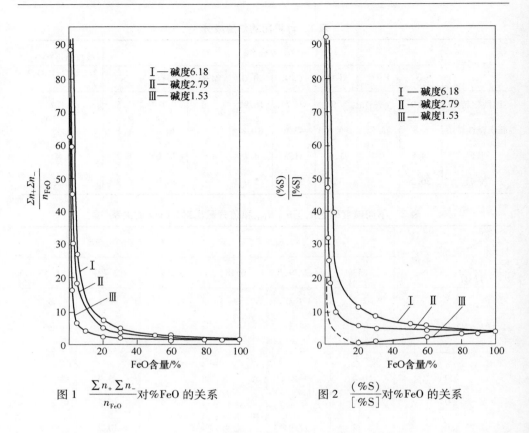

图 1 $\dfrac{\Sigma n_+ \Sigma n_-}{n_{FeO}}$ 对 %FeO 的关系 　　　图 2 $\dfrac{(\%S)}{[\%S]}$ 对 %FeO 的关系

增加时，$\Sigma n_+ \Sigma n_- / n_{FeO}$ 值的下降率是不一致的。当 FeO 含量<2%～3%时，每增加 0.5%FeO 使 $\Sigma n_+ \Sigma n_- / n_{FeO}$ 值大大地下降；当 FeO 含量 >10%～20%时，$\Sigma n_+ \Sigma n_- / n_{FeO}$ 值下降率较缓；而当 FeO 含量 >50%～60%时，$\Sigma n_+ \Sigma n_- / n_{FeO}$ 值下降率更缓，几乎接近一常数而不变。

(2) 比较图1内，Ⅰ、Ⅱ、Ⅲ三条线可以看出，当氧化铁含量相同时，碱度越大，则 $\Sigma n_+ \Sigma n_- / n_{FeO}$ 值越大，这说明炉渣的碱度增高对脱硫有利。

(3) 研究脱硫作用或硫分配比时，在一定炉渣碱度下，不能孤立地只考虑氧化铁的作用，必须把铁液内硫的活度系数 f_S 和渣液内硫离子的活度系数 $\gamma_{S^{2-}}$ 全面地考虑。随着成分的不同，生铁的 $f_S = 3 \sim 6$。由式（6）表示的 $\gamma_{S^{2-}}$ 是按平炉类型的炉渣推出，在本计算中假定也适用于碱性化铁炉、碱性转炉，甚至低碱度类型的炉渣。$\gamma_{S^{2-}}$ 对硫分配比有很大影响。它可抵消一部分，甚至大部分 $\Sigma n_+ \Sigma n_- / n_{FeO}$ 值对脱硫的有利作用。对高碱度的炉渣，随着氧化铁含量的增加，一般来讲，$\Sigma n_+ \Sigma n_- / n_{FeO}$ 值是下降的。但对低碱度的炉渣，随着氧化铁含量的增加，$\Sigma n_+ \Sigma n_- / n_{FeO}$，因之硫分配比在一定范围内是可能上升的，计算中碱度1.53的炉渣（图2曲线Ⅲ）便是如此，但在这种情况下，纵然硫分配比随氧化铁含量的增加而上升,但它的最高值也不超过氧化铁炉渣的 3.6,此值远较高碱度炉渣的硫分配比为小。

明显地可以看出,根据炉渣完全离子化理论和活度系数的引用,氧化铁对脱硫的作用可以较全面地予以解释,不同学者对不同氧化铁含量对脱硫作用的不同估价,正如文献提出的四种分歧的结论,实际上不是相互矛盾,而是有内在联系的,它们说明氧化铁在不同条件下显示出不同的作用,这些不同作用都可用式(8)统一地表示出来。

上面提到的 Борнацкий[7] 的理论计算采用了同样的方法,但他未计算含 20%~100%FeO 的炉渣,也未采用 $\gamma_{S^{2-}}$;因之,他未发现 $\gamma_{S^{2-}}$ 对低碱度渣在一定范围内有促使硫分配比升高的影响。

显然,图 2 中曲线 III 指出硫分配比有一最低值,计算证明,出现最低值的碱度随炉渣成分不同而不同,同时相当于最低值的 %FeO 也随炉渣成分而异,进一步的计算指出,曲线 III 在高氧化铁区域尚出现一"驼峰",但此驼峰值大于 3.6 不多,所以一般来说,可认为低碱度渣的硫分配比是逐步上升到 3.6 的(参阅表 3 以纯 $CaO\text{-}FeO\text{-}SiO_2$ 三元系炉渣计算的结果)。

表 3　纯 $CaO\text{-}FeO\text{-}SiO_2$ 炉渣的硫分配比 (%S)/[%S]

%FeO	碱　度		
	2.0	2.5	2.8
1	19.2	—	—
5	2.8	8.4	12.7
10	1.2	5.2	6.2
20	—	3.8	5.1
30	—	3.5	4.4
40	—	3.5	4.1
60	2.6	3.8	4.4
70	3.6	4.2	4.4
80	3.9	4.0	4.1
90	3.7	3.7	3.8
100	3.6	3.6	3.6

对炼钢工艺来说,增加炉渣氧化铁含量意味着提高金属铁的消耗。因之,某些炼钢工作者企图用提高炉渣氧化铁含量以增加脱硫作用是值得商榷的。

在二氧化硅以 SiO_4^{4-} 形态存在于渣中的假定下,式(8)不能应用于高氧化硅含量的炉渣。当 $SiO_2>30\%$ 时，则 $\sum SiO_4^{4-}>1$ 而 $N_{O^{2-}}$ 成负值，再用式（6）以计算 $\gamma_{S^{2-}}$ 即不合理。图 2 曲线Ⅲ不能用式（8）计算的部分用虚线示意地表示出来。对于一般含二氧化硅高于 30% 的高炉渣，如欲利用式（8）计算硫分配比，则必须先确定二氧化硅在渣中的存在形态（可能是更复杂的硅酸根离子，铝酸根离子也类似），再求出相当于式（6）$\gamma_{S^{2-}}$ 对存在的阴离子的关系。这些问题有待进一步的研究，但从理论上推导，利用式（8）计算高炉渣的硫分配比的可能性是存在的。

式（8）符合于平衡状态，对平炉纯沸腾末期的炉液，式（8）基本上是适用的[7]。作者利用一些碱性化铁炉和侧吹碱性转炉实际操作的数据根据式（8）进行计算（见表 4 及表 5），得出下列结论：（1）碱性化铁炉脱硫反应可认为达到平衡，这和高炉内脱硫反应达不到平衡的情况不同[14]；（2）侧吹碱性转炉实际的硫分配比低于计算值，这可能是由于脱硫反应未达平衡，而更主要的原因是由于气化去硫反应的存在[15]。

式（8）的 L_S 与温度有关，纯氧化铁渣硫分配比与温度的关系，根据较近的资料[16]，由式（9）表示：

$$\lg\frac{(\%S)}{[\%S]} = \frac{2620}{T} - 0.827 \tag{9}$$

由式（4）换算得出：

$$\lg L_S = \frac{2620}{T} - 2.476 \tag{10}$$

式（10）指出 L_S 随温度上升而下降，但根据 Борнацкий[7]的式（11）：

$$\lg L_S = -\frac{3160}{T} + 0.46 \tag{11}$$

则 L_S 随温度上升而增加，这种分歧尚待于进一步的研究。由于 f_S 及 $\gamma_{S^{2-}}$ 或多或少地也与温度有关，由式（8）推论硫分配比与温度的关系更显复杂。至于增高温度引起动力学方面脱硫的影响，以及三氧化铁能帮助化渣，从而使脱硫反应易于进行，这里不加讨论。

结 论

理论计算证明，由炉渣完全离子化理论导出的硫分配比公式：

$$\frac{(\%S)}{[\%S]} = \frac{2.56 f_S}{\gamma_{S^{2-}}} \times \frac{\sum n_+ \sum n_-}{n_{FeO}}$$

在引用活度系数的条件下，可以较全面地解释文献中氧化铁对脱硫的四种不同评价，即：

（1）纯氧化铁渣可以脱硫。
（2）平炉渣氧化铁量在 3%~70% 范围变化对硫分配比无影响。
（3）电炉（或高炉）渣氧化铁含量越少，则硫分配比越高。

表 4 碱性化铁炉脱硫资料

来源	炉号	铁液温度/°C	生铁配入 S/%	铁液成分/%					炉渣成分/%							(%S)/[%S]	
				C	Mn	Si	P	S	CaO	MgO	MnO	ΣFeO	SiO₂	Al₂O₃	P₂O₅	S	实际	计算
上海[11]	3-198	1390	0.66	3.40	0.39	1.90	0.305	0.055	44.73	18.17	0.34	2.60	27.0	6.50	1.0	1.45	26.3	36.3
	3-199	1400	0.66	3.08	0.21	0.92	0.39	0.19	47.08	16.44	0.51	2.20	25.40	6.40	1.0	1.94	10.2	41.6
	3-205	1390	0.96	2.64	0.25	1.78	0.45	0.24	40.26	16.44	0.63	0.80	31.10	6.80	1.0	3.74	15.5	22.6
	3-18A-210	1375	0.96	2.54	0.27	1.55	0.42	0.16	45.54	13.55	0.37	0.45	31.00	4.80	0.58	3.71	22.8	41.4
	3-21A-213	1340	1.76	2.68	0.22	1.90	0.36	0.103	45.17	15.12	0.46	1.60	28.70	5.40	1.0	4.30	41.6	33.0
	2-8-82-1	1325	0.58	3.31	1.12	1.70	0.34	0.012	48.45	17.12	0.48	1.40	23.70	4.80	—	1.49	124.0	95.0
美国[13]	5	—	—	3.00(估计)	—	—	—	0.047	41.2	19.6	1.2	1.8	28.6	6.5	—	0.56	11.9	19.4
	6	—	—	—	—	—	—	0.030	32.5	28.6	0.6	2.9	26.0	9.2	—	1.26	42.0	21.6
	7	—	—	—	—	—	—	0.032	40.3	23.2	0.6	2.0	25.8	6.5	—	3.08	96.0	50.3
	8	—	—	—	—	—	—	0.010	62.4	9.0	0.6	0.4	18.8	12.5	—	2.50	250.0	268.0

表 5 侧吹碱性转炉脱硫资料

来源	炉号	试样	铁液成分/%					炉渣成分/%							(%S)/[%S]	
			C	Mn	Si	P	S	CaO	MgO	MnO	ΣFeO	SiO₂	Al₂O₃	P₂O₅	S	实际	计算
唐山[13]	5-4148	不留渣法，氧化渣	0.05	0.18	—	0.024	0.038	46.00	5.31	6.97	23.40	13.06	1.20	3.60	0.203	5.3	7.3
	1-4637	不留渣法，扒渣	3.33	0.36	0.14	0.605	0.057	49.80	6.80	5.81	5.20	25.86	2.97	2.86	0.287	4.3	9.0
	3-836	留渣双渣法，混合渣	3.45	0.62	0.90	0.224	0.064	48.36	7.91	5.64	12.50	18.58	2.13	4.81	0.202	4.6	19.5
上海[11]	9-822	留渣法，吹炼31min后样	0.54	0.31	0.02	0.051	0.098	47.19	6.40	6.26	13.08	8.76	—	16.85	0.105	1.1	8.0
	9-822	留渣法，吹炼36min后样	0.10	0.22	0.10	0.065	0.063	47.02	8.50	5.68	16.04	8.20	—	18.27	0.180	2.8	6.8
	9-823	不留渣法，吹炼28min后样	0.07	0.14	0.02	0.058	0.071	46.20	5.20	3.55	24.60	6.00	—	12.65	0.300	4.2	7.6
	9-823	不留渣法，吹炼32min后样	0.04	0.10	0.08	0.050	0.053	42.20	6.00	3.42	31.60	5.40	—	10.85	0.214	4.0	7.4
唐山[13]	3-548	后吹试验，扒渣法，第一次样	0.06	—	—	0.046	0.055	27.80	7.74	0.47	48.80	6.64	2.86	4.49	0.172	3.1	4.8
	3-2531	留渣法，扒渣后第一次样	0.07	—	—	0.013	0.053	38.66	9.00	0.70	34.20	12.28	0.84	3.09	0.122	2.3	6.0
	3-2532	留渣法，扒渣后第二次样	0.04	—	—	0.019	0.058	38.99	4.87	0.35	41.50	8.54	1.76	2.91	0.202	3.5	6.0

（4）对低碱度炉渣在一定碱度范围内，增加氧化铁含量能使硫分配比加大。

上列四种分歧的结论不是相互矛盾的，而是有内在联系的。它们说明氧化铁在不同条件下显示不同的作用，而这些不同的作用都可用上列公式统一地来表示。

由于假定二氧化硅以 SiO_4^{4-} 形态存在于炉渣中，这个公式不能应用于酸性渣，同时硫分配比与温度的关系尚有待于进一步的阐明。

采用碱性化铁炉实际操作的数据进行计算，得知碱性化铁炉内的脱硫反应达到平衡。侧吹碱性转炉由于有气化去硫反应的存在，实际的硫分配比则较计算值为小。

参 考 文 献

[1] Bardenheuer P, Geller W. Mitt K. with. Inst. Eisenforsch, 1934, 16: 77.
[2] Grant N J, Chipman J. Trans. AIME, 1946, 167: 134.
[3] Hatch G G, Chipman J. Tran. AIME, 1949, 185: 274.
[4] Rocca R, Grant N J, Chipman J. Trans. AIME, 1951, 191: 319.
[5] Bishop H L, Lander H N, Grant N J, Chipman J. Trans. AIME, 1956, 206: 862.
[6] Самарин А М, Шварцман Л А, Темкин М. Ж. физ. химии, 1946, 20: 111.
[7] Борнацкнй И И. 平炉钢脱硫. 胡可夫译. 北京: 冶金工业出版社, 1955: 8, 11~12, 108.
[8] Самарин А М, Шварцман Л А. Изв. АН СССР. Отд. техн. н. 1948（9）: 1457.
[9] Темкин М. Ж. физ. химии, 1946, 20: 105.
[10] 魏寿昆. 活度在冶金物理化学中的应用. 北京: 中国工业出版社, 1964.
[11] 上海冶金工业局. 全国转炉会议资料汇编 第一辑. 上海: 上海科技出版社, 1959: 53~57, 80~81.
[12] Carter S F. Trans. Am. Foundrymen's Soc., 1953, 61: 52.
[13] 作者个人搜集资料.
[14] Filer E W, Darken L S. Trans AIME, 1952, 194: 253.
[15] 魏寿昆, 王鑑, 王光雍, 万天骥. 北京钢铁学院1952~1962年论文集, 炼钢部分, 北京钢铁学院, 1962: 1~20.
[16] Томилин И А Шварцман Л А. Изв. АН СССР, Отд. техн. н., 1956（10）: 122.

高炉型渣脱硫的离子理论[*]

魏寿昆

摘　要：根据新提出的碱度指标，对低碱度高炉型渣的离子结构作了假定。利用前文提出的、适用于各种碱性炼钢型渣的硫分配比公式：

$$\frac{(\%S)}{[\%S]} = \frac{32 L_S f_S}{\gamma_{Fe^{2+}} + \gamma_{S^{2-}}} \times \frac{\sum n_+ \sum n_-}{n_{FeO}}$$

对 Hatch 和 Chipman、Taylor 和 Stobo 以及 Куликов 等的饱和碳铁液和高炉型渣平衡实验数据进行处理分析，得出：

$$\lg \gamma_{Fe^{2+}} \gamma_{S^{2-}} = -53.5 N_{O^{2-}} + 2.12$$

进一步计算证明了该硫分配比公式也适用于一般的低碱度渣。对碱性炼钢型渣适用的、关于 FeO 含量对脱硫作用的规律证明了也适用于低碱度渣。将 f_S 与 $(\%S)/[\%S]$ 合并考虑，则 $(\%S)/a_S$ 基本上不受温度和铁液成分的影响。

1　引言

在前文[1]内，本文作者根据由炉渣完全离子化理论导出的硫分配比公式[❶]：

$$\frac{(\%S)}{[\%S]} = \frac{32 L_S f_S}{\gamma_{Fe^{2+}} + \gamma_{S^{2-}}} \times \frac{\sum n_+ \sum n_-}{n_{FeO}} \qquad (1)$$

通过引用活度系数的理论计算，证明了该公式适用于各种炼钢型炉渣，并能全面地解释前人关于炉渣 FeO 含量对脱硫作用的四种分歧意见。但该公式不能应用于 SiO_2 含量一般 >30% 的低碱度炉渣，特别是碱度较低的高炉型渣。本文企图证明该公式也可适用于高炉型渣，并对可能适用的条件进行探索性的尝试。

文献上对高炉型渣的实际脱硫分配比的计算公式有所报道[2,3]。由于高炉内硫在渣铁液间的分配远未达到平衡[4,5]，不少人进行了高炉型渣（包括二元系及三元系渣）和饱和碳铁液间硫分配的研究[4~13]。表 1 综合了近年来的研究结果。可以看出，高炉型渣的脱硫作用都根据分子理论来解释。Frohberg[18] 曾企图利用离子理论研究高炉型渣的脱硫作用，但迄未能给出计算硫分配比的公式。

---

[*]　原刊于《金属学报》，1966，9：127~141。

[❶]　前文内，以 2.56 代 $32 L_S$，因 $L_S = 0.08$；根据 Самарин 等的假定，$\gamma_{Fe^{2+}} = 1$，因之略去 $\gamma_{Fe^{2+}}$。

表 1 高炉型渣的脱硫

类别	来源	公式	备注
高炉操作	Гольдштейн[2]	$\dfrac{(\%S)}{[\%S]} = 10.9[\%Si]\dfrac{(\%CaO)}{(\%SiO_2)} + 21.7\dfrac{(\%CaO)}{(\%SiO_2)} + 1.8[\%Si] - 16.2$	适用于平炉生铁: Si 0.5%~1.1%; Mn 1.2%~1.8%
	Гольдштейн[2]	$\dfrac{(\%S)}{[\%S]} = 36.9[\%Si]\dfrac{(\%CaO)}{(\%SiO_2)} - 27.3[\%S] + 6.7$	适用于铸造生铁: Mn 0.5%~0.8%
	Воскобойников[3]	$\dfrac{(\%S)}{[\%S]} = 98B^2 - 160B + 72 - [0.6(\%Al_2O_3) - 0.012(\%Al_2O_3)^2 - 4.032]B^4$ $B = \dfrac{(\%CaO) + (\%MgO) + (\%MnO)}{(\%SiO_2)}$	适用于 1450℃
饱和碳铁液的平衡实验	Hatch 和 Chipman[4]	$\dfrac{(\%S)}{[\%S]} = f(剩余碱)$ $剩余碱 = (n_{CaO} + \dfrac{2}{3}n_{MgO}) - (n_{SiO_2} + n_{Al_2O_3})$	100g 炉渣的摩尔数
	邹元爔[14]	$\dfrac{(\%S)}{[\%S]} = f(B)$ $B = \dfrac{(\%CaO) + (\%MgO)}{(\%SiO_2) + 0.6(Al_2O_3)\left[\dfrac{(\%CaO) + (\%MgO)}{(\%SiO_2)} - 1.19\right]}$	
	Куликов 等[9]	$\dfrac{(\%S)}{[\%S]} = 17.4B^{1.45}$ $B = \dfrac{(\%CaO) + 0.5(\%MgO) + (\%MnO) - 1.75(\%S)}{(\%SiO_2) + 0.6(\%Al_2O_3)\left[\dfrac{(\%CaO) + 0.5(\%MgO) + (\%MnO) - 1.75(\%S)}{(\%SiO_2)} - 1.19\right]}$	
	Куликов[15]	$\lg\dfrac{(\%S)}{[\%S]} = \lg f_S - \lg p_{CO} + 2.55B - 2.07$ $B = \dfrac{(\%CaO) + \alpha(\%MgO) + 2(\%MnO)}{(\%SiO_2) + 0.6(\%Al_2O_3)\left[\dfrac{(\%CaO) + \alpha(\%MgO)}{(\%SiO_2)} - 1.19\right]}$ $\alpha = \dfrac{1.84(\%SiO_2) - 0.9(\%CaO)}{(\%SiO_2) + 0.9(\%MgO)}$	p_{CO} = CO 的分气压（大气压）

续表 1

类别	来源	公式	备注
饱和碳铁液的平衡实验	Schenck、Frohberg 和 Gammal [10~13]	$\lg\dfrac{(\%S)}{a_S} = \left\{\left[0.6 - 0.31\left(\dfrac{n_{CaO}}{n_{SiO_2}}\right)^2\right] \cdot \dfrac{n_{Al_2O_3}}{n_{SiO_2}} + \dfrac{n_{CaO}}{n_{SiO_2}}\right\}^2 +$ $0.06\left(\dfrac{n_{CaO}}{n_{SiO_2}}\right)^2 - 0.65$	$\dfrac{n_{CaO}}{n_{SiO_2}} < 1.39$ (Al_2O_3 呈碱性;1550℃)
		$\lg\dfrac{(\%S)}{a_S} = \left[\dfrac{n_{CaO}}{0.8n_{Al_2O_3} - 1.55n_{Al_2O_3} \cdot \left(\dfrac{n_{SiO_2}}{n_{CaO}}\right)^2 + n_{SiO_2}}\right]^2 +$ $0.06\left(\dfrac{n_{CaO}}{n_{SiO_2}}\right)^2 - 0.65$	$\dfrac{n_{CaO}}{n_{SiO_2}} > 1.39$ (Al_2O_3 呈碱性;1550℃)
	Kalyanram、MacFarlane 和 Bell [16]	$\lg\dfrac{(\%S)}{a_S} = 1.3\left[\dfrac{(\%CaO) + 0.7(\%MgO)}{0.94(\%SiO_2) + 0.18(\%Al_2O_3)}\right] - 0.43$	处理 Hatch 和 Chipman 的数据
	Giedroyc、Mcphail 和 Mitchell [17]	$\lg\dfrac{(\%S)}{a_S} = 1.58\left[\dfrac{(\%CaO) + (\%MgO)}{(\%SiO_2) + (\%Al_2O_3)}\right] + 0.19$	处理 Hatch 和 Chipman 的数据

本文利用 Hatch 和 Chipman[4]、Taylor 和 Stobo[8]以及 Куликов、Кожевников 和 Цылев[9]等平衡实验(实验条件:$a_C = 1$,$p_{CO} = 1$ 大气压)的数据进行处理和分析。

2 理论分析和计算

前文[1]曾指出,无论脱硫反应按式(2)或式(3)进行

$$[FeS] = (Fe^{2+}) + (S^{2-}) \tag{2}$$

$$[S] + (O^{2-}) = (S^{2-}) + [O] \tag{3}$$

硫分配比均按式(1)计算。对含碳的铁液来说,脱硫反应:

$$[S] + (O^{2-}) + [C] = (S^{2-}) + \{CO\} \tag{4}$$

由于式(4)可看作按两步进行:

$$[S] + (O^{2-}) = (S^{2-}) + [O]$$

$$[C] + [O] = \{CO\}$$

不难推导，式（4）的硫分配比也可用式（1）计算。同样地也可以证明，任何有元素 M 参与的脱硫反应

$$x[M] + y[S] + y(O^{2-}) = y(S^{2-}) + (M_xO_y)$$

它的硫分配比也用式（1）计算。

对碱性炼钢型渣，液中有 SiO_4^{4-}、AlO_3^{3-} 及 PO_4^{3-} 等酸根离子，Самарин 和 Шварцман[19] 曾得到下列关系：

$$\lg\gamma_{Fe^{2+}}\gamma_{S^{2-}} = 1.52\sum N_{SiO_4^{4-}} - 0.17 \tag{5}$$

但当渣中 SiO_2 含量>30%时，渣中硅酸根离子结构有所变化，式（5）不适用。他们曾假定 $\gamma_{Fe^{2+}} = 1$，但此假定并无必要。

式（4）的平衡常数是：

$$K_4 = \frac{\gamma_{S^{2-}} N_{S^{2-}} p_{CO}}{f_S[\%S]\gamma_{O^{2-}} N_{O^{2-}} a_C} \tag{6}$$

在平衡实验的条件下，$p_{CO} = 1$ 大气压，$a_C = 1$，所以有：

$$\frac{(\%S)}{f_S[\%S]} = 32K_4 n_{O^{2-}}\left(\frac{\gamma_{O^{2-}}}{\gamma_{S^{2-}}}\right) \tag{7}$$

式中，$n_{O^{2-}}$ 代表 100g 炉渣氧离子的摩尔数。

在碱性渣的范围内，Самарин 等[19] 曾证明 $\gamma_{S^{2-}} = \gamma_{O^{2-}}$。Frohberg[18] 在处理他人数据时也证实了：

$$\frac{\gamma_{O^{2-}}}{\gamma_{S^{2-}}} = 1$$

当 $N_{SiO_4^{4-}} < 0.7$ 时可求出：

$$\lg\gamma_{Fe^{2+}}\gamma_{O^{2-}} = \lg\gamma_{Fe^{2+}}\gamma_{S^{2-}} = 1.21 N_{SiO_4^{4-}} \tag{8}$$

当炉渣含 SiO_2 过高时，$\gamma_{O^{2-}}$ 与 $\gamma_{S^{2-}}$ 不能相等的原因，实质上是由于炉渣已进入酸性范围，原假定的酸根离子结构已不符合实际情况。但如果对含 SiO_2 高的低碱度渣的酸根离子作出合理的假定，我们可以认为 $\gamma_{O^{2-}}$ 与 $\gamma_{S^{2-}}$ 仍是相等的。

根据 $\gamma_{O^{2-}}$ 和 $\gamma_{S^{2-}}$ 相等的假定，式（7）变为：

$$\frac{(\%S)}{[\%S]} = 32 f_S K_4 n_{O^{2-}} \tag{9}$$

根据已知的硫分配比和由表2计算出的 K_4 值，用式（9）可以计算 $n_{O^{2-}}$。如果能计算出炉渣内的 $\sum n_+$ 和 $\sum n_-$，则即可由式（1）计算 $\gamma_{Fe^{2+}}\gamma_{S^{2-}}$ 并试找后者的规律。关键在于如何确定低碱度的高炉型渣的离子结构。

表 2　K_4 的热力学计算数据

$L_S = \dfrac{\gamma_{Fe^{2+}} N_{Fe^{2+}} \gamma_{S^{2-}} N_{S^{2-}}}{f_S [\%S]}$;	$\lg L_S = -\dfrac{920}{T} - 0.5784$ [18]
$L_O = \dfrac{\gamma_{Fe^{2+}} N_{Fe^{2+}} \gamma_{O^{2-}} N_{O^{2-}}}{f_O [\%O]}$;	$\lg L_O = \dfrac{5762}{T} - 2.439$ [20]
$[S] + (O^{2-}) = (S^{2-}) + [O]$　　(3); 　$K_3 = \dfrac{L_S}{L_O}$, 所以　$\lg K_3 = -\dfrac{6682}{T} + 1.861$	
	$\Delta F_3^\circ = 30580 - 8.52T$
$[C]_\% + [O]_\% = \{CO\}$;	$\Delta F^\circ = -5350 - 9.48T$ [21]
$C_{石墨} = [C]_\%$;	$\Delta F^\circ = 5100 - 10.00T$ [21]
所以　$[C]_饱 + [O]_\% = \{CO\}$;	$\Delta F^\circ = -250 - 19.48T$
$[C]_饱 + [S] + (O^{2-}) = (S^{2-}) + \{CO\}$　　(4);	$\Delta F_4^\circ = 30330 - 28.00T$
	所以　$\lg K_4 = -\dfrac{6628}{T} + 6.119$

前人[18,22~24]对液态硅酸盐的结构进行了不少的研究和分析。根据 CaO-Al_2O_3-SiO_2 三元系相图的分析和高炉型渣的岩相研究[25]，随着碱度的不同，高炉型渣可能有 Ca_2SiO_4（硅酸二钙）、$Ca_3Si_2O_7$（钙方柱石）、$CaSiO_3$（硅灰石）、$Ca_2Al_2SiO_7$（铝方柱石）、$CaAl_2Si_2O_8$（钙斜长石）等组成物。采用文献[4,8,9] 的实验数据进行反复分析计算，经找出，对高炉型渣的离子结构作下列的假定较为合理：（1）炉渣所含的硅酸根离子随碱度不同逐步变化；（2）Al_2O_3 一般和 SiO_2 及 O^{2-} 生成 $Al_2SiO_7^{4-}$，在碱度较低时，可同时生成 $Al_2Si_2O_8^{2-}$，而在碱度较高时，也可同时生成 AlO_3^{3-}。表 3 示出假定的高炉型渣离子结构和碱度的关系。

表 3　高炉型渣结构和碱度的关系

碱　度		碱度指标	炉渣内存在的负离子
$\dfrac{\%CaO + \%MgO}{\%SiO_2 + \%Al_2O_3}$	$\dfrac{n_{CaO} + n_{MgO}}{n_{SiO_2} + n_{Al_2O_3}}$	$\dfrac{\sum b - 2n_{Al_2O_3}}{n_{SiO_2} - n_{Al_2O_3}}$	(O^{2-}、S^{2-} 除外)
>1.2	>1.6	>2.0	SiO_4^{4-}、$Al_2SiO_7^{4-}$、AlO_3^{3-}
1.0~1.2	1.4~1.6	1.5~2.0	SiO_4^{4-}、$Si_2O_7^{6-}$、$Al_2SiO_7^{4-}$
0.8~1.0	1.0~1.4	1.0~1.5	$Si_2O_7^{6-}$、$Si_3O_9^{6-}$、$Al_2SiO_7^{4-}$
<0.8	<1.0	<1.0	$Si_3O_9^{6-}$、$Al_2SiO_7^{4-}$、$Al_2Si_2O_8^{2-}$

由于以重量百分数或以摩尔数计算的碱度和假定的离子结构不大严格地符合（见表5），特提出一个新的"碱度指标"，即 $(\sum b - 2n_{Al_2O_3})/(n_{SiO_2} - n_{Al_2O_3})$（式中 $\sum b = n_{CaO} + n_{MgO} + n_{MnO} + n_{FeO}$）。此指标得自于下列的离子生成反应：

$$SiO_2 + Al_2O_3 + 2O^{2-} = Al_2SiO_7^{4-} \qquad (n_{O^{2-}} : n_{SiO_2} = 2:1)$$

$$SiO_2 + 2O^{2-} \rightleftharpoons SiO_4^{4-} \quad (n_{O^{2-}} : n_{SiO_2} = 2:1)$$

$$2SiO_2 + 3O^{2-} \rightleftharpoons Si_2O_7^{6-} \quad (n_{O^{2-}} : n_{SiO_2} = 1.5:1)$$

$$3SiO_2 + 3O^{2-} \rightleftharpoons Si_3O_9^{6-} \quad (n_{O^{2-}} : n_{SiO_2} = 1:1)$$

$$2SiO_2 + Al_2O_3 + O^{2-} \rightleftharpoons Al_2Si_2O_8^{2-} \quad (n_{O^{2-}} : n_{SiO_2} = 0.5:1)$$

碱度指标的分子($\Sigma b - 2n_{Al_2O_3}$)代表生成 $Al_2SiO_7^{4-}$ 后所余的 O^{2-} 摩尔数,而分母($n_{SiO_2} - n_{Al_2O_3}$)代表生成 $Al_2Si_2O_7^{4-}$ 后所余的 SiO_2 的摩尔数,如该指标大于2,则除再能生成 SiO_4^{4-} 外,尚有余 O^{2-} 以生成 AlO_3^{3-};如指标在1.5~2之间,则所余的 SiO_2 不能全部生成 SiO_4^{4-},将有一部分生成 $Si_2O_7^{6-}$;如指标为1~1.5,则剩余的 SiO_2 能生成 $Si_2O_7^{6-}$ 和 $Si_3O_9^{6-}$;如指标小于1,则除生成 $Si_3O_9^{6-}$ 外,尚能生成 $Si_2O_5^{2-}$,但后者不能单独存在,而与 Al_2O_3 生成 $Al_2Si_2O_8^{2-}$。

炉渣的离子结构已定,即可依据式(1)计算 $\gamma_{Fe^{2+}}$ 和 $\gamma_{S^{2-}}$。f_S 根据文献[26]求出,并用文献[27]推荐的方法修正。求炉渣各组成物的摩尔数 n 时,先由炉渣原分析的%S 求出%CaS,并减去与此%CaS 相当的%CaO(对炼钢型渣,由于所含的%S 很少,可不作此种计算),将炉渣各组成物的和按比例调整到100%后再求出各组成物的 n。L_S 和 K_4 均按1500℃时的数值计算:$L_S = 0.08$,而 $K_4 = 240$。

表4给出一个计算 $\gamma_{Fe^{2+}}$、$\gamma_{S^{2-}}$ 的实例。表5给出29个实验数据的计算总结果。

表4 炉号 H-48 的计算实例

炉 号	铁水成分,%	炉渣				
		组成物	原分析,%	改正后,%	调整后,%	摩尔数 n, mol
H-48[4] 1500℃	C 4.76 Si 1.19 S 0.006	CaO	39.46 - 4.34	35.12	34.96	0.6234
		MgO	16.55	16.55	16.47	0.4086
		SiO_2	31.18	31.18	31.04	0.5166
		Al_2O_3	12.00	12.00	11.94	0.1171
		FeO	0.030	0.030	0.030	0.00042
		S	2.48	—	—	
		CaS	— 5.58	5.58	5.56	0.0771
				100.46	100.00	

$$f_S = 6.3 \quad SiO_2 + Al_2O_3 + 2O^{2-} = Al_2SiO_7^{4-}; \quad n_{Al_2SiO_7^{4-}} = 0.1171$$

$$\frac{(\%S)}{a_S} = 66 \quad \quad \Sigma b = n_{CaO} + n_{MgO} + n_{FeO} = 1.0324$$

$$n_{O^{2-}} = \frac{66}{32 \times 240} = 0.0086 \quad \quad \frac{\Sigma b - 2n_{Al_2O_3}}{n_{SiO_2} - n_{Al_2O_3}} = \frac{0.7982}{0.3995} = 1.998$$

续表 4

$$\frac{\%CaO + \%MgO}{\%SiO_2 + \%Al_2O_3} = 1.19 \quad\quad SiO_2 + 2O^{2-} = SiO_4^{4-}; \quad 令 x = n_{SiO_4^{4-}}$$

$$2SiO_2 + 3O^{2-} = Si_2O_7^{6-}$$

$$\frac{n_{CaO} + n_{MgO}}{n_{SiO_2} + n_{Al_2O_3}} = 1.63 \quad\quad 0.7982 - [2x + \frac{3}{2}(0.3995 - x)] = 0.0086$$

所以 $x = n_{SiO_4^{4-}} = 0.3808$

$n_{Si_2O_7^{6-}} = \frac{1}{2}(0.3995 - 0.3808) = 0.0093$

$\sum n_+ = \sum b + n_{CaS} = 1.1095$

$\sum n_- = n_{O^{2-}} + n_{Al_2SiO_7^{4-}} + n_{SiO_4^{4-}} + n_{Si_2O_7^{6-}} + n_{S^{2-}} = 0.5929$

$N_{O^{2-}} = \frac{0.0086}{0.5929} = 0.0145$

$\gamma_{Fe^{2+}}\gamma_{S^{2-}} = \frac{2.56 \times 1.1095 \times 0.5929}{66 \times 0.00042} = 61$

表5 高炉型渣 $\gamma_{Fe^{2+}}\gamma_{S^{2-}}$ 计算的总结果

来源	炉号	$t/℃$	f_S	$\frac{(\%S)}{a_S}$	$\frac{\%CaO + \%MgO}{\%SiO_2 + \%Al_2O_3}$	$\frac{n_{CaO} + n_{MgO}}{n_{SiO_2} + n_{Al_2O_3}}$	$\frac{\sum b - 2n_{Al_2O_3}}{n_{SiO_2} - n_{Al_2O_3}}$	$n_{O^{2-}}$	$n_{SiO_4^{4-}}$	$n_{Si_2O_7^{6-}}$
Hatch 等[4]	H-48	1500	6.3	66	1.19	1.63	2.00	0.0086	0.3808	0.0093
	H-44	1500	6.3	62	1.15	1.43	1.64	0.0081	0.1136	0.1675
	H-39	1500	6.0	37	1.03	1.39	1.61	0.0048	0.0928	1.1758
	H-51	1500	6.3	36	1.08	1.29	1.41	0.0047	—	0.2011
	H-41	1500	8.9	21	0.86	1.19	1.40	0.0027	—	0.1356
	H-56	1500	9.3	43	1.08	1.31	1.40	0.0056	—	0.2045
	H-40	1500	13.3	18	1.06	1.25	1.31	0.0023	—	0.1825
	H-47	1500	8.1	27	0.84	1.09	1.22	0.0035	—	0.0635
	H-34	1500	6.5	34	0.89	1.10	1.20	0.0044	—	0.0725
	H-55	1500	11.2	24	0.81	1.09	1.17	0.0031	—	0.0651
	H-37	1500	10.8	13	0.85	1.07	1.12	0.0017	—	0.0570
	H-31	1500	7.9	11	0.70	0.99	0.98	0.0014	—	—
	H-54	1500	18.6	10	0.71	0.96	0.93	0.0013	—	—
	H-49	1500	12.4	7.8	0.65	0.93	0.81	0.0010	—	—

续表 5

来源	炉号	$n_{Si_3O_9^{6-}}$	$n_{Al_2Si_2O_8^{2-}}$	$n_{Al_2SiO_7^{4-}}$	$n_{AlO_3^{3-}}$	$\sum n_+$	$\sum n_-$	$N_{O^{2-}}$	$\gamma_{Fe^{2+}}+\gamma_{S^{2-}}$	$\lg\gamma_{Fe^{2+}}+\gamma_{S^{2-}}$
Hatch 等[4]	H-48	—	—	0.1171	—	1.1095	0.5929	0.0145	61	1.7853
	H-44	—	—	0.1055	—	1.0182	0.4640	0.0174	41	1.6128
	H-39	—	—	0.1216	—	1.0381	0.4721	0.0102	79	1.8976
	H-51	0.0317	—	0.1008	—	0.9547	0.3883	0.0121	49	1.6902
	H-41	0.0234	—	0.1856	—	0.9325	0.4289	0.0063	174	2.2405
	H-56	0.0386	—	0.0702	—	1.0269	0.4704	0.0119	41	1.6128
	H-40	0.0804	—	0.0568	—	0.9842	0.4016	0.0057	141	2.1492
	H-47	0.0592	—	0.2059	—	0.8509	0.3997	0.0088	53	1.7243
	H-34	0.0816	—	0.1659	—	0.8685	0.3943	0.0112	61	1.7853
	H-55	0.0915	—	0.1549	—	0.9269	0.4589	0.0068	50	1.6990
	H-37	0.1251	—	0.1401	—	0.8771	0.3728	0.0045	178	2.2504
	H-31	0.0832	0.0033	0.2522	—	0.8066	0.3879	0.0036	114	2.0569
	H-54	0.1394	0.0152	0.1469	—	0.8748	0.4493	0.0029	72	1.8573
	H-49	0.0855	0.0271	0.2290	—	0.7909	0.3910	0.0026	282	2.4502

来源	炉号	$t/℃$	$\dfrac{(\%S)}{a_S}$	$\dfrac{\%CaO+\%MgO}{\%SiO_2+\%Al_2O_3}$	$\dfrac{n_{CaO}+n_{MgO}}{n_{SiO_2}+n_{Al_2O_3}}$	$\dfrac{\sum b-2n_{Al_2O_3}}{n_{SiO_2}-n_{Al_2O_3}}$	$n_{O^{2-}}$	$n_{SiO_4^{4-}}$	$n_{Si_2O_7^{6-}}$
Taylor 等[8]	59	1500	9.3	24	0.94	1.14	1.25	0.0031	—	0.1050
	43	1500	6.0	16	0.89	1.11	1.24	0.0021	—	0.0835
	53	1500	5.6	21	0.89	1.10	1.19	0.0027	—	0.0727
	44	1500	5.4	10	0.83	1.04	1.09	0.0013	—	0.0329
	12	1500	5.6	3.1	0.82	1.00	1.02	0.00040	—	0.0072
	9	1500	5.3	2.8	0.61	0.76	0.55	0.00036		

来源	炉号	$n_{Si_3O_9^{6-}}$	$n_{Al_2Si_2O_8^{2-}}$	$n_{Al_2SiO_7^{4-}}$	$n_{AlO_3^{3-}}$	$\sum n_+$	$\sum n_-$	$N_{O^{2-}}$	$\gamma_{Fe^{2+}}+\gamma_{S^{2-}}$	$\lg\gamma_{Fe^{2+}}+\gamma_{S^{2-}}$
Taylor 等[8]	59	0.0744	—	0.1445	—	0.8854	0.3822	0.0081	22	1.3424
	43	0.0637	—	0.1815	—	0.8592	0.3822	0.0055	20	1.3010
	53	0.0854	—	0.1663	—	0.8637	0.3813	0.0071	48	1.6812
	44	0.1053	—	0.1830	—	0.8305	0.3710	0.0035	26	1.4150
	12	0.1475	—	0.1586	—	0.8224	0.3543	0.0011	57	1.7559
	9	0.1162	0.1025	0.0990	—	0.7022	0.3708	0.0010	183	2.2625

续表 5

来源	炉号	$t/℃$	f_S	$\dfrac{(\%S)}{a_S}$	$\dfrac{\%CaO + \%MgO}{\%SiO_2 + \%Al_2O_3}$	$\dfrac{n_{CaO} + n_{MgO}}{n_{SiO_2} + n_{Al_2O_3}}$	$\dfrac{\sum b - 2n_{Al_2O_3}}{n_{SiO_2} - n_{Al_2O_3}}$	$n_{O^{2-}}$	$n_{SiO_4^{4-}}$	$n_{Si_2O_7^{6-}}$
Куликов 等[9]	15	1460	5.2	77	1.30	1.81	2.47	0.0100	0.3873	—
	14	1460	5.5	64	1.28	1.77	2.44	0.0083	0.3740	—
	12	1480	5.7	86	1.31	1.75	2.36	0.0112	0.3747	—
	13	1460	5.7	42	1.17	1.61	2.04	0.0055	0.3872	—
	30	1500	5.7	35	1.15	1.57	2.03	0.0046	0.3696	—
	28	1500	5.7	27	1.17	1.54	1.88	0.0035	0.3020	0.0530
	26	1500	6.3	23	1.09	1.48	1.87	0.0030	0.2698	0.0526
	29	1500	5.1	32	1.07	1.44	1.76	0.0042	0.2052	0.0988
	27	1500	5.7	21	1.07	1.42	1.72	0.0027	0.1706	0.1181

来源	炉号	$n_{Si_3O_9^{6-}}$	$n_{Al_2Si_2O_8^{2-}}$	$n_{Al_2SiO_7^{4-}}$	$n_{AlO_3^{3-}}$	$\sum n_+$	$\sum n_-$	$N_{O^{2-}}$	$\gamma_{Fe^{2+}}\gamma_{S^{2-}}$	$\lg\gamma_{Fe^{2+}}\gamma_{S^{2-}}$
Куликов 等[9]	15	—	—	0.0847	0.0847	1.1425	0.6204	0.0161	6.6	0.8195
	14	—	—	0.0962	0.0918	1.1222	0.6061	0.0137	19	1.2788
	12	—	—	0.0989	0.0740	1.0992	0.5887	0.0190	5.5	0.7404
	13	—	—	0.1316	0.0074	1.0836	0.5611	0.0098	15	1.1761
	30	—	—	0.1406	0.0042	1.0670	0.5548	0.0083	17	1.2304
	28	—	—	0.1249	—	1.0582	0.5253	0.0067	58	1.7634
	26	—	—	0.1480	—	1.0297	0.5067	0.0059	64	1.8062
	29	—	—	0.1389	—	1.0302	0.4884	0.0086	36	1.5563
	27	—	—	0.1396	—	1.0124	0.4659	0.0058	52	1.7160

作 $\lg\gamma_{Fe^{2+}}\gamma_{S^{2-}}$ 对 $N_{O^{2-}}$ 图（图 1），经用最小二乘法处理，得一直线关系式（相关系数 $r = 0.62$）：

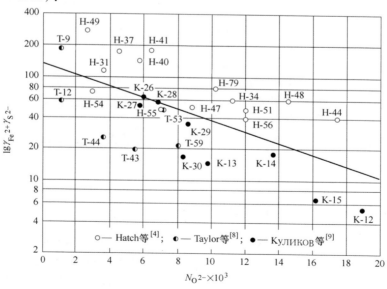

图 1　$\lg\gamma_{Fe^{2+}}\gamma_{S^{2-}}$ 与 $N_{O^{2-}}$ 的关系

$$\lg\gamma_{Fe^{2+}}\gamma_{S^{2-}} = -53.5 N_{O^{2-}} + 2.12 \qquad (10)$$

将 $N_{O^{2-}}$ 分成五组，取平均值。同样地取相应的 $\lg\gamma_{Fe^{2+}}\gamma_{S^{2-}}$ 平均值（见表6和图2），再用最小自乘法处理，得出下列直线关系式（相关系数 $r = 0.94$）：

$$\lg\gamma_{Fe^{2+}}\gamma_{S^{2-}} = -55.6 N_{O^{2-}} + 2.14 \qquad (11)$$

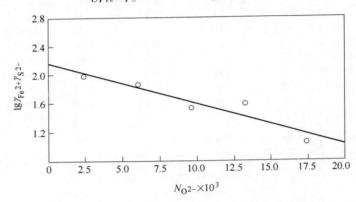

图2 $\lg\gamma_{Fe^{2+}}\gamma_{S^{2-}}$ 与 $N_{O^{2-}}$ 的关系（综合成组后）

表6 $\lg\gamma_{Fe^{2+}}\gamma_{S^{2-}}$ 与 $N_{O^{2-}}$ 的关系（综合成五组）

$N_{O^{2-}}$ 的范围	炉 号	平均值	
		$\lg\gamma_{Fe^{2+}}\gamma_{S^{2-}}$	$N_{O^{2-}}$
0~0.004	T-9、T-12、H-49、H-54、T-44、H-31	1.9663	0.0024
0.004~0.008	H-37、T-43、H-40、K-27、K-26、H-41、K-28、H-55、T-53	1.8452	0.0060
0.008~0.012	T-59、K-30、K-29、H-47、K-13、H-39、H-34、H-56	1.5406	0.0096
0.012~0.016	H-51、K-14、H-48	1.5837	0.0134
0.016~0.020	K-15、H-44、K-12	1.0576	0.0175

3 讨论

3.1 经验式（10）和式（11）的确实性问题

由于 $N_{O^{2-}}$ 的数值很小，式（10）和式（11）基本上是相同的。从图1可以看出，式（10）的数据点比较分散，相关系数为0.62。图2指出式（11）的直线关系较明显，相关系数为0.94。但式（10）和式（11）究竟哪一个更较确实，这需要对相关系数加以检验。表7[28]给出检验的结果。可以明显地看出，式

（10）的置信概率大于 99.9%；也就是说，$\lg\gamma_{Fe^{2+}}\gamma_{S^{2-}}$ 和 $N_{O^{2-}}$ 的非线性关系估计在 1000 个试样中至多只有一个。而式（11）中二者的非线性关系估计在 100 个试样中则可有 1~2 个。两个式子的直线相关都有很大的显著性。但两式比较，式（10）则更较确实，因此，在下面讨论中都以式（10）为准。必须指出，考虑相关系数大小时，必须同时考虑试样数目的多少，孤立地只以相关系数作出评价会导致错误的结论。

表7　式（10）及式（11）的置信概率

公式	试样数目	自由度	相关系数	显著性水准	置信概率/%
（10）	29	27	0.62	<0.001	>99.9
（11）	5	3	0.94	0.01~0.02	98~99

某些数据，例如 H-49 的 $\gamma_{Fe^{2+}}\gamma_{S^{2-}}=282$，有些太高。比较 H-49 和 H-55 的铁渣成分和硫分配比，可以看出，H-49 的 $\gamma_{Fe^{2+}}\gamma_{S^{2-}}$ 值太高，原因之一可能是试样尚未达到平衡。由于国际上尚无分离铁粒和渣样的标准分析方法，FeO 欠准确的分析结果，特别是三种来源的试样，也能导致数据的分歧。比较 T-43 和 T-53 的铁渣成分和硫分配比，可以看出，较高的 FeO 分析值导致较低的 $\gamma_{Fe^{2+}}\gamma_{S^{2-}}$ 值。由于数据处理采用较多的式样，可以认为过高过低数值的作用最终相互抵消。

最后，式（10）的线性关系是否确实主要取决于炉渣结构的假定是否妥当和正确。本文研究是一种尝试，炉渣结构的确定有待今后研究。

3.2　经验式（10）能否扩大应用于一般的低碱度炉渣问题

本文作者[1]曾证实碱性化铁炉渣间硫分配达到平衡，当时采用碱度较高的炉渣进行计算。表 8 给出一些碱性化铁炉低碱度操作硫分配比计算。计算时除将炉渣的 S 折合成 CaS 而减去与 CaS 相当的 CaO 并调整到组成物之和为 100% 外，依据碱度指标 $(\sum b - 2n_{Al_2O_3})/(n_{SiO_2} - n_{Al_2O_3})$ 先定出炉渣的离子结构。欲确定不同硅酸根离子的比例，又依据表 5 内碱度指标与硅酸根离子的摩尔数的关系制成表 9~表 12，并绘成图 3~图 6。确定了不同离子的比例之后，即可计算 $n_{O^{2-}}$、$N_{O^{2-}}$ 及 $\gamma_{Fe^{2+}}\gamma_{S^{2-}}$，从而计算硫分配比。表 9 和表 10 给出的硅酸根离子的比例和碱度指标的关系比较严格而合理，因为渣内所有的 Al_2O_3 均按结合成 $Al_2SiO_7^{4-}$ 处理。表 11 只给出一近似关系，因为这里 Al_2O_3 不仅结合为 $Al_2SiO_7^{4-}$，而又有一部分结合为 $Al_2Si_2O_8^{2-}$，这样便和 $(\sum b - 2n_{Al_2O_3})/(n_{SiO_2} - n_{Al_2O_3})$ 的指标不大相适应。在 $(\sum b - 2n_{Al_2O_3})/(n_{SiO_2} - n_{Al_2O_3}) > 2$ 的炉渣（表 12），渣内的 O^{2-} 已超过与 SiO_4^{4-} 及 $Al_2SiO_7^{4-}$ 相结合的量，Al_2O_3 必然有一部分与 O^{2-} 结合成 AlO_3^{3-}，这种炉渣逐步过渡到只有 SiO_4^{4-} 及 AlO_3^{3-} 的炼钢型碱性渣，因之选用另一碱度指标 $(\sum b - 3n_{Al_2O_3})/n_{SiO_2}$ 以求出 $n_{SiO_4^{4-}}/(n_{SiO_4^{4-}} + n_{Al_2SiO_4^{4-}})$ 的比例。当 $n_{Al_2SiO_4^{4-}} = 0$，也即

$n_{SiO_4^{4-}} = 100\%$ 时，此指标 $(\sum b - 3n_{Al_2O_3})/n_{SiO_2} \approx 2$（近于2是因为尚有微量的 $n_{O^{2-}}$）。

表8 低碱度操作的碱性化铁炉脱硫资料

来源	炉号	铁水成分/%					f_S	炉渣成分/%							$\dfrac{(\%S)}{[\%S]}$	
		C	Mn	Si	P	S		CaO	MgO	SiO$_2$	Al$_2$O$_3$	MnO	FeO	S	实际	计算
上海冶金局中心实验室[19]	5-1A	2.78	0.60	0.64	0.225	0.331	2.6	20.03	19.44	40.76	12.30	0.32	1.90	1.18	3.6	1.0
	5-24	3.40	0.28	1.40	0.460	0.153	4.0	32.49	17.20	32.42	12.40	0.11	0.90	1.90	12.4	16
	4-32A	2.65	0.32	1.84	0.545	0.323	3.2	34.32	16.66	37.40	7.68	1.03	1.05	3.26	10.1	2.9
	4-34A	2.69	0.26	2.28	0.520	0.459	3.6	33.96	14.39	38.80	9.12	1.08	0.80	3.51	7.7	12
Carter[30]	3	—	—	—	—	0.061	4.0（估算）	30.5	22.0	34.3	9.4	1.7	1.7	0.39	6.4	3.1
	4	—	—	—	—	0.064	4.0（估算）	35.0	16.8	35.1	7.8	1.7	2.2	0.46	7.2	7.3

表9 碱度指标 $(\sum b - 2n_{Al_2O_3})/(n_{SiO_2} - n_{Al_2O_3})$ 为 1.5~2.0 时硅酸根离子的比例

来 源	炉 号	$\dfrac{\sum b - 2n_{Al_2O_3}}{n_{SiO_2} - n_{Al_2O_3}}$	$\dfrac{n_{SiO_4^{4-}}}{n_{SiO_4^{4-}} + n_{Si_2O_7^{6-}}} \times 100/\%$
Hatch 等[4]	H-48	2.00	97.6
Куликов 等[9]	28	1.88	85.1
	26	1.87	83.7
	29	1.76	67.5
	27	1.72	59.1
Hatch 等[4]	H-44	1.64	40.4
	H-39	1.61	34.5

表8指出式（1）和式（10）基本上可以适用于低碱度的碱性化铁炉渣，这里铁水不一定为碳所饱和。计算出的硫分配比和实际值在工艺操作的变化范围内基本上是一致的，说明低碱度操作时化铁炉的铁渣间的硫分配达到平衡。

表10　碱度指标 $(\sum b - 2n_{Al_2O_3})/(n_{SiO_2} - n_{Al_2O_3})$ 为 1.0~1.5 时硅酸根离子的比例

来源	炉号	$\dfrac{\sum b - 2n_{Al_2O_3}}{n_{SiO_2} - n_{Al_2O_3}}$	$\dfrac{n_{Si_2O_7^{6-}}}{n_{Si_2O_7^{6-}} + n_{Si_3O_9^{6-}}} \times 100/\%$
Hatch 等[4]	H-51	1.41	86.4
	H-41	1.40	85.3
	H-56	1.40	84.1
	H-40	1.31	69.4
Taylor 等[8]	59	1.25	58.7
	43	1.24	56.7
Hatch 等[4]	H-47	1.22	51.8
	H-34	1.20	47.0
Taylor 等[8]	53	1.19	46.0
Hatch 等[4]	H-55	1.17	41.6
	H-37	1.12	31.3
Taylor 等[8]	44	1.09	23.8
	12	1.02	4.7

表11　碱度指标 $(\sum b - 2n_{Al_2O_3})/(n_{SiO_2} - n_{Al_2O_3})$ <1 时硅酸根离子的比例

来源	炉号	$\dfrac{\sum b - 2n_{Al_2O_3}}{n_{SiO_2} - n_{Al_2O_3}}$	$\dfrac{n_{Si_3O_9^{6-}}}{n_{Si_3O_9^{6-}} + n_{Al_2Si_2O_8^{2-}}} \times 100/\%$
Hatch 等[4]	H-31	0.98	96.2
	H-54	0.93	90.2
	H-49	0.81	76.0
Taylor 等[8]	9	0.55	53.1

表12　碱度指标 $(\sum b - 2n_{Al_2O_3})/(n_{SiO_2} - n_{Al_2O_3})$ >2 时硅酸根离子的比例

来源	炉号	$\dfrac{\sum b - 2n_{Al_2O_3}}{n_{SiO_2} - n_{Al_2O_3}}$	$\dfrac{\sum b - 3n_{Al_2O_3}}{n_{SiO_2}}$	$\dfrac{n_{SiO_4^{4-}}}{n_{SiO_4^{4-}} + n_{Al_2SiO_7^{4-}}} \times 100/\%$
Куликов 等[9]	15	2.47	1.48	82.0
	14	2.44	1.40	79.6
	12	2.36	1.40	79.1
	13	2.04	1.25	74.5
	30	2.03	1.18	72.4

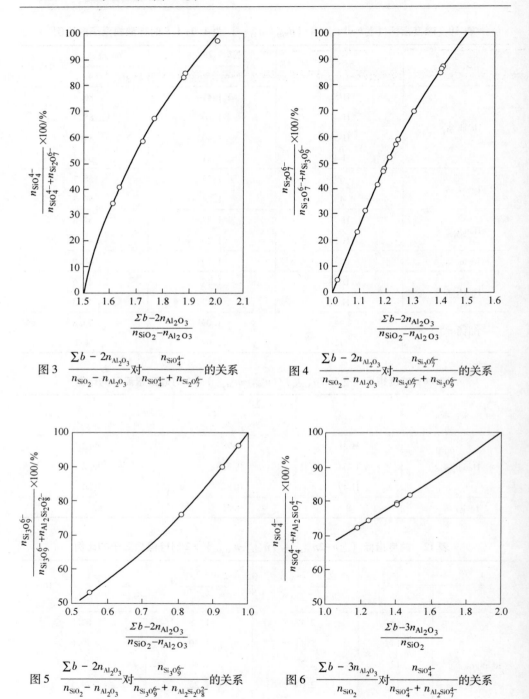

图 3 $\dfrac{\Sigma b - 2n_{Al_2O_3}}{n_{SiO_2} - n_{Al_2O_3}}$ 对 $\dfrac{n_{SiO_4^{4-}}}{n_{SiO_4^{4-}} + n_{Si_2O_7^{6-}}}$ 的关系

图 4 $\dfrac{\Sigma b - 2n_{Al_2O_3}}{n_{SiO_2} - n_{Al_2O_3}}$ 对 $\dfrac{n_{Si_2O_7^{6-}}}{n_{Si_2O_7^{6-}} + n_{Si_3O_9^{6-}}}$ 的关系

图 5 $\dfrac{\Sigma b - 2n_{Al_2O_3}}{n_{SiO_2} - n_{Al_2O_3}}$ 对 $\dfrac{n_{Si_3O_9^{6-}}}{n_{Si_3O_9^{6-}} + n_{Al_2Si_2O_8^{2-}}}$ 的关系

图 6 $\dfrac{\Sigma b - 3n_{Al_2O_3}}{n_{SiO_2}}$ 对 $\dfrac{n_{SiO_4^{4-}}}{n_{SiO_4^{4-}} + n_{Al_2SiO_7^{4-}}}$ 的关系

3.3 FeO 含量对低碱度酸性渣脱硫作用的影响

选用表 8 炉号 4-34A，调整后的炉渣成分（%）为：

CaO	MgO	SiO$_2$	Al$_2$O$_3$	MnO	FeO	CaS
27.85	14.40	38.83	9.13	1.08	0.80	7.91

计算时将 FeO 的百分数改变，而其他组成物的比例维持不变。为了摆脱生铁成分对脱硫的影响，特计算 $(\%S)/a_S$。表 13 给出计算的结果。可以看出，FeO 的降低大大地提高了 $(\%S)/a_S$，而随着 FeO 的提高，$(\%S)/a_S$ 达到一最低值，当 FeO 再度增高，则 $(\%S)/a_S$ 开始重新增加，到 FeO 为 30%~40% 时，由于 O^{2-} 的增加，该炉渣已进入碱性渣的范围，$\gamma_{Fe^{2+}}\gamma_{S^{2-}}$ 需用式（5）计算，此后 $(\%S)/a_S$ 仍逐渐增加，最终到纯 FeO 时达到 3.6。这样，在前文[1]得出的对炼钢型碱性渣关于 FeO 脱硫作用的规律完全适用于低碱度的酸性渣。

表 13 炉渣不同氧化铁含量对 $(\%S)/a_S$ 的影响

FeO/%	$n_{O^{2-}}$	$n_{Si_2O_7^{6-}}$	$n_{Al_2SiO_7^{4-}}$	$n_{Si_3O_9^{6-}}$	$n_{SiO_4^{4-}}$	$n_{AlO_3^{3-}}$	Σn_+	Σn_-	$\gamma_{Fe^{2+}}\gamma_{S^{2-}}$ 按式(10)计算	$\gamma_{Fe^{2+}}\gamma_{S^{2-}}$ 按式(5)计算	$\dfrac{(\%S)}{a_S}$
0.05	0.006	0.1287	0.0902	0.1011	—	—	0.9869	0.4371	20.5	—	77
0.10	0.0071	0.1287	0.0901	0.1011	—	—	0.9872	0.4375	17.8	—	45
0.80	0.0053	0.1391	0.0896	0.0927	—	—	0.9897	0.4364	29.5	—	3.4
2.0	0.0016	0.1581	0.0885	0.0779	—	—	0.9947	0.4342	83.8	—	0.47
10	0.0001	0.2447	0.0812	—	0.0156	—	1.0272	0.4410	128	—	0.06
20	0.0042	0.0668	0.0723	—	0.3151	—	1.0678	0.5468	51.2	—	0.10
30	0.0104	—	0.0274	—	0.4286	0.0718	1.1074	0.6155	16.4	—	0.25
40	0.1401	—	—	—	0.3902	0.1084	1.1493	0.7049	—	8.1	0.45
60	0.5555	—	—	—	0.2606	0.0724	1.2295	0.9327	—	2.4	1.5
80	0.9737	—	—	—	0.1305	0.0360	1.3107	1.1622	—	1.1	3.1
100	1.3918	—	—	—	—	—	1.3918	1.3918	—	1.0	3.6

注：炉号 4-34A：$\dfrac{\%CaO+\%MgO}{\%SiO_2+\%Al_2O_3}=0.88$；$\dfrac{n_{CaO}+n_{MgO}}{n_{SiO_2}+n_{Al_2O_3}}=1.16$。

3.4 温度对硫分配比的影响

式（1）的 L_S 与温度的关系在文献 [1] 上存有分歧。本研究选用 Frohberg[18] 的公式，根据后者温度对 L_S 的影响很小，可以认为在 1500~1600℃ 范围内，L_S 基本上守常。同时，$\gamma_{Fe^{2+}}\gamma_{O^{2-}}$（也即 $\gamma_{Fe^{2+}}\gamma_{S^{2-}}$）在 1540~1800℃ 范围内也

不随温度而改变[18]。在高炉操作和石墨坩埚的平衡实验中，随着温度的增高，铁液内 Si 量增加，虽然含 C 量略为减小，但总结果是 f_S 加大；也就是增高温度则 $(\%S)/[\%S]$ 加大。如按 $(\%S)/a_S$ 计算，则温度和铁水成分的影响均可不加考虑。

4 结论

通过前文的研究和本文的尝试，可以认为，由炉渣完全离子化理论导出和引用活度系数的硫分配比公式：

$$\frac{(\%S)}{[\%S]} = \frac{32 L_S f_S}{\gamma_{Fe^{2+}} \gamma_{S^{2-}}} \times \frac{\sum n_+ \sum n_-}{n_{FeO}}$$

是一个较综合的公式。它适用于含任何碳量的铁液（包括饱和碳铁液）和任何成分的熔渣（无论是高碱度或低碱度，高 FeO 或低 FeO）。

高炉型渣随着新提出的碱度指标 $(\sum b - 2 n_{Al_2O_3})/(n_{SiO_2} - n_{Al_2O_3})$ 的不同，可能有 SiO_4^{4-}、$Si_2O_7^{6-}$、$Si_3O_9^{6-}$、$Al_2SiO_7^{4-}$、$Al_2Si_2O_8^{2-}$ 及 AlO_3^{3-} 等酸根离子，而

$$\lg \gamma_{Fe^{2+}} \gamma_{S^{2-}} = -53.5 N_{O^{2-}} + 2.12$$

FeO 含量对脱硫作用的规律在低碱度高炉型渣是和碱性炼钢型渣完全一致。

由于 L_S 随温度变化不大，而 $\gamma_{Fe^{2+}} \gamma_{S^{2-}}$ 也不受温度影响，由上列公式计算出的 $(\%S)/f_S[\%S]$ 或 $(\%S)/a_S$ 基本上不受温度的影响，同时也不受铁液成分的影响。

参 考 文 献

[1] 魏寿昆. 金属学报，1964，7：157.

[2] Гольдщтейн Н Л. Сталъ，1953 (7)：598.

[3] Воскобойников В Г. Сталъ，1955 (7)：583.

[4] Hatch G G, Chipman J. Trans. AIME, 1949, 185：274, 831.

[5] Filer E W, Darken L S. Trans. AIME, 1952, 194：253.

[6] Grant N J, Kalling U, Chipman J. Trans. AIME, 1951, 191：666.

[7] Grant N J, Troili O, Chipman J. Trans. AIME, 1951, 191：672.

[8] Taylor J, Stobo J J. J. Iron Steel Inst. (London), 1954, 178：360.

[9] Куликов И С, Кожевников И Ю, Цылев Л М. Современные проблемы металлургин, Изд. Акад. наук СССР, 1958：149~162.

[10] Schenck H, Frohberg M G, Gammal T El. Arch. Eisenhuttenw, 1960, 31：11.

[11] Schenck H, Frohberg M G, Gammal T El. Arch. Eisenhuttenw, 1960, 31：471.

[12] Schenck H, Frohberg M G, Gammal T El. Arch. Eisenhuttenw, 1961, 32：63.

[13] Schenck H, Frohberg M G, Gammal T El. Arch. Eisenhuttenw, 1962, 33: 589.
[14] 邹元爔. 金属学报, 1956, 1: 127, 143.
[15] Куликов И С. Сталь, 1961 (11): 972.
[16] Kalyanram M R, MacFarlane T G, Bell H B. J. Iron Steel Inst. (London), 1960, 195: 58.
[17] Giedroyc V, McPhail A N, Mitchell J G. J. Iron Steel Inst. (London), 1964, 202: 11.
[18] Frohberg M G. Arch. Eisenhuttenw, 1961, 32: 597.
[19] Самарин А М, Щварцман Л А. Изб. АН СССР, Отд. техн. н, 1948, 9: 1457.
[20] Gokcen N A. Trans. AIME, 1956, 206: 1558.
[21] AIME. Electric Furnace Steelmaking (Interscience), 1963, 2: 133~134.
[22] Bockris J O'M, Kitchener J A, Ignatowicz S, et al. Trans. Faraday Soc., 1952, 48: 75.
[23] Bockris J O'M, Mackenzie J D, Kitchener J A. Trans. Faraday Soc., 1955, 51: 1734.
[24] Bockris J O'M, Tomlinson J W, White J L. Trans. Faraday Soc., 1956, 52: 299.
[25] Велянкйн Д С, Иванов В В, Лапин В В. Петрография технического камия, Изд. Акад. Наук СССР, 1952: 320~344.
[26] Sherman C W, Chipman J. Trans. AIME, 1952, 194: 597.
[27] Morris J P, Buehl R C. Trans. AIME, 1950, 188: 371; 另见文献 [21], 1963, 2: 122.
[28] AIME. Electric Furnace Steelmaking (Intersscience), 1963, 2: 455 (Tab. 22-12).
[29] 上海市冶金工业局中心试验室, 中国科学院冶金陶瓷研究所. 全国转炉会议资料汇编, 第1辑. 上海: 上海科技出版社, 1959: 28~78.
[30] Carter S F. Trans. Am. Foundrymen's Soc, 1953, 61: 52.

从炉渣离子理论计算的硫分配比来看攀钢钒钛铁矿中 TiO_2 的属性[*]

魏寿昆

摘　要：采用攀钢高炉炉渣一个典型数据，按炉渣离子理论对硫分配比进行计算。TiO_2 作为酸性氧化物，硫分配比为 9.1，TiO_2 作为碱性氧化物，则硫分配比为 60.7。高炉生产实践得到的硫分配比一般不大于 10，因之可以相信，TiO_2 在高炉渣内呈酸性，不可能是碱性氧化物。

对攀钢钒钛铁矿，计算自熔性的碱度应采用 $CaO\% + MgO\%/SiO_2\% + Al_2O_3\% + TiO_2\%$，因而该矿是酸性矿，不可能是半自熔性矿。

1　攀钢钒钛铁矿中 TiO_2 的属性

攀钢钒钛铁精矿含 TiO_2 太高（约 13.5%），不能单独用来炼铁。配用 15% 普通矿后，高炉炉渣 TiO_2 含量仍高达 24%，此渣脱硫能力很差，硫分配比一般不大于 10，而普通高炉炉渣的硫分配比通常为 30~60。TiO_2 在脱硫作用中呈酸性，抑属碱性，通过炉渣离子理论的计算可以阐明。

根据炉渣离子理论，脱硫由下列反应表示：

$$[S] + (O^{2-}) = (S^{2-}) + [O]$$

炉渣内 O^{2-} 离子越多，则脱硫作用越大。碱性氧化物是 O^{2-} 离子的提供者，而酸性氧化物则是 O^{2-} 离子的消耗者。TiO_2 若是碱性，则能促进脱硫；但若呈酸性，则将显著降低脱硫的作用。

硫分配比计算公式的推导见文献 [1]：

$$\frac{(\%S)}{[\%S]} = \frac{32 L_S f_S}{\gamma_{Fe^{2+}} \gamma_{S^{2-}}} \times \frac{\sum n_+ \sum n_-}{n_{FeO}}$$

式中　L_S——FeS 分配反应的平衡常数；

f_S——铁液中 S 的活度系数；

$\gamma_{Fe^{2+}}$——炉渣中 Fe^{2+} 离子的活度系数；

---

[*]　原刊于《钢铁钒钛》，1988（4）：1~3，65。

$\gamma_{S^{2-}}$——炉渣中 S^{2-} 离子的活度系数；

$\sum n_+$——100g 炉渣所含的正离子摩尔数的总和；

$\sum n_-$——100g 炉渣所含的负离子摩尔数的总和；

n_{FeO}——100g 炉渣所含的 FeO（即 Fe^{2+}）摩尔数。

表 1 列出攀钢高炉炉渣一个典型实例的组成及 100g 炉渣含有氧化物的摩尔数。

表 1 炉渣组成及 100g 炉渣中氧化物的摩尔数

氧 化 物	重量百分数/%	分子量	100g 中的摩尔数 n_i/mol
SiO_2	25.5	60.09	0.4244
TiO_2	24.0	79.90	0.3004
Al_2O_3	16.6	101.96	0.1628
CaO	24.4	56.08	0.4351
MgO	7.2	40.31	0.1786
FeO	1.6	71.85	0.0223
S	0.7	32.06	0.0218
	100.0		

1.1 TiO_2 呈酸性

Al_2O_3 在高炉渣内的离子，根据 $CaO-Al_2O_3-SiO_2$ 相图[2]，一般以 $Al_2SiO_7^{4-}$ [钙铝黄长石（gehlenite）$2CaO \cdot SiO_2 \cdot Al_2O_3$ 生成] 离子存在；参考 $CaO-TiO_2-SiO_2$ 相图[3]，TiO_2 在高炉渣中可以 $TiSiO_5^{2-}$ [楔石（sphene）$CaO \cdot TiO_2 \cdot SiO_2$ 生成] 及 TiO_3^{2-} [钙钛矿（perofskite）$CaO \cdot TiO_2$ 生成] 等离子存在。

$$SiO_2 + Al_2O_3 + 2O^{2-} = Al_2SiO_7^{4-}$$

由于 $n_{Al_2O_3} = 0.1628$，因之生成的 $n_{Al_2SiO_7^{4-}} = 0.1628$，而消耗的 $n_{O^{2-}} = 0.3256$，消耗的 $n_{SiO_2} = 0.1628$。剩余的 $n_{SiO_2} = 0.4244 - 0.1628 = 0.2616$。

$$SiO_2 + TiO_2 + O^{2-} = TiSiO_5^{2-}$$

根据上式生成的 $n_{TiSiO_5^{2-}} = 0.2616$，消耗的 $n_{O^{2-}} = 0.2616$，消耗的 $n_{TiO_2} = 0.2616$。剩余的 $n_{TiO_2} = 0.3004 - 0.2616 = 0.0388$。

$$TiO_2 + O^{2-} = TiO_3^{2-}$$

因之，生成的 $n_{TiO_3^{2-}} = 0.0388$，消耗的 $n_{O^{2-}} = 0.0388$。

提供的 $\sum n_{O^{2-}} = 0.4351 + 0.1786 + 0.0223 = 0.6360$。消耗的 $\sum n_{O^{2-}} = 0.3256 + 0.2616 + 0.0388 = 0.6260$。所以炉渣中的 $n_{O^{2-}} = 0.6360 - 0.6260 = 0.0100$。

$$\sum n_- = n_{Al_2SiO_7^{4-}} + n_{TiSiO_5^{2-}} + n_{TiO_3^{2-}} + n_{O^{2-}} + n_{S^{2-}}$$
$$= 0.1628 + 0.2616 + 0.0388 + 0.0100 + 0.0218 = 0.4950$$

$$\sum n_+ = \sum n_{O^{2-}} = 0.6360$$

根据作者的研究[4]，有下列的经验公式：

$$\lg \gamma_{Fe^{2+}} \gamma_{S^{2-}} = -53.5 N_{O^{2-}} + 2.12$$

$$N_{O^{2-}} = \frac{n_{O^{2-}}}{\sum n_-} = \frac{0.0100}{0.4950} = 0.0202$$

$$\lg \gamma_{Fe^{2+}} \gamma_{S^{2-}} = 1.0390; \quad \gamma_{Fe^{2+}} \gamma_{S^{2-}} = 10.95$$

则 $L_S = 0.08^{[5]}$。

按 $C = 4.0\%$、$e_S^C = 0.11$ 计算，铁水其他杂质元素忽略不计，则 $f_S = 2.75$。故

$$\frac{(S\%)}{[S\%]} = \frac{32 \times 0.08 \times 2.75}{10.95} \times \frac{0.6360 \times 0.4950}{0.0223} = 9.1$$

1.2 TiO_2 呈碱性

碱性氧化物 TiO_2 提供 Ti^{4+} 及 $2O^{2-}$，因之，由表 1 数据可知 $n_{Ti^{4+}} = 0.3004$，$n_{O^{2-}} = 0.6008$。

和上面一样，生成的 $n_{Al_2SiO_7^{4-}} = 0.1628$，消耗的 $n_{O^{2-}} = 0.3256$，消耗的 $n_{SiO_2} = 0.1628$，剩余的 SiO_2 生成 SiO_4^{4-}（$2CaO \cdot SiO_2$ 产物）：

$$SiO_2 + 2O^{2-} \rightleftharpoons SiO_4^{4-}$$

剩余的 $n_{SiO_2} = 0.4244 - 0.1628 = 0.2616$，因之生成的 $n_{SiO_4^{4-}} = 0.2616$，消耗的 $n_{O^{2-}} = 0.5232$。

提供的 $\sum n_{O^{2-}} = 2 \times 0.3004 + 0.4351 + 0.1786 + 0.0223 = 1.2368$。消耗的 $\sum n_{O^{2-}} = 0.3256 + 0.5232 = 0.8488$。因此炉渣中的 $n_{O^{2-}} = 1.2368 - 0.8488 = 0.3880$。

$$\sum n_- = n_{Al_2SiO_7^{4-}} + n_{SiO_4^{4-}} + n_{O^{2-}} + n_{S^{2-}} = 0.1628 + 0.2616 + 0.3880 + 0.0218 = 0.8342$$

$$\sum n_+ = n_{Ti^{4+}} + n_{Ca^{2+}} + n_{Mg^{2+}} + n_{Fe^{2+}} = 0.3004 + 0.4351 + 0.1786 + 0.0223 = 0.9364$$

由于碱性渣，$\lg \gamma_{Fe^{2+}} \gamma_{S^{2-}}$ 有另一关系式[6,7]：

$$\lg \gamma_{Fe^{2+}} \gamma_{S^{2-}} = 1.53 \sum N_{SiO_4^{4-}} - 0.17$$

$$\sum N_{SiO_4^{4-}} = \frac{n_{Al_2SiO_7^{4-}} + n_{SiO_4^{4-}}}{\sum n_-} = \frac{0.1628 + 0.2616}{0.8342} = 0.5088$$

则

$$\lg \gamma_{Fe^{2+}} \gamma_{S^{2-}} = 0.6085$$

$$\gamma_{Fe^{2+}} \gamma_{S^{2-}} = 4.059$$

$$\frac{(\%S)}{[\%S]} = \frac{32 \times 0.08 \times 2.75}{4.059} \times \frac{0.9364 \times 0.8342}{0.0223} = 60.7$$

上列计算结果可归纳为表 2。

表 2 100g 炉渣含有的离子的摩尔数

离子		TiO_2 呈酸性	TiO_2 呈碱性
正离子	Ca^{2+}	0.4351	0.4351
	Mg^{2+}	0.1786	0.1786
	Fe^{2+}	0.0223	0.0223
	Ti^{4+}	—	0.3004

续表 2

离子		TiO_2 呈酸性	TiO_2 呈碱性
负离子	$Al_2SiO_7^{4-}$	0.1628	0.1628
	$TiSiO_5^{2-}$	0.2616	—
	TiO_3^{2-}	0.0388	—
	SiO_4^{4-}	—	0.2616
	O^{2-}	0.0100	0.3880
	S^{2-}	0.0218	0.0218
硫分配比		9.1	60.7

由上面计算清楚地看出，只有 TiO_2 呈酸性时，计算的脱硫能力才符合生产实际。因此，TiO_2 不可能是碱性氧化物，而是呈酸性。

2 关于攀枝花钒钛铁矿的自熔性问题

一般普通铁矿的自熔性按碱度公式 $\dfrac{\%CaO+\%MgO}{\%SiO_2+\%Al_2O_3}$ 计算，碱度>0.8，称为自熔性；碱度 = 0.5~0.8，称为半自熔性；碱度<0.5，称为酸性。

攀枝花矿含 TiO_2 比较高，后者除对护炉起到一定的好作用外，给高炉冶炼带来一系列的难以克服的困难（如不能单独用攀矿炼铁等）。从攀矿的自熔性考虑，必须加入 TiO_2 一项。TiO_2 从理论上和实践上都被证明是酸性的。碱度公式应写为 $\dfrac{\%CaO+\%MgO}{\%SiO_2+\%Al_2O_3+\%TiO_2}$ 方为合理。严格地讲，TiO_2 和 SiO_2 相比，其酸性较弱，适宜加一校正系数 $R(R<1)$。但由于比 SiO_2 酸性弱的 Al_2O_3 向来不加系数，整个碱度公式为简便计算，都以不加系数为好。

攀钢高炉渣内含有少量的低价钛氧化物，如 TiO 或 Ti_2O_3。它们是碱性氧化物，是在冶炼过程中不希望产生的氧化物，但由于它们的量很少（最多不过3%），可忽略不计。

按提出的碱度公式，攀枝花矿应是酸性矿，而不是半自熔性矿。

参 考 文 献

[1] 魏寿昆. 冶金过程热力学. 上海：上海科学技术出版社，1980：239.
[2] Sims C E. Electric Furnace Steelmaking. Interscience Publishers，vol. 2，1963：243.
[3] Sims C E. ibid，vol. 2，1963：244.
[4] 魏寿昆. 金属学报，1966，9：127~141.
[5] 魏寿昆. 冶金过程热力学. 上海：上海科学技术出版社，1980：232，235.
[6] 魏寿昆. ibid：233.
[7] 魏寿昆. 金属学报，1964，7：157~164.

The Ionization Theory of Desulphurization of the Panzhihua Blast Furnace Slag[*]

Wei Shoukun

Abstract: With the proper choice of the ion species present in the Panzhihua blast furnace slag, the sulphur partition ratio was calculated based on the ionization theory of the slag. It can be concluded that TiO_2 reacts as an acidic oxide and the Panzhihua ore or concentrate can not be treated as a semi-self-fluxing ore.

Keywords: blast furnace slag, desulphurization, sulphur partition ratio

1 Introduction

The titaniferous vanadio-magne titedeposit in the Panzhihua area is characterized by its high Ti content (Table 1)[1]. Direct smelting of the iron ore orconcentrate has caused troubles in the blast furnace operation, such as the high viscous nature of the slag, the non-coalescence and non-separation of the hot metal droplets in the slag, and the accumulation of heavy in soluble deposit on the inner walls or the hearth. All these would result in the irregular operation of the blast furnace. Recent practice of charging 10% ~ 15% ordinary ore together with sintered concentrate has made much improvement and achieved smooth operation of the furnace. But owing to the strongly acidicnature of the slag the sulphur partition ratio has not been able to reach a value of more than ten. It is the purpose of the present paper to make a tentative analysis of the ionization theory of desulphurization of the slag, the sulphur partition ratio being quantitatively calculated with use of the empirical formula for $\gamma_{Fe^{2+}}\gamma_{S^{2-}}$, which had been elaborated by the auther and already reported elsewhere[2,3].

2 Theoretical

Based on the ionization theory, desulphurization of hot metal takes place according to the following reaction:

$$[S] + (O^{2-}) \rightleftharpoons (S^{2-}) + [O]$$

---

[*] 原刊于 Chin. J. Met. Sci. Technol, 1989, 5: 313~318.

Increase in the number of moles of O^{2-} in the slag will promote the transfer of [S] from the metal to the slag as S^{2-}. It is well-known that basic oxides serve as donators for O^{2-} ions. while acidic oxides consume O^{2-} ions to form complex anions and conse quently will lessen the Sulphur partition ratio. Temkin and Samarin[4,5] have first derived the formula for calculating the sulphur partition ratio for steel making slags based on the ionization theory of the slag as follows:

$$\frac{(\%S)}{[\%S]} = \frac{32 L_S f_S}{\gamma_{Fe^{2+}} \gamma_{S^{2-}}} \frac{\sum n_+ \sum n_-}{n_{FeO}} \tag{1}$$

where L_S——the distribution coefficient of S, equal to 0.08;

f_S——the activity coefficient of S;

$\sum n_+$——total sum of moles of cations;

$\sum n_-$——total sum of moles of antions;

$\gamma_{Fe^{2+}}, \gamma_{S^{2-}}$——the activity coefficient of Fe^{2+} and S^{2-} respectively.

Samarin[5] has also ascertained the empirical relation

$$\lg \gamma_{Fe^{2+}} \gamma_{S^{2-}} = 1.53 \sum_{SiO_4^{4-}} - 0.17 \tag{2}$$

which holds true for $SiO_2 < 30\% (wt\%)$.

In the year of sixties, the author has proved that Eq. (1) is a most general and universal formula for partition of sulphur, which could be applied for molten iron with any C content, and slags with either low or high FeO content, as well as slags with either high basicity for the steel making slags or low basicity for the blast furnace slags. Details of derivation of Eq. (1), especially for hot metal, can be found in Ref. [3].

With the proper assumption and grouping of complex anions, and using composition data of slag and hot metal from different kinds of blast furnace operation as found in the literature, the author has worked out the following empirical formula similar to Eq. (2)[2]:

$$\lg \gamma_{Fe^{2+}} \gamma_{S^{2-}} = -53.5 N_{O^{2-}} + 2.12 \tag{3}$$

which holds true for $N_{O^{2-}} = 0 \sim 0.02$. Since $\sum N_{SiO_4^{4-}} + N_{O^{2-}} + N_{S^{2-}} = 1$, for correlation, Eq. (2) can be rewritten as follows:

$$\lg \gamma_{Fe^{2+}} \gamma_{S^{2-}} = -1.53 N_{O^{2-}} + 1.36 - 1.53 N_{S^{2-}} \tag{4}$$

Calculation was made with 2 slags from the Panzhihua Iron and Steel Co. with composition given in Table 3[6]. Since TiC and TiN are insoluble in the hot metal, their percentages are not used in the calculation. The metallic iron Fe_M supposed to be their on droplets in the slag, is deducted from the total iron content Fe_t to obtain the difference used for calculating the percentage of FeO present in the slag. Ti_xO_y is assumed to contain 50% TiO and 50% Ti_2O_3. Then all the percentages of the constituents of the slag are calculated to the full 100% basis; this is necessary, because it is required to know the number of moles n of each constituent present in 100g slag. These values are shown in Tables 2 and 4.

Table 1 Typical composition of Panzhihua raw ore and concentrate

Item	Fe_t	FeO	SiO_2	Al_2O_3	CaO	MgO	MnO	TiO_2	V_2O_5	Cr_2O_3	Co	Ni	Cu	S	P
ore	31.80	—	21.15	9.05	6.53	6.36	0.28	10.25	0.32	0.029	0.018	0.013	0.022	0.60	0.02
concentrate	51.56	30.51	4.64	4.69	1.57	3.91	0.33	12.73	0.564	0.032	0.02	0.013	0.020	0.532	0.0045

Table 2 Composition of slag No. 1 as oxides and ions (Fraction x of $TiSiO_5^{2-}$ in mol% = 34.7)

	Oxides		Ions				
					TiO_2 as acidic oxide		TiO_2 as basic oxide
Constituent	wt%	No. of moles per 100g slag	Constituent	wt%	No. of moles per 100g slag	wt%	No. of moles per 100g slag
CaO	25.10	0.4475	Ca^{2+}	17.94	0.4475	17.94	0.4475
MgO	9.69	0.2404	Mg^{2+}	5.84	0.2404	5.84	0.2404
FeO	1.15	0.0160	Fe^{2+}	0.89	0.0160	0.89	0.0160
TiO	0.94	0.0147	Ti^{2+}	0.70	0.0147	0.70	0.0147
Ti_2O_3	0.94	0.0065	Ti^{3+}	0.62	0.0130	0.62	0.0130
SiO_2	23.68	0.3941	Ti^{4+}	—	—	14.35	0.2995
Al_2O_3	13.84	0.1357	$Al_2SiO_7^{4-}$	26.33	0.1357	26.33	0.1375
TiO_2	23.93	0.2995	$TiSiO_5^{2-}$	16.21	0.1039	—	—
V_2O_5	0.36	0.0020	TiO_3^{2-}	18.76	0.1956	—	—
S	0.37	0.0115	SiO_4^{4-}	—	—	23.80	0.2584
	100.00		$Si_3O_9^{6-}$	11.76	0.0515		
			VO_4^{3-}	0.47	0.0040	0.47	0.0040
			O^{2-}	0.11	0.0070	8.69	0.5432
			S^{2-}	0.37	0.0115	0.37	0.0115
				100.00		100.00	
Actual(%S)/[%S] = 6.6			$(\%S)/[\%S]_{calculated}$ = 6.6			$(\%S)/[\%S]_{calculated}$ = 143.3	

炉渣离子理论 · 131 ·

Table 3 Original composition of slag and hot metal

Item	CaO	MgO	SiO$_2$	Al$_2$O$_3$	TiO$_2$	Ti$_x$O$_y$	V$_2$O$_5$	TiC	TiN	Fe$_t$	Fe$_M$	S	(%S)/[%S]
Slag No. 1	25.4	9.80	23.96	14.00	24.21	1.90	0.37	0.33	0.40	4.80	3.90	0.37	6.6
Slag No. 2	25.07	9.17	23.43	14.00	24.12	2.06	0.36	0.15	0.19	3.69	2.94	0.42	7.2

Item	C	Si	V	Ti	S	P
Hot metal No. 1	4.0①	0.089	0.310	0.150	0.056	0.053
Hot metal No. 2	4.0①	0.099	0.360	0.150	0.058	0.039

①assumed.

Table 4 Composition of slag No. 2 as oxides and ions (Fraction x of 'TiSiO$_5^{2-}$' in mol% = 38.55)

	Oxides		Ions				
					TiO$_2$ as acidic oxide	TiO$_2$ as basic oxide	
Constituent	wt%	No. of moles per 100g slag	Constituent	wt%	No. of moles per 100g slag	wt%	No. of moles per 100g slag
CaO	25.17	0.4488	Ca^{2+}	17.99	0.4488	17.99	0.4488
MgO	9.21	0.2285	Mg^{2+}	5.55	0.2885	5.55	0.2285
FeO	0.97	0.0135	Fe^{2+}	0.75	0.0135	0.75	0.0135
TiO	1.03	0.0161	Ti^{2+}	0.77	0.0161	0.77	0.0161
Ti$_2$O$_3$	1.03	0.0072	Ti^{3+}	0.69	0.0144	0.69	0.0144
SiO$_2$	23.53	0.3916	Ti^{4+}	—	—	14.52	0.3031
Al$_2$O$_3$	14.06	0.1379	Al$_2$SiO$_7^{4-}$	26.76	0.1379	26.76	0.1379
TiO$_2$	24.22	0.3031	TiSiO$_5^{2-}$	18.22	0.1168	—	—
V$_2$O$_5$	0.36	0.0020	TiO$_3^{2-}$	17.87	0.1863	—	—
S	0.42	0.0131	SiO$_4^{4-}$	10.41	0.0456	23.36	0.2537
			Si$_3$O$_9^{6-}$	—	—	—	—
			VO$_4^{3-}$	0.46	0.0040	0.46	0.0040
			O^{2-}	0.11	0.0067	8.73	0.5455
			S^{2-}	0.42	0.0131	0.42	0.0131
	100.00			100.00		100.00	
Actual(%S)/[%S]=7.2			(%S)/[%S]$_{calculated}$=7.2			(%S)/[%S]$_{calculated}$=171.3	

The key to the success of ionization theory of slag lies in the properchoice of the ion species in the slag. Perusal of the phase diagram of the $CaO\text{-}TiO_2\text{-}SiO_2$ system[7] shows the presence of titanite $CaO \cdot TiO_2 \cdot SiO_2$ and perofskite $CaO \cdot TiO_2$. Anions of $TiSiO_5^{2-}$ and TiO_3^{2-} are thus assumed to be present in the slag. According to Ref. [2], $Al_2SiO_7^{4-}$ and $Si_3O_9^{6-}$ are also assumed.

Reactions for the formation of complex anions are:

$$SiO_2 + Al_2O_3 + 2O^{2-} = Al_2SiO_7^{4-}$$
$$SiO_2 + TiO_2 + O^{2-} = TiSiO_5^{2-}$$
$$TiO_2 + O^{2-} = TiO_3^{2-}$$
$$3SiO_2 + 3O^{2-} = Si_3O_9^{6-}$$
$$V_2O_5 + 3O^{2-} = 2VO_4^{3-}$$

The Al_2O_3 forms $Al_2SiO_7^{4-}$ with SiO_2. A fraction x of TiO_2 forms $TiSiO_5^{2-}$ with SiO_2, while the remaining portion of TiO_2 forms TiO_3^{2-}. The remaining part of SiO_2 forms $Si_3O_9^{6-}$. The following relations are helpful for calculation:

$$\sum n_+ = n_{CaO} + n_{MgO} + n_{FeO} + n_{TiO} + 2n_{Ti_2O_3}$$
$$n_{O^{2-}}\text{-sup plied} = \sum n_+ + n_{Ti_2O_3}$$
$$n_{O^{2-}}\text{-consumed} = n_{SiO_2} + n_{Al_2O_3} + 3n_{V_2O_5} + (1-x)n_{TiO_2}$$
$$n_{O^{2-}}\text{ present in the slag} = \sum n_+ + n_{Ti_2O_3} - n_{SiO_2} - n_{Al_2O_3} - 3n_{V_2O_5} - (1-x)n_{TiO_2}$$
$$\sum n_- = \sum n_+ + n_{Ti_2O_3} + n_S + (2/3)xn_{TiO_2} - (2/3)n_{SiO_2} - (1/3)n_{Al_2O_3} - n_{V_2O_5}$$

Values of f_S are calculated conventionally with values of e_S^j taken from Ref. [8]. For both cases $f_S = 2.69$.

3 Results and discussion

Results of calculation of slags No. 1 and No. 2 with the sulphur partition ratio calculated in coincidence with that found in actual practice are also given in Tables 3 and 4, the weight percentages of differentions being also listed in the same tables. To study the effect of x upon the sulphur partition ratio calculation was made with different values of x with results shown in Tables 5 and 6. Although the basicity of these two slags differs not much, it can be shown that variance in basicity does cause a change in the fraction of $TiSiO_5^{2-}$. To give a deeper insight into this effect, the same way of calculation was carried out upon the slag No. 3 from the previous paper[9]. Table 7 gives its composition (hot metal with 4% C, $f_S = 2.75$), and Table 8 gives the results of calculation. Column 7 in Table 8 gives a case where $n_{Si_3O_9^{6-}} = 0$, and $xn_{TiO_2} = n_{SiO_2} - n_{Al_2O_3}$, which has been discussed in detail in Ref. [9]. Comparative values of x for the three slags with the same sulphur partition ratio of 7 are put together in Table 9. It could be obviously shown that the higher the basicity was, the less would be the percentage of, $TiSiO_5^{2-}$ and the bigger the proportion of TiO_3^{2-}.

It might be of interest to treat TiO_2 as a basic oxide, Al_2O_3 behaving as $Al_2SiO_7^{4-}$ as before, to make a calculation of the sulphur partition ratio. Since the slag in this case would be strongly basic, Eq. 2 should be used for the evaluation of $\gamma_{Fe^{2+}}\gamma_{S^{2-}}$. As shown in Tables 3 and 4, the calculated partition ratio for slag No. 1 equals 143, while that for slag No. 2 equals 171. These values of sulphur partition ratio would be too high to be realizable at any blast furnace under any operational conditions. Hence it can be concluded that it is absurd to treat TiO_2 as a basic oxide.

Table 5　Calculated (%S)/[%S] values for slag No. 1

%$TiSO_5^{2-}$	32	33	34.5	34.7	34.77	35	36
$n_{O^{2-}}$	-0.0011	0.0019	0.0064	0.0070	0.00724	0.0079	0.0109
$N_{O^{2-}}$	—	0.0038	0.0126	0.0138	0.0142	0.0155	0.0213
(%S)/[%S]	—	1.9	5.7	6.6	7.0	8.2	—

Table 6　Calculated (%S)/[%S] values for slag No. 2

%$TiSiO_5^{2-}$	36	37	38	38.50	38.55	39	40
$n_{O^{2-}}$	-0.0010	0.0020	0.0051	0.0066	0.0067	0.0081	0.0111
$N_{O^{2-}}$	—	0.0039	0.0100	0.0129	0.0131	0.0158	0.0216
(%S)/[%S]	—	2.3	4.9	7.0	7.2	10.0	—

Table 7　Composition of slag No. 3

Constituent	wt%
CaO	24.4
MgO	7.2
FeO	1.6
SiO_2	25.5
Al_2O_3	16.6
TiO_2	24.0
S	0.7
	100.0

Table 8 Calculated $(\%S)/[\%S]$ values for slag No. 3

%TiSiO$_5^{2-}$	83.5	85	86.5	86.76	86.8	87.08	87.5
$n_{O^{2-}}$	-0.0008	0.0037	0.0082	0.0090	0.0091	0.0100	0.0112
$N_{O^{2-}}$	—	0.0076	0.0167	0.0183	0.0185	0.0202	0.0227
$(\%S)/[\%S]$	—	1.9	5.9	7.0	7.4	9.1	—

Table 9 Effect of slag basicity upon the fraction x of TiSiO$_5^{2-}$ for the $(\%S)/[\%S]=7$

Item	Basicity		x/mol%
	$\dfrac{\%CaO+\%MgO}{\%SiO_2+\%Al_2O_3}$	$\dfrac{\%CaO+\%MgO}{\%SiO_2+\%Al_2O_3+\%TiO_2}$	(Fraction of TiSiO$_5^{2-}$)
slag No. 1	0.93	0.57	34.77
slag No. 2	0.91	0.56	38.50
slag No. 3	0.75	0.48	86.76

The acidic nature of the highly titaniferous slag could not bring about an effective desulphurization of the hot metal. For making low-S steels, a pretreatment of hot metal with desulphurizing agent would be quite desirable. To assure smooth operation of the blast furnace, one of the measures is to run the furnace at rather low temperature. The low Si-content of the hot metal and its pre-removal of V would give an end-temperature of the blown steel too low for continuous casting. Extra means of raising the temperature of the steel to the proper temperature for continuous casting would be indispensable. This might be done by blowing coal powder into the bath in the converter to increase the heat source or by external heating of the steel in the ladle during the secondary refining.

4 Concluding remark

From the point of view of desulphurization, the TiO$_2$ in the Panzhihua ore or its concentrate can only be an acidic oxide. Calculation of basicity for the ore or concentrate according to $(\%CaO+\%MgO)/(\%SiO_2+\%Al_2O_3+\%TiO_2)$ would be adequate and highly desirable. Owing to the acidic nature of the TiO$_2$, the Panzhihua ore or concentrate can never be considered as semi-self-fluxing.

References

[1] Publications from Panzhihua Iron and Steel Co. (in Chinese)

[2] Wei Shoukun. Acta Metall. Sin. ,1966,9:127. (in Chinese)
[3] Wei Shoukun. Physical Chemistry of Metallurgical Processes, Shanghai Science and Technology Press,1980:227,239. (in Chinese)
[4] Samarin A M,Shwartsman L A,Temkin M. Zh . Fiz. Khim. ,1946,20:114. (in Russian)
[5] Samarin A M,Shwartsman L A . Akad. Nauk SSSR,Zzv. OTN,1948(9):1457. (in Russian)
[6] Wang Xiqing. Personal communication.
[7] DeVries R C,Roy R,Osborn E F. J. Amer. Ceram. Soc. ,1955,38. 158;Sims C E. Electric Furnace Steelmaking,Interscience Publ. ,vol. 2,1963:244.
[8] Sigworth G K,Elliott J F. Metal Science,1974,8:298.
[9] Wei Shoukun. Gang Tie Fan Tai,1988(4):1. (in Chinese)

Some Advances on the Theoretical Research of Slag[*]

Jiang Guochang　Xu Kuangdi　Wei Shoukun

Abstract: This pasper is a review and a comment regarding slag models and optical basicity. It was thought to be better to establish slag model and basicity concept based on the cell structure of slag.

Keywords: slag model, basicity, cell structure

1　Introduction

It has been long, that the development of metallurgical technology relyed upon a lot of experiment-sieving procedures. The development of materials was in a similar situation. During many years it always originated from a "flavouring" experiment in a laboratory small furnace. Nevertheless, in recent years the invention of new materials changes its way to rely on the so called composition design. Certainly, the principle of theoretical estimation in advance should be also abided for the creation of a new metallurgical technology.

To perform such a theoretical estimation the support of a thermodynamic data base, particularly the systematical activity data of melt components in various cases is needed first of all. Based on the now available activity data of metal for ironmaking and steelmaking, it is, at least, possible to carry out many estimations, even if some interesting problems as component activity in concentrated alloys and the relationship between activity and microstructure of liquid metal, are waiting to be studied. However, this is not the case for slag. There are many instances, that many reliable estimations are failure due to lack of slag data. Perhaps, this is why many metallurgists nowadays focus their attention on the research regarding slag.

The International Conferences on Molten Salgs and Fluxes held in 1980, 1984, 1988 collect and exhibit the results of this field. Moreover, as glass and magma are silicate similar to molten slag, a number of glass-chemists and geochemists run some correspond-

---

[*]　原刊于 ISIJ International, 1993, 33(1): 20~25.

ing studies with almost the same approaches. It is doubtlessly worth while for a metallurgist to know something about their contributions. This paper intends to give a brief review concerning some results of these three fields.

2 A review on slag modelling

The process of gradually understanding slag is just the process of developing of various slag models. Based on Mysen's opinion,[1] it is reasonable to divide the slag models into three categories:

(1) The physical parameter model.

The essence of this kind of slag model is an inference on microstructure of slag from the measured physical properties. For example, the bonding energy of Si-O bond is estimated according to the activation energy of viscous flow and based on the presumed relationship between microstructure and the physical properties.

(2) The thermodynamic parameter model.

The ultimate aim of these models is to deduce an equation group from the available thermodynamic experimental data, and by means of this equation group to evaluate interpolatingly and/or extrapolatingly. The foundation of these models is a set of presumed microstructural units. And the relationship between these units and experimental data is stipulated to follow the basic principle of thermodynamics.

(3) The structural model.

According to this approach the properties are related to the microstructure which is either measured by some of advanced instruments or determined along the theoretical approach of chemical bonding.

In ferrous metallurgy, the well known molecular theory, the complete ion theory, the molecular-ion coexisted theory and the regular solution theory, all are thermodynamic parameter model. Apart from these, the sub-regular solution model which is recently used in slag research as well as the latest developed models of Gaye and Pelton also belong this category. Among the mentioned three categories, the progression of this category is the fastest. The following is a brief comment on the models involving in this category.

2.1 Regular solution model

Since Lumsden adopted regular solution model in his slag investigation 30 years ago, this model has been used quite well in equilibrium calculation for oxidation slag. Particularly, for the redox equilibrium of iron in slag and the partition of O, Mn, P between slag and metal. The calculated results were found in harmony with measurements in a comparatively large composition region. [2~4]

Nowadays, the most contributions to the use of regular solution model in slag are given by Ban-ya. As he indicated, the basic point of this model is that O^{2-} anions compose of the main microscopic lattice in slag. Various cations distribute in the gaps among them to form different kinds of cells (i-O-j). Here, i and j denote cations. The equation for the calculation of activity coefficients, in fact, is an extension of the quadratic form from Darken. However, many α_{ij} parameters including in this equation were evaluated by Ban-ya[2]. Furthermore, an assumed pure liquid is taken as the standard in regular solution model. Ban-ya pointed out a conversion of standard must be included in the equation. [2]

On the other hand, so far the following weak points still limit the use of this model in slag.

(1) Generally the deviation of calculation from measurement is in a degree of 10%. [3] Eventhough the error of this kind of experiment perhaps arrives 10%~30%[2], it seems that some effects arrise from the model. The Darken's quadratic form modified by Ban-ya for activity coefficient evaluation is

$$RT\ln\gamma_i = \sum_j \alpha_{ij}X_j^2 + \sum_j \sum_k (\alpha_{ij} + \alpha_{ik} + \alpha_{jk})X_jX_k + \Delta G \quad i \neq j \neq k \quad (1)$$

According to the theory of regular solution[5], $\alpha_{ij} = \frac{1}{2}ZN_0(2U_{ij} - U_{ii} - U_{jj})$. Here, bonding energy of cells U_{ij}, U_{ii}, U_{jj} all are assumed to be independent on slag composition. So as the approach of Darken and Ban-ya, some α_{ij} could be evaluated pursuanting to the data of binary and ternary systems. Considering that actually there are interaction between different kinds of cells, the bonding energy in binary or ternary systems should be not the same of that in multipul component system. Thereby it is difficult to keep the evaluated α_{ij} in constant. Moreover, Ji Chunlin pointed out that, at least, in some cases $\alpha_{jk} \neq \alpha_{kj}$[6].

(2) As Ban-ya's idea, one of the premises of regular solution model is an assumption that the coordination number of cations is constant. In contrast, this is possible only if it is referred to an intermediate concentration region. Hence the component which content is lower than 5% is neglected by Ban-ya in his evaluation of γ_i for multipul component systems. [2] Perhaps, this is just the reason to explain why regular solution model is not so suitable in simulation the behaviour of reduction slag.

(3) Till now based on regular solution model it is impossible to treat the behaviour of slag containing S^{2-}, F^-, Cl^-, which are used more and more in ironmaking and steelmaking.

2.2 Sub-regular solution model

By means of sub-regular solution model, Shim published his results concerning a_{FeO} of steelmaking slag[7] and $Fe_tO-Na_2O-SiO_2$ slag.[8] The following is the basic equation of this model.

$$RT\ln\gamma_i = \sum_j \left[\varepsilon_{ij}X_j^3 + (\vartheta_{ij} + \xi_{ij}T)X_j^2\right] + \\
\sum_j \sum_{k>j} \left[(\varepsilon_{ij} + \varepsilon_{ik})X_jX_k(X_j + X_k + 0.25) + \right.\\
\left(\vartheta_{ij} + \vartheta_{ik} - \frac{1}{2}\vartheta_{jk} - \frac{1}{2}\vartheta_{kj}\right)X_jX_k + \\
\left.\left(\xi_{ij} + \xi_{ik} - \frac{1}{2}\xi_{jk} - \frac{1}{2}\xi_{kj}\right)TX_jX_k\right] + \\
\sum_j \sum_{k>j} \sum_{l>k} (\varepsilon_{ij} + \varepsilon_{ik} + \varepsilon_{il})X_jX_kX_l \quad (2)$$

The approach of evaluation ε, ϑ and ξ parameters is quite similar to that used in regular solution model. The merit of this model essentially superior than regular solution model is that the bonding energy is assumed to be changeable in some degree to follow the variation of composition. The a_{FeO} of Shim[7] is obviously closed to the experiment than that given by regular solution model.

The following equation was used by Jiang Guochang to calculate the component activities of molten C-Fe-X (X=Cr, Ni, Si, Mn) alloy[9]

$$RT\ln\gamma_i = \sum_j \sum_k A_{jk} Y_j Z_k \quad (3)$$

Here, Y and Z are variables of composition. The evaluation of A_{jk} parameters through value fitting is based on the thermodynamic data of the boundary of liquid region. It is thought, that this approach should also be possibly useful for slag simulation. The problem is that some of the data is either in shortage or no generally accepted value. Thereby, the process to evaluate A_{jk} parameters is just the process to distinguish and judge which are the reliable boundary data.

2.3 Molecular-ion coexisted model

In China, the leading exponent of molecular-ion coexisted model is Zhang Jian. Based on the concept of "effective concentration", the basic assumption of this model, the activity of Al_2O_3 in basic slag used for steel refining was evaluated.[10] In contrast, regular solution model has not given the value of $\alpha_{Ca^{2+}-Al^{3-}}$ and $\alpha_{Mg^{2+}-Al^{3-}}$. So far the molecular-ion coexisted model was also claimed to be useful for the description of the oxidational ability of CaO-MgO-Fe$_2$O$_3$-SiO$_2$ melted slag[11] as well as the partition of sulphur between this slag and molten steel.[12]

The so called associated solution model[13] or two-sublattice model with hypothetical vacancies[14] are based on the similar assumption as the approach of Zhang. However, this category does not to be widely accepted in metallurgical field. On the other hand, several researchers belonging to other specialities work with the similar approach. For instance, Hastie et al.[15] and Bottinga et al.[16] all take the idea that it is not necessary to declare whether the presumed microstructural units are ion or molecular. And they all choose some of the intermediate compounds of the concerned systems as microstructural unit. In addition, it is thought by Hastie that various microstructural units mix together ideally, so component activity equivalent to the concerned unit content which is namely the "effective concentration" defined by Zhang. Hastie's approach has been applied to simulate a system of 8 components including elements such as Na, Al, B, Cl, Br. The difference between them is: Zhang aims at the behaviour under steelmaking temperature and so calculates directly based on equilibrium constant under 1873K. On the contrary, minimizing free energy of the system is the way of other two research groups. This makes an important achievement in description of phase diagram.

2.4 Pelton's model

This is a modified quasi-chemical approximation model based on Yokokawa-Niwa's model[17] which is one of the classical and standard ways in latticelike modelling of slag. So far it was used in simulation the equilibrium between SiO-CaO-MgO-MnO-FeO-Na$_2$O slag vs. solid oxide or molten metal or gaseous phase.[18] Its feature is as follows:

(1) Take various i-O-j cells including the cells of $i=j$ as microstructural units.

(2) During mixing the following reaction happens

$$(i\text{-O-}i) + (j\text{-O-}j) = 2(i\text{-O-}j) \tag{4}$$

The relative amount of cells is determined by the variation of free energy of the mentioned reaction. Namely,

$$\frac{Y^2_{(i\text{-O-}i)}}{Y_{(i\text{-O-}i)}Y_{(j\text{-O-}j)}} = 4\exp[-2W_{ij}/(ZRT)] = K \tag{5}$$

where Y denotes mole fraction of cells; Z denotes coordination number; $W_{ij} = (\omega_{ij} - \eta_{ij}T)$, ω_{ij} denotes the variation of mole enthalpy, η_{ij} denotes the variation of mole non-configurational entropy. Both of them are related to slag compositron in a polynominal form. The parameters involved in these polynominals are evaluated through value fitting with the published thermodynamic data. Thereby, the equilibrium constant of the mentioned cell reaction is changeable to follow the variation of slag composition.

In fact, the equation used to determine the mole fraction of cells is just the characteristics of quasi-chemical approximation model. The Pelton's modification is that to involve the variation of non-configurational entropy.

(3) Through the so called asymmetrical approximation the W_{ij} deduced from various $MO-SiO_2$ and Na_2O-SiO_2 was used to describe $M^*O-M^{**}O-SiO_2$ ternary system. An additional item taken from ternary data is needed only in several cases. [19]

2.5 Gaye's model

As the further development of Kapoor-Frohberg's Model, [20] Gaye's Model has been widely adopted, at least, in Europe. Provided, at present, this is the most successful thermodynamic parameter model. According to the paper published in 1984, a slag containing 6 oxides can be treated. [21] Moreover, in 1990, the result of the cooperation of IRSID with NSC was published. [22] It extends the application to the slags in which multi-anions as O^{2-}, S^{2-}, F^- are coexisted, and to the slag containing P_2O_5.

This is also a modified quasi-chemical approximation model. Similar to Pelton's approach, it just uses binary system parameters to simulate multipul component systems. The difference from Pelton's Model is that not only the formation energy W_{ij}^A of i-A-j cell but also the interaction energy E_{ij}^A and E_i^{AB} between various cells, where A and B denote O^{2-} or S^{2-} or $(2F)^{2-}$, are taken into account. The binary parameters involved in this model were suggested to be a linear function of slag composition and to be independent on temperature. In slag two kinds of sub-lattices were assumed. One is composed of anions, the another is composed of cations. Various cation entities distribute on the cation sub-lattice. This is an assumption to deal with P_2O_5 and Al_2O_3. For example, it is presumed that the cation entities of P_2O_5 are PO^{3+} and P^{5+}. But, as pointed out by Gaye himself, dealing with Al_2O_3 with this approach needs further improvement.

It is worthwhile to note, that dealing with C_S data of SiO_2-Al_2O_3-Fe_2O_3-FeO-MnO-MgO-CaO slag by means of Gaye's model is much superior than the treatment based on optical basicity[23]. Moreover, a number of researchers use Gaye's Model to estimate C_S, [23] the melting point[24] of some slags and the modification of non-metallic inclusions. [25]

2.6 Polymer model

Masson is known widely due to his great contribution[26] to polymer model. In metallurgical field this model has been highlighted. A plenty of metallurgists study slag behaviour along this model. As they trust that the slag microstructure should be a polymer. And anyhow, in description of polymerization of some binary silicate in part of their liquid region the adoption of polymer model is successful. On the contrary, this model was criticized a lot by geochemists. As Mysen[1] and Bottinga[16] indicated:

(1) The experiment, polymer model depends on the results of which as evident is not reliable.

(2) Generally, it is just possible to deal with binary silicate of either chain-like or branched-chain-like microstructure. Namely, it is limited in the region of mole fraction of SiO_2 is lower than 0.5. It seems to be difficult to deal with the slag containing some elements the valency as well as the coordination number of which is changeable.

(3) The following polymeraization reaction is assumed as the foundation of polymer model

$$SiO_4^{4-} + SiO_{3n+1}^{2(n+1)-} = Si_{n+1}O_{n+4}^{2(n+2)-} + O^{2-} \qquad (6)$$

This reaction claims that the mole fraction of O^{2-} should be larger if the polymerization degree becomes larger. Nevertheless, the real mole fraction of O^{2-} is always a small quantity. And, the equilibrium constant of this polymerization reaction becomes larger to follow the increase of temperature. This claims that an increase of polymerization degree results from an elevation of temperature. Obviously this could not be in reality.

Magma which is the objective of geochemists is a silicate based material as slag. So it should be a polymer. The microstructural unit of magma is considered as a ring, which is composed of 3 to 6 monomers,[1] by geochemists. At present, it seems no theoretical model is available to describe this kind of ring-microstructure.

3　A review on slag basicity

In spite of that basicity is a concept frequently used in metallurgy, it entered into the hall of modern theory just after the appearance of optical basicity. It was first suggested by Duffy and Ingram in their research on glass-chemistry.[27] Together with them, Sommerville made efforts to introduce this concept to metallurgists.[28] Later on, Nakamura et al. suggested a new optical basicity theory based on photoacoustic spectroscope and average electron density.[29] In 1987, Jiang Guochang introduced the concept of optical basicity in a national seminar.[30]

Nowadays optical basicity has been utilized by a number of metallurgists to sum up various capacity data of many slags as well as the solubility of (MgO) and redox equilibrium of (Fe_tO).[31] Intrinsically, the reason is that the chemical characteristics of slag has to be concerned if one refers to slag-metal reaction. However, basicity is just the brief and practical index reflecting chemical characteristics of slag. This function is not replaceable with slag modelling.

On the other hand, some weak points were found during utilizing optical basicity:

(1) Various cells as i-O-j and i-O-i are taken essentially to be microstructural units in optical basicity. But for calculating optical basicity of a practical slag, the basicity characteristics of oxide molecular in stead of that of cells is utilized.

(2) Neither Duffy's approach nor Nakamura's approach is ideal in elucidation completely the influence of "environment" on anion activity. The comparison of bonding energy of Al-O in variant MO-Al_2O_3-SiO_2 systems (where MO = Na_2O, CaO, MgO) is a typical example. Appen[32] and Varshal[33] indicated that of Al-O in Na_2O system is the strongest one, and that in MgO is the weakest one. This is, however, in contrast to optical basicity. Yokokawa also doubted the basicity characteristics of Al_2O_3 given by the theory.[34] The characteristics of Na_2O needs to be improved further yet.[31]

(3) For transition metal oxides, the characteristics given by Duffy et al. is successful in description of phosphorus capacity only. On the contrary, that given by Nakamura et al. is successful just in simulation of surphur capacity. It seems that a transition metal oxide should have a variable characteristics.

4 Discussion

(1) In history, molecular theory, ion theory and molecular-ion coexisted model were used to describe slags. The later does not get wide support in metallurgists. One of the possible reason is that some moleculars used as microstructural unit in this model could not be detected with experiment. On the other hand, it was understood earlier that the ions in slag are far from the completely independent state. Experiments indicated all the ions are involved in this or that kind of cell. As already mentioned, many modern slag modelling take various cells as the microstructural units. Hence, it is reasonable to say this is the Cell Theory era.

(2) In reality, slag should be a polymer made up of many cells. Unfortunately, even the real polymerization feature of CaO-Al_2O_3-SiO_2 slag so far does not be understood thoroughly. Nevertheless, the polymerization degree might effect directly on physical properties, such as viscosity. For simulating chemical activation of slag it seems fairly to undertake study based on cell characteristics. This is also the view point of Bottinga.[16]

The so called cell characteristics is essentially two items. Namely, the relative amount of cells and the bonding energy of them. In Gaye's Model and Pelton's Model, these two items determine the mixing free energy of the slag exactly. In fact, in a slag system the variation of composition is just the variation of the amount of cells with different bonding energy. One of the contributions of molecular-ion coexisted model is it indicates that, the thermodynamic activity of component does not have a meaningful relationship with the apparent concentration, it is a function of the relative amount of concerned microstructural unit supposed to be mixing together with others ideally. According to optical basicity, the basicity characteristics(B) given for compounds is intrinsically just an extent of bonding intensity of some cells. Thereby, its stipulation of taking oxide compound as calculation unit is a weak point which must neglect or screen the influence of other cells. The "bonding basicity" concept suggested by Jiang is a step aiming at modify on this weak point.[35] It is thought that the bonding basicity($B_{ij}Y_{ij}$) of cell(i-O-j) should have essentially a relationship with the thermodynamic activity of related component.

(3) Even it is limited in the term of cell characteristics, several topics need to be talked over further.

1) The existing state of sulphur is changeable to follow the variation of oxygen pressure.[5] If sulphur in oxidational slag is contained in cell(Ca-O-S), in reductional slag it may become to (Ca-S-Ca) or (Ca-S-Si). Similarly, during dephosphorization with reduction slag, phosphorus might also be involved in cell(Ca-P-Ca) and so on. Provided, from this point it is able to sum up the capacity data of both of oxidational and reductional slags.

2) Alumina is a typical amphoteric oxide, the coordination number of it is variable. And through the formation of complex, cell(i-O-j) becomes to ($i-O<^k_j$)[27]. In addition, different cations could have different effect on this triple-bonded oxygen anion. Perhaps, these are the grasps to understand the behaviour of alumina and so on.

3) As a typical transition metal oxide, the behaviour of Fe_tO is difficult to be described satisfactorily, at present, with optical basicity. The complexity of this description concerns the redox equilibrium together with the amphoteric feature of Fe_2O_3. Recently, it has been found that some kinds of metallic bonds might affect this oxide also.[36]

(4) The research on cell structure depends upon the progression of modern detection instruments. Meanwhile, it cannot help paying special attention to the improvement of preparation technology for slag samples. A typical instance is that, the experiments regarding activity of iron oxide are always performed under a lower temperature due to the strong corrosivity of the slags.

Considering that the experiments so far are not so reliable, it is essential to approach the reality from both sides: theoretical way and experimental way, and to rely upon the agreement of them. Nowadays some estimations about the energy characteristics of cells by means of quantum chemistry and so on seems to be noticeable.

(5) For metallurgists it is too early to develop an actural structural model. The thermodynamic parameter models are more practical. Whether these models are successful or not, it all depends on whether they are able to sum up reasonably the known thermodynamic data. And it does not concern the assumed microstructural units whether they are of reality or not. From this point, the molecular-ion coexisted model should have its suitable status. But, in every sense, it is not to say the assumed microstructural units are proved to be of acturality, pursuanting to that the model can describe the concerned activity. If a thermodynamic parameter model is used widely and its microstructural units for various slag systems are assumed not to follow the desire of getting agreement with experimental results, but in accordance with a seriously universal stipulation, then the assumption of this model might close to reality in a closer degree. From this point, the molecular-ion coexisted model is not so satisfied.

The molecular-ion coexisted model usually takes some unstable compounds as microstructural units, and then subjects criticism. From the view point of cell theory, these moleculars factually hint the clusters of some cells, and different molecular relates to different cell cluster. Eventhough, the existance of unstable compounds under high temperature is not easy to acquire evidence, the existence of many cells is now proved. In the liquid region of $CaO-Al_2O_3$ (1873K), the cell reaction which governing the thermodynamics of this system is

$$(Ca-O-Ca) + (Al-O-Al) = 2(Ca-O-Al) \tag{7}$$

The formation free energy of unstable compound $3CaO-Al_2O_3$, but not of other stable compound, was choosen by Jiang to calculate equilibrium constant of the cell reaction. [35] This is a reasonable selection, because only $3CaO-Al_2O_3$ contains just cell of (Ca-O-Al). This unstable compound was also taken as microstructural unit in molecular-ion coexisted model, but this was only for getting agreement with experimental results. [10]

After all, a thermodynamic parameter model may acquire more reliability if its microstructural units are more closer to reality. Perhaps, from this view point to undertake the next improvement of Gaye's model is beneficial. The amphoteric behaviour of alumina and the variation of environment of S^{2-} have not been dealt with quantitatively in Gaye's model. And, provided, for P_2O_5 and so on one have to take double bonding into account.

(6) To compare the approaches of Pelton, Gaye and we ourselves, Eq. (5) can be useful in binary system. According to either Pelton or Gaye, the energy parameters for calculation of K are evaluated through value fitting of thermodynamic data. As my opinion, however,

$$K = f(B_{ij}, G^m) \tag{8}$$

and then the data of micro structure as bonding energy can be adopted. For lime-alumina-silica ternary system, the governing equation suggested by ourselves is[37]

$$Y_{12}^{\alpha} Y_{13}^{\beta} Y_{14}^{\delta} Y_{24}^{\nu} / Y_{44}^{\sigma} Y_{11}^{\tau} Y_{22}^{\nu} Y_{33}^{l} = f^{*} (B_{ij}, G^{m}) \qquad (9)$$

Here, $\alpha, \beta, \delta, \nu, \sigma, \tau, v, l$ all are stoichiometric parameters in cell mixing reaction. Subscripts 1, 2, 3, 4 denotes Si, Al_a (acid Al^{3+}), Al_b (basic Al^{3+}), Ca respectively, 12 denotes Si-O-Al cell and so on.

5 Conclusion remarks

This pasper is a review and a comment regarding slag models and optical basicity. It is possible to say, undertaking the research from the following two aspects might be beneficial. One is to elucidate the cell structure depending on both experiment and theory. Based on this, the precise evaluation of basicity characteristics in various cases can be given. The another is establishment of a model to calculate the relative amount of various cells, taking the influence of the variation of coordination number and chemical valency into account. And finally to correct the research by experiment.

Acknowledgements

This research is supported by The National Natural Science Foundation of China.

We are very grateful to Professor Zhang Jian for his systematical introduction of the Molecular-Ion Coexisted Model.

References

[1] Mysen B O. Structure and properties of silicate melts, Elsevier Sci. Publishers B. V. , 1988:48, 79,116.

[2] Ban-ya S, Hino M . Tetsu-to-Hagane, 1988,74(9):1701.

[3] Hino M, Kikuchi I, Fujisawa A, Ban-ya S. Proc. of 6th IISC Vol. 1, ISIJ, Tokyo, 1990:264.

[4] Nagabayashi R, Hino M, Ban-ya S. Proc. of 3rd Int. Conf. on Molten Slags and Fluxes, Glasgow, U. K. , June, 1988:24.

[5] Richarsaon F D. Physico-Chemistry of Melts in Metallurgy, Academic Press Inc. , London, 1974:116.

[6] Qi Guojun, Ji Chunlin. Proc. of 5th National Conf. on Metallurgical Physico-Chemistry, China, 1984:227.

[7] Dow-Bin Hyun, Jae-Dong Shim. Trans. Iron Steel Inst. Jpn, 1988, 28:736.

[8] Dow-Bin Hyun, Jae-Dong Shim. Proc. of 6th IISC, Vol. I, ISIJ, Tokyo, 1990:177.

[9] Jiang Guochang, Zhang Xiaobing, Xu Kuangdi. Acta Metall. Sin. , 1992,28B:240.

[10] Zhang Jian, Wang Chao. Proc. of National Academic Symp. on Special Steel Melting, China,

1986:1.
- [11] Zhang Jian, Wang Chao, Tong Fusheng. A Calculation Model on the Oxidational Power of a Multi-Component Molten Slag, Private Communication.
- [12] Zhang Jian, Wang Chao, Tong Fusheng. Proc. of The Wei Shoukun Symp. , China, 1990:57.
- [13] Sharma R C, Chang Y A. Metall. Trans. , 1979, 10B:103.
- [14] Hillert M, Sundman B, Wang X Z. Metall. Trans, 1990, 21B:303.
- [15] Hastie J W, Bonnell D W, Plante E R. Proc. of 3rd Intern. Conf. on Molten Slags and Fluxes, Glasgow, U. K. June, 1988:254.
- [16] Bottinga Y, Weill D F, Richet P. Thermodynamics of Minerals and Melts, ed. by Newton R C, Springer-Verlag, N. Y. , Heidelberg, Berlin, 1981:207.
- [17] Yokokawa T, Niwa K. Trans. Jpn. Inst. Met. , 1969, 10:3, 81.
- [18] Pelton A D, Eriksson G, Blander M. Proc. of 3rd Int. Conf. on Molten Slags and Fluxes, Glasgow, U. K. , June, 1988:66.
- [19] Pelton A D, Blander M. Metall. Trans. , 1986, 17B:805.
- [20] Kapoor M L, Frohberg M G. Proc. of Symp. on Chemical Metallurgy of Iron and Steel, Sheffield, U. K. , 1971:17.
- [21] Gaye H, Welfringer J. Proc. of 2nd Intern. Symp. on Metallurgical Slags and Fluxes, Lake Tahoe, Nevada, Nov. , 1984:257.
- [22] Lehmann J, Gaye H, Yamda W, Matsumiya T. Proc. of 6th IISC, Vol. 1, ISIJ, Tokyo, 1990:256.
- [23] Saint-Jours C, Allibert M. Proc. of 3rd Int. Conf. on Molten Slags and Fluxes, Glasgow, U. K. , June, 1988:65.
- [24] Boom R, Deo B, Van Der Knoop W, Mensonides F, Van Unen G. Proc. of 3rd Intern. Conf. on Molten Slags and Fluxes, Glasgow, U. K. , June, 1988:273.
- [25] Kay D A R, Jiang Junpu. Proc. of 3rd Int. Conf. on Molten Slags and Fluxes, Glasgow, U. K. , June, 1988:263.
- [26] Masson C R. Proceedings of Symp. on Chemical Metallurgy of Iron & Steel, London, U. K. , 1971:3.
- [27] Duffy J A, Ingram M D. J. Non-Cryst. Solids, 1976, 21:373.
- [28] Sosinsky J, Sommerville I D. Metall. Trans. , 1986, 17B:331.
- [29] Nakamura T, Ueda Y, Toguri J M. J. of J/Af, 1986, 50:456.
- [30] Jiang Guochang. J. of Shanghai University of Technology, 1989, 10(3):257.
- [31] Sommerville I D. Proc. of Technological Advances in Metallurgy, Lulea, Sweden, Sept. , 1986:13.
- [32] Appen A A. Glass Chemistry(Chinese Version) , Publisher of Civil Eng. of China, 1974:268.
- [33] Varshal V G. Izv. A. N. SSSR Inorg. Materials, 1972(5):934.
- [34] Yokokawa T, Maekawa T, Uchida N. Bull. JIM, 25(1986), 3.
- [35] Jiang Guochang, Xu Kuangdi. Proc. of 6th IISC, Vol. 1, ISIJ, Tokyo, 1990:240.
- [36] Ingram M D. Proc. of 3rd Int. Conf. on Molten Slags and Fluxes, Glasgow, U. K. , June, 1988:166.
- [37] Jiang Guochang, Zhang Xiaoping. Proc. of 4th Int. Conf. on Molten Slag and Fluxes, ISIJ, Tokyo, 1992.

选择性氧化理论

XUANZEXING YANGHUA LILUN

炼钢过程中铁液内磷、碳等元素氧化的热力学[*]

魏寿昆

采用已知的活度数据,通过三种理论方法的计算,即(1)氧化物分解压力 P_{O_2};(2)铁液内与各元素平衡的[O]量;(3)氧化的反应自由能 ΔF。对铁液内磷、碳及其他元素氧化的顺序进行了分析和研究。文中对 Карнаухов 及 Морозов 计算 P_{O_2} 不正确之处,及对 Левин 计算铁液内磷、碳平衡的[O]量欠妥之处提出了讨论。理论计算证明:(1)脱磷是一界面反应,没有适当成分的炉渣是不可能进行的;(2)迅速造成高氧化钙及高氧化铁炉渣,在较低温度有可能在托马斯法吹炼中使磷与碳同时氧化;(3)增加一氧化碳的分压能使碳的氧化受到抑制而使磷先于碳被氧化。

炼钢过程中铁液内元素,随着炼钢方法的不同,有不同的氧化顺序。一般来讲,硅锰先于磷而氧化。对碱性底吹转炉炼钢,碳先于磷而氧化。但对于电炉及碱性平炉炼钢,则磷先于碳而氧化。对氧气顶吹转炉、Kaldo 转炉及卧式 Rotor 转炉,则磷与碳同时氧化,且随着操作工艺的变化,人们可改变磷和碳的氧化速度。对我国的侧吹碱性转炉,磷与碳也同时氧化,关于元素氧化顺序的理论,Карнаухов[1]首先提出利用氧化物的分解压力解释元素氧化的顺序,但这种方法只适用于纯物质的反应,而不适用于有溶液参加的反应。Морозов[2]引用元素溶于铁液的标准自由能来纠正上列的计算方法,但他未引用铁液及炉渣各组成物的活度系数。Левин[3]计算铁液内与各元素平衡的[O]量,而以它评定碳及磷的氧化顺序,但他的计算结果值得商讨,同时他也忽略了溶液各组成物的活度系数。由于分解压力及平衡[O]量都是根据反应达到平衡状态而计算的,而大多数的反应达不到平衡,因之反应自由能 ΔF 应该是评定氧化顺序最好的准则,但文献上缺少对 ΔF 系统的计算。本文旨在采用已知的活度数据,通过热力学计算,对铁液内主要元素的氧化反应作进一步的研究,分析并讨论氧化顺序的机理。

---

[*] 原刊于《金属学报》,1964,7:240~249。
　本文注释:采用液态 SiO_2 的生成自由能,因 Elliott 计算 a_{SiO_2} 是采用液态 SiO_2 为标准状态。编者注释:文中自由能数值单位为 cal,1cal = 4.184J。

1 计算中采用的热力学数据

$Si_{(液)} + O_2 = SiO_{(液)}$; $\Delta F^\circ = -223800 + 46.08T$[4]

$[Mn] + [O] = MnO_{(液)}$; $\Delta F^\circ = -58400 + 25.98T$[4,5] (1% 标准状态)

$2Fe_{(液)} + O_2 = 2FeO_{(液)}$; $\Delta F^\circ = -113800 + 23.64T$[4]

$[C] + [O] = CO$; $\Delta F^\circ = -5350 - 9.48T$[4] (1% 标准状态)

$\frac{2}{5}P_{2(气)} + O_2 = \frac{2}{5}P_2O_{5(液)}$; $\Delta F^\circ = -146700 + 48.40T$[6]

$Si_{(液)} = [Si]$; $\Delta F^\circ = -28500 - 6.09T$[4] (1% 标准状态)

$Mn_{(液)} = [Mn]$; $\Delta F^\circ = -9.11T$[4,5] (1% 标准状态)

$C_{(石墨)} = [C]$; $\Delta F^\circ = 5100 - 10.0T$[4] (1% 标准状态)

$\frac{1}{2}P_{2(气)} = [P]$; $\Delta F^\circ = -29250 - 4.55T$[6] (1% 标准状态)

$\frac{1}{2}O_{2(气)} = [O]$; $\Delta F^\circ = -28000 - 0.69T$[4] (1% 标准状态)

a_{FeO} 自 Turkdogan 和 Pearson[7] 的三元等活度曲线图读出;

a_{SiO_2} 根据 Elliott[8] 的三元等活度系数曲线图计算;

a_{MnO} 根据 Bishop、Grant 和 Chipman[9] 的三元等活度系数曲线图计算;

$a_{P_2O_5}$ 根据 Turkdogan 和 Pearson[6] 的方法计算;

f_{Si}、f_{Mn}、f_C、f_P 及 f_O 计算方法见文献[10]。

计算时考虑托马斯法炼钢的初期阶段,选定生铁成分(%)为:

C	Si	Mn	P
4	0.6	1	2

初期生成的炉渣成分(mol%)为:

CaO	MnO	FeO	SiO$_2$	P$_2$O$_5$
40	20	5	33	2

(均按摩尔百分数计算)

以空气吹炼,空气入铁液的总压力为 1.3atm[11]。

2 气-液相氧化反应(直接氧化反应)

气-液相的氧化反应存在于底吹顶吹及吹转炉的炼钢过程中。这些反应生成的氧化物的分解压力 P_{O_2},可用式(2)表示:

$$[A] + O_2 = (AO_2) \tag{1}$$

$$P_{O_2} = \frac{a_{AO_2}}{a_A} P_{O_2}^o \tag{2}$$

式（2）中的 $P_{O_2}^o$ 为纯氧化物 AO_2 的分解压力。式（2）可换写为[10]：

$$P_{O_2} = \frac{\gamma_{AO_2} N_{AO_2}}{\gamma_A^o f_A \frac{55.85}{M_A}} \times \frac{P_{O_2}^o}{\frac{[\%A]}{100}} \tag{3}$$

式中　M_A——元素 A 的分子量；

γ_A^o——元素 A 在无限稀溶液时按 Raoult 定律计算的活度系数。

在计算分解压力时，Карнаухов[1] 忽略了式（3）中的 $\dfrac{\gamma_{AO_2} N_{AO_2}}{\gamma_A^o f_A \dfrac{55.85}{M_A}}$ 一项，而 Морозов[2] 则忽略了式（3）中的 $\dfrac{\gamma_{AO_2} N_{AO_2}}{f_A}$ 一项，所以他们的计算都有一定的误差。

式（3）又可换写为：

$$\lg P_{O_2} = \frac{\Delta F^o}{4.575T} + \lg A_{O_2} - \lg f_A - \lg [\%A] \tag{4}$$

式（4）中的 ΔF^o 是以元素 A 的 1% 溶液为标准状态时式（1）的反应标准自由能。

对碳的氧化反应：

$$2[C] + O_2 = 2CO \tag{5}$$

$$\lg P_{O_2} = \frac{\Delta F^o}{4.575T} + 2\lg P_{CO} - 2\lg f_C - 2\lg [\%C] \tag{6}$$

式（6）中的 ΔF^o 是以碳的 1% 溶液为标准状态时式（5）的反应标准自由能，而 P_{CO} 是生成物 CO 平衡时的分压力。

氧化物的分解压力越小，则该氧化物越稳定，也即该元素被氧化的可能性越大。因之，以式（4）和式（6）计算出的 P_{O_2}，可评定元素氧化的顺序。

式（4）和式（6）分别是按式（1）和式（5）在平衡状态时计算的 P_{O_2}。如按反应物的原始状态和生成物的终结状态来考虑，反应生成的自由能 ΔF 是评定反应发生可能性大小最好的准则。ΔF 负值越大，该反应发生的可能性也越大，这样，氧化顺序按氧化反应的 ΔF 来评列。

对式（1）：

$$\Delta F = \Delta F^\circ + RT\ln \frac{\gamma_{AO_2} N_{AO_2}}{f_A [\%A] P'_{O_2}} \tag{7}$$

对式 (5):

$$\Delta F = \Delta F^\circ + RT\ln \frac{P'^2_{CO}}{f_C^2 [\%C]^2 P'_{O_2}} \tag{8}$$

式 (7) 和式 (8) 的 ΔF° 是反应标准自由能, 标准状态为: 元素的 1% 溶液、1atm 气体和纯氧化物或饱和氧化物的炉渣。P'_{O_2} 是反应物 O_2 最初的实际压力, 而 P'_{CO} 是生成物 CO 最后的实际压力。

假定上列选定成分的生铁及炉渣处于平衡状态, 按式 (4) 和式 (6) 计算出的 P_{O_2} 见图 1。假定选定成分的生铁及炉渣分别代表反应物的原始状态及生成物的终结状态, 按式 (7) 和式 (8) 计算出的 ΔF 见图 2。为比较起见, 图 2 仍列出两种另外情况的自由能, 即 (1) 标准自由能 ΔF°; (2) 由选定成分的生铁及 1.3atm 生成纯氧化物或饱和氧化物炉渣的自由能 ΔF。

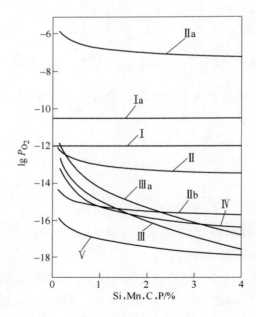

图 1　分解压力 P_{O_2} (1300℃)

Ⅰ—Fe; Ⅰa—Fe (按 $2Fe_{(液)} + O_2 = 2FeO_{(液)}$, $\Delta F^\circ = -113800 + 23.64T$ 计算);

Ⅱ—P; Ⅱa—P (按 $4/5[P] + O_2 = 2/5P_2O_{5(饱)}$, $\Delta F^\circ = -126800 + 49.79T$ 计算);

Ⅱb—P (铁酸钙炉渣); Ⅲ—C ($P_{CO} = 0.43$atm); Ⅲa—C ($P_{CO} = 1$atm); Ⅳ—Mn; Ⅴ—Si

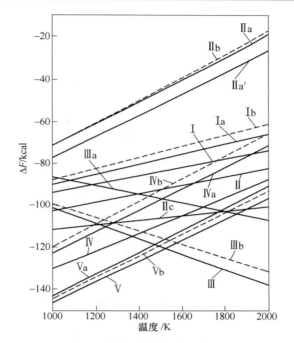

图 2 气体直接氧化反应的 ΔF

Ⅰ—$2[Fe] + O_2 = 2(FeO)$，$\Delta F = -113800 + 19.92T$；Ⅰa—$2[Fe] + O_2 = 2FeO_{(液)}$，$\Delta F^o = -113800 + 23.64T$；Ⅰb—$2[Fe] + O_2 = 2FeO_{(液)}$，$\Delta F = -113800 + 26.32T$；Ⅱ—$4/5[P] + O_2 = 2/5(P_2O_5)$，$\Delta F^o = -123300 + 20.40T$；Ⅱa—$4/5[P] + O_2 = 2/5 P_2O_{5(液)}$，$\Delta F^o = -123300 + 52.04T$；Ⅱa′—$4/5[P] + O_2 = 2/5 P_2O_{5(气)}$，$\Delta F^o = -126800 + 49.79T$；Ⅱb—$4/5[P] + O_2 = 2/5 P_2O_{5(液)}$，$\Delta F^o = -123300 + 52.39T$；Ⅱc—$4/5[P] + O_2 = 2/5(P_2O_5)$，$\Delta F = -123300 + 11.34T$(铁酸钙炉渣)；Ⅲ—$2[C] + O_2 = 2CO$，$\Delta F = -66700 - 35.61T(P'_{CO} = 0.43atm)$；Ⅲa—$2[C] + O_2 = 2CO$，$\Delta F^o = -66700 - 20.34T$；Ⅲb—$2[C] + O_2 = 2CO$，$\Delta F = -66700 - 32.19T(P'_{CO} = 1atm)$；Ⅳ—$2[Mn] + O_2 = 2(MnO)$，$\Delta F = -172800 + 42.39T$；Ⅳa—$2[Mn] + O_2 = 2MnO_{(液)}$，$\Delta F^o = -172800 + 50.58T$；Ⅳb—$2[Mn] + O = 2MnO_{(液)}$，$\Delta F = -172800 + 53.60T$；Ⅴ—$[Si] + O_2 = (SiO_2)$，$\Delta F = -195300 + 49.32T$；Ⅴa—$[Si] + O_2 = SiO_{2(液)}$，$\Delta F^o = -195300 + 52.17T$；Ⅴb—$[Si] + O_2 = SiO_{2(液)}$，$\Delta F = -195300 + 51.17T$

3 液-液相的氧化反应（间接氧化反应）

铁液内元素氧化也可通过液-液相反应进行。炉渣内的氧化铁溶于铁液，再与各元素起氧化作用。Левин[3]利用托马斯转炉的实际操作数据，作出[%O]对[%C]或[%P]的曲线图（图3）：

$$\lg K_P = \lg \frac{a_{4CaO \cdot P_2O_5}}{[\%P]^2 [\%O]^5 a_{CaO}^4} = \frac{71667}{T} - 28.73 \tag{9}$$

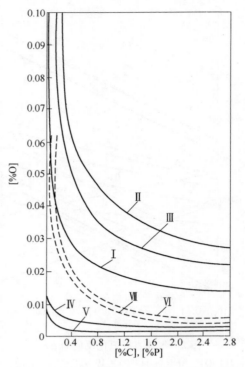

图 3 [%O] 与 [%C] 或 [%P] 的关系
（托马斯转炉——Левин）

Ⅰ—P, 1600℃, lg$A$=1；Ⅱ—P, 1500℃, lg$A$=4.8；Ⅲ—P, 1400℃, lg$A$=6.8；
Ⅳ—P, 1300℃, lg$A$=3.6；Ⅴ—P, 1200℃, lg$A$=3.2；Ⅵ—C, 1600℃；Ⅶ—C, 1300℃

令 $A = \dfrac{a^4_{CaO \cdot P_2O_5}}{a^4_{CaO}}$，他发现在吹炼第一期 lg$A$=3，第二期末 lg$A$ 加大到 7，而在后吹期末则 lg$A$=1。根据 [%O] 的大小来评定碳和磷氧化的顺序，而此顺序和 lg$A$ 值有关。

Левин 采用的 K_P 系根据 Winkler 和 Chipman[12] 的数据，后者未采用活度系数，所以 lg$A = N_{4CaO \cdot P_2O_5}/N^4_{CaO}$。作者根据托马斯转炉吹炼一些实际操作又重新进行了计算，但并未发现不同吹炼期 lg$A$ 的规律性。同时 Winkler 和 Chipman 对炉渣分子作出不合理的假定，而对生铁又未考虑 f_P 的因素，因之，Левин 计算的结论值得进一步地商讨。

采用上面选定成分的生铁和炉渣，利用已知的活度系数对达到平衡时的 [%A]×[%O] 值进行计算，结果见表1。式（7）和式（8）的 P'_{O_2} 换以 f_O[%O]，即可计算间接氧化反应的 ΔF（图4）。[%O] 可由已知的炉渣氧化铁活度及温度用式（10）估计：

$$[\%O] = a_{FeO} \times [\%O]_{饱} \qquad (10)$$

计算时采用1300℃时的 $[\%O]_{饱}$ 值。

表1 间接氧化反应的 $[\%A] \times [\%O]$ 值

反 应	温度/℃	$[\%A][\%O]$ 值
$[Si] + 2[O] = (SiO_2)$	1300	$[Si]^{0.5}[O] = 10^{-4} \times 0.3$
$[C] + [O] = CO$	1300	$[C][O] = P_{CO} \times 10^{-4} \times 1.6$
$4/5[P] + 2[O] = 2/5(P_2O_5)$	1300	$[P]^{0.6}[O] = 10^{-8} \times 3.0$
$4/5[P] + 2[O] = 2/5(P_2O_5)$	1300	$[P]^{0.4}[O] = 10^{-4} \times 2.4$(铁酸钙炉渣)
$[Si] + 2[O] = (SiO_2)$	1500①	$[Si]^{0.5}[O] = 10^{-8} \times 0.3$
$[C] + [O] = CO$	1500	$[C][O] = P_{CO} \times 10^{-4} \times 2.0$
$4/5[P] + 2[O] = 2/5(P_2O_5)$	1500	$[P]^{0.6}[O] = 10^{-2}$

① 计算时仍采用原生铁成分；如考虑到铁液内杂质在此温度已大量减少，实际的 $[\%A] \times [\%O]$ 值要大于表内的计算值。

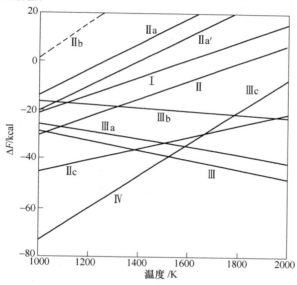

图4 间接氧化反应的 ΔF

I—$[Fe] + [O] = (FeO)$, $\Delta F = -57800 + 36.92T$;
II—$4/5[P] + 2[O] = 2/5(P_2O_5)$, $\Delta F = -67440 + 37.10T$; IIa—$4/5[P] + 2[O] = 2/5 P_2O_{5(气)}$, $\Delta F^o = -67440 + 53.18T$; IIa'—$4/5[P] + 2[O] = 2/5 P_2O_{5(气)}$, $\Delta F^o = -70940 + 50.93T$;
IIb—$4/5[P] + 2[O] = 2/5 P_2O_{5(液)}$, $\Delta F = -67440 + 69.18T$; IIc—$4/5[P] + 2[O] = 2/5(P_2O_5)$,
$\Delta F = -67440 + 22.73T$(铁酸钙炉渣); III—$[C] + [O] = CO$, $\Delta F = -10700 - 18.74T(P'_{CO} = 0.43atm)$;
IIIa—$[C] + [O] = CO$, $\Delta F = -10700 - 15.04T(P_{CO} = 1atm)$; IIIb—$[C] + [O] = CO$,
$\Delta F = -10700 - 6.25T(P'_{CO} = 10atm)$; IIIc—$[C] + [O] = CO$, $\Delta F = -10700 - 3.52T(P_{CO} = 20atm)$;
IV—$[Si] + 2[O] = (SiO_2)$, $\Delta F = 139300 + 66.11T$

4 计算结果的分析和讨论

4.1 Карнаухов 和 Морозов 法计算 P_{O_2} 的误差

设 $\Delta F^o_{纯}$ 为反应式（1）当元素 A 以纯物质为标准状态时的标准自由能；

ΔF^o 为反应式（1）当元素 A 以 1%溶液为标准状态时的标准自由能；

ΔF_A 为元素 A 溶于铁液内生成 1%溶液的标准自由能；

$P_{O_2(K)}$ 为根据 Карнаухов 法求出的氧分压；

$P_{O_2(M)}$ 为根据 Морозов 法求出的氧分压。

则不难看出：

$$\Delta F^o = \Delta F^o_{纯} - \Delta F^o \tag{11}$$

根据式（4）：

$$\lg P_{O_2} = \frac{\Delta F^o_{纯}}{4.575T} - \frac{\Delta F^o_A}{4.575T} + \lg a_{AO_2} - \lg f_A - \lg[\%A] \tag{12}$$

$$\lg P_{O_2(K)} = \frac{\Delta F^o_{纯}}{4.575T} - \lg[\%A] + \lg 100 \tag{13}$$

$$\lg P_{O_2(M)} = \frac{\Delta F^o_{纯}}{4.575T} - \frac{\Delta F^o_A}{4.575T} - \lg[\%A] \tag{14}$$

表 2 给出对选定的生铁及炉渣成分和指定的空气吹炼情况，根据式（12）~式（14）计算的 $\Delta \lg P_{O_2}$ 值（1300℃）。

表 2 不同方法计算的 $\Delta \lg P_{O_2}$ 值（1300℃）

元 素	$\lg P_{O_2} - \lg P_{O_2(K)}$	$\lg P_{O_2} - \lg P_{O_2(M)}$
Si	+1.9	-1.4
Mn	+0.4	-2.4
C	+2.1	-2.5
P	-5.2	-7.3
Fe	-1.4	-1.4

可以看出，对 Si、Mn 及 C，Карнаухов 法使 P_{O_2} 变小，而 Морозов 法则使 P_{O_2} 值变大。对铁则二法均使 P_{O_2} 值变大，以 P_{O_2} 值在 $10^{-12} \sim 10^{-18}$ 的范围而论，这些差别并不为大。但对磷则二法均提出一个结论，即磷的 P_{O_2} 加大颇多，以致磷的 $\lg P_{O_2}$ 曲线超过铁的 $\lg P_{O_2}$ 曲线之上（参阅图 1），也即使得磷在铁液内虽有炉渣生成不能先于铁而被氧化。这个结论是不合理的。下面将谈到，用 ΔF 法计算证明，只有在无炉渣生成时则磷不能先于铁而被氧化。这说明不考虑炉渣的活度可能导致错误的结论。Карнаухов 在原计算中，因当时缺乏炼钢氧化反应的 ΔF^o 值（更缺乏相应的活度值），采用反应的生成热用近似方法以计算 $\lg P_{O_2}$，因之，他

的全部计算结果已无参考价值。Морозов 采用另一自由能数据：

$$\frac{4}{5}[P] + O_2 + \frac{6}{5}CaO_{(固)} = \frac{2}{5}(Ca_3P_2O_8), \quad \Delta F^o = -202000 + 61.24T$$

使 $a_{Ca_3P_2O_8} = 1$，$f_P = 1$，得到磷在炉渣生成时能先于铁而氧化的结论，因之，他未发现不用炉渣活度系数能导致错误结论的缺点。

4.2 炼钢过程中杂质元素氧化顺序的分析和讨论

从上面计算的结果来看，无论用哪一种方法，所得元素氧化的顺序基本上是一致的。凡有最小的氧分解压力 P_{O_2}、最小的 [%A]×[%O] 值或最小的反应自由能 ΔF（最大的负值），该元素则优先氧化。分解压力和 [%A]×[%O] 值两方法适用于某一定温度下的平衡状态，而 ΔF 则适用于不同温度范围内的不平衡状态，ΔF 的方法更有普遍应用的意义，例如从图 2 和图 4 可以读出，对该成分生铁的托马斯法吹炼，碳焰开始上升的温度约为 1520K。这基本上与实际工艺操作的情况相符合。

对托马斯法吹炼，空气吹入铁液，元素氧化反应系直接氧化或间接氧化，可视熔池内元素分布情况及供氧强度而定。从热力学观点来看，当硅、铁、锰、碳及磷原子同时存在或当供氧强度不足时，吹入的空气最先氧化硅和锰。由于铁液内铁原子数远较其他元素的原子数为多，当风嘴较远区域的硅及锰由于扩散迟缓而赶不到风嘴附近时，铁被大量氧化，生成的氧化铁溶于铁液为 [O]，此 [O] 再扩散到有其他元素的区域，因而形成间接氧化。由此可见，转炉吹炼过程中，直接氧化及间接氧化反应均可能发生。文献 [13] 认为转炉中只能有间接氧化反应是不够全面的。随着温度的上升，在硅、锰全部或大部被氧化掉后，碳开始大量氧化。直到碳焰下降之后，大部的磷方能氧化，比较图 2 中的曲线 Ⅰ、Ⅱ 及 Ⅱb，得出一个重要的结论：磷的氧化必须有炉渣生成，也就是说，磷的氧化是在铁液及渣液面进行，如果有氧气参加，则磷的氧化在气体、铁液和渣液三者的界面进行。在后吹之前，加入的石灰有 40%~50%[14,15]不溶解，炉渣系含氧化铁不高的酸性渣（特别在吹炼初期），对去磷不利。Lellep[16] 曾进行过试验，在吹炼过程中加入事先制备好的铁酸钙，则磷可提前氧化（图 5）。试以同样成分的生铁，而炉渣成分（均系摩尔百分数）为：

CaO	FeO	SiO$_2$	P$_2$O$_5$
48	48	2	2

再进行计算，图 1 和图 2 分别绘出的 P_{O_2}（曲线 Ⅱb）及 ΔF（曲线 Ⅱc）均给 Lellep 试验以理论上的证实。图 1 的曲线 Ⅱb 及 Ⅲ 指出，当 [C] 降低到 1% 时，和 [C] 平衡的 O 的 $P_{O_2} = 10^{-15}$ atm，但与同样 O 量平衡的 [P] 值是 0.5%，说明由于铁酸钙炉渣的存在，磷已先于碳而氧化。

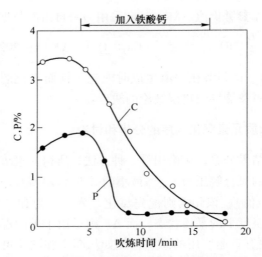

图 5　加入铁酸钙时，托马斯吹炼的碳、磷氧化情况（Lellep）

由于铁液成分不均匀及间接氧化反应的存在，在吹炼的第一期碳有可能被氧化一小部分，而在碳焰下降之前，铁及磷也均可能被氧化一小部分；在后期吹炼，由于铁液内杂质元素减少，铁液内 [O] 量大增[17]，炉渣内氧化铁量也激增，在适当情况下炉渣可一度起分层现象[18,19]，有时由于磷的猛烈氧化，炉渣内的氧化锰被还原，造成锰的"驼峰"现象[20]。

必须特别指出：磷的间接氧化反应也必须有碱性炉渣存在，否则不能进行，而高氧化钙和高氧化铁的炉渣对去磷更为有利（比较图4中的曲线Ⅰ、Ⅱ、Ⅱb及Ⅱc）。这更足以说明脱磷是一界面反应。

平炉铁液内元素氧化都通过 [O]，氧通过炉渣进行传递。磷能先于碳氧化，一方面由于炉渣含高氧化铁，而另一方面更主要的是碳的氧化有被抑制现象。CO 气泡如能在炉底生成，则 P_{CO} 至少大于 1atm[21]；CO 如在铁液内生成，P_{CO} 可大至于无限大。CO 的生成与炉底粗糙状态及铁液内有无固体颗粒有关。Rellermeyer 等[22]估计平炉铁液内 P_{CO} 可高至 30atm。这样，从表1和图4可以看出，碳可以被抑制而后于磷氧化。当然，石灰沸腾有助于碳的氧化，所以平炉内碳和磷基本上是同时氧化的。

顶吹转炉（LD 法）随着操作制度的不同而有不同的氧化顺序[23~27]。当氧气压力高或氧气吹嘴距离炉液近时，氧气直接与铁液接触，并促使铁滴四飞，有利于碳的氧化。由于炉渣与铁液不接触，磷的氧化被抑制。当氧气压力较低或氧气吹嘴距离炉液较远时，氧气直接氧化炉渣，再传递于铁液内，这就有利于去磷。但适当的炉渣搅拌，使得磷能由铁液下层扩散到渣面附近，对去磷仍系必要。对高磷生铁，通过吹入石灰粉提前造渣（LD-AC 或 OLP 法），以及多次造渣均能得到低磷优质钢，如同时适当控制碳的氧化速度，则可进行高拉碳操作。

Kaldo 及 Rotor 法通过炉子的转动速度可以控制气-渣及渣-铁液反应，Rotor 又利用两个吹嘴调整氧化速度，二法都能根据需要调整操作，使得磷先于或同时与碳氧化。

对磷的间接氧化，作者采用托马斯转炉[20]、顶吹转炉[19]以及后期炉渣分层[18]的一些实际数据进行 ΔF 的计算，得到：

$$\Delta F = -67440 + (29.0 - 35.5)T$$

证明上列由选定成分的炉渣计算所得的结论可以应用到一般的实际操作中。

最后尚须作一补充说明，本文计算磷氧化的 ΔF° 均以纯液态 P_2O_5 为准，实际上在炼钢温度 P_2O_5 早已气化。参照下列数据[28]：

$$P_{2(气)} + 2\frac{1}{2}O_2 = P_2O_{5(气)}, \quad \Delta F^\circ = -375500 + 115.38T \tag{15}$$

得出：

$$\frac{4}{5}[P] + O_2 = \frac{2}{5}P_2O_{5(气)}, \quad \Delta F^\circ = -126800 + 49.79T \tag{16}$$

$$\frac{4}{5}[P] + 2[O] = P_2O_{5(气)}, \quad \Delta F^\circ = -70940 + 50.93T \tag{17}$$

图 1 中曲线 IIa 是由式（16）计算出的 $\lg P_{O_2}$ 曲线，它远在纯氧化铁分解的 $\lg P_{O_2}$ 线之上，说明直接产生气态 P_2O_5 不可能。图 2 中曲线 IIa′ 及图 4 中曲线 IIa′ 分别表示式（16）及式（17）的 ΔF°，同样地证明磷的氧化无炉渣不能进行，可以看出，无论采用液态抑或气态 P_2O_5 进行计算，所得的结论一致。

5 结论

（1）本文所用三种方法计算出的元素氧化顺序基本上是一致的，但反应自由能 ΔF 法有更普遍的意义，因为它反映不同温度范畴不平衡状态下反应发生的顺序。

（2）氧化气体吹入铁液内，氧化反应可以由两种机构进行：1）氧气直接与杂质元素化合；2）由于铁液内铁原子数很多，当在扩散不好的情况下，氧气先与铁生成 [O]，后者再与杂质元素起反应。

（3）无论是气体氧直接氧化，抑或通过 [O] 间接氧化，去磷是一界面反应，没有适当成分的炉渣是不可能进行的。

（4）磷与碳的氧化顺序决定于两个因素：1）有无适当成分的炉渣存在；2）碳氧化反应是否被抑制，迅速造成适当成分（高 CaO 和高 a_{FeO}）的炉渣，在较低温度，有可能在托马斯转炉使磷与碳同时氧化。平炉磷能先于碳而氧化，主要由于碳氧化有被抑制的现象。

（5）通过改变工艺操作，如吹炼方式、底吹侧吹或顶吹、吹入气体的压力、吹嘴与炉渣面的距离、造渣方式、炉液转动以及温度控制等，可以改变并控制磷与碳的氧化顺序。

参 考 文 献

[1] Карнаухов М М. Металлургии стали. ОНТИ, 1933~1934, 1: 22~26.
[2] Морозов А М. Современный мартеновский процесс. Металлургиздат, 1961: 108~109.
[3] Левин С Л. Изв. высш. уцебн. заведений. Черн. Металлургия, 1958 (2): 76.
[4] AIME. Electric Furace Steelmaking. Interscience, vol. 2, 1963: 128~137.
[5] Philbrook W O, Bever M B. Basic Open Hearth Steelmaking. AIME, 1951, 2nd ed: 638.
[6] Turkdogan E T, Pearson J. J. Iron Steel Inst, London, 1953, 175: 398.
[7] Turkdogan E T, Pearson J. J. Iron Steel Inst, London, 1953, 173: 217.
[8] Eliot J F. Trans. AIME, 1955, 203: 485.
[9] Bishop H L, Grant N J, Chipman J. Trans. AIME, 1958, 212: 890.
[10] 魏寿昆. 活度在冶金物理化学中的应用. 北京: 中国工业出版社, 1964.
[11] Schenck H. Physikalische Chemie der Eisenhüttenprozesse. Verlag von J. Springer, 1934. Bd. 2: S82.
[12] Winkler T B, Chipman J. Trans. AIME, 1946, 167: 111.
[13] Ростовцев С Т. 北京钢铁学院物化教研组译. 冶金过程理论. 北京: 冶金工业出版社, 1956: 580.
[14] Bardenheuer P, Thanheiser G. Stahl Eisen, 1934, 112: 725.
[15] Bading W. Stahl Eisen, 1936: 409.
[16] Lellep O. Bericht Mexico, 1941; Bulle G. Stahl Eisen, 1951: 1442; Bulle G. Iron & Coal Trades Review, 1952, 164 (4385): 921.
[17] Fischer W A, Straube H. Stahl Eisen, 1960: 1194.
[18] vom Ende H, Mahn G. Stahl Eisen, 1960: 136.
[19] vom Ende H, Mahn G. Stahl Eisen, 1961: 641.
[20] Eichel K. Das Basische Windfrischverfahren. Verlag Technik, 1952, S. 108: 115~117.
[21] Schenck H. Physikalische Chemie der Eisenhüttenprozesse. Verlag von J. Springer, 1934, Bd. 2, S 94.
[22] Rellermeyer H, Knüppel H, Sittard J. Stahl Eisen, 1957: 1296.
[23] Kosnider H, Neuhaus H, Schenck H. Stahl Eisen, 1957: 1277.
[24] Springorum F A, Spcith K G, Därmann O, von Ende H. Stahl Eisen, 1957: 1284.
[25] Hcischkeil W, Kootz T. Stahl Bisen, 1959: 205.
[26] Kootz T. J. Iron Steel Ins. , London, 1960, 196: 253.
[27] Pearson J. Iron & Coal Trades Revierw, 1960, 181 (4824): 1407.
[28] Elliott J F, Gleiser M. Thermochemistry for Steelmaking. Addison-Weslcy, 1960, 1: 193.

镍锍选择性氧化的热力学及动力学[*]

魏寿昆　洪彦若

摘　要：利用氧气吹炼镍锍直接得金属镍，其关键在于去硫保镍。本文利用选择性氧化原理，提出氧化转化温度的概念。热力学分析指出，去硫保镍的条件是：

（1）镍锍熔体用 O_2 开吹的温度必须超过该组成硫、镍氧化的转化温度；对含硅 20%～25%的镍硫，其开吹温度不能低于 1350～1400℃。

（2）随着熔体中硫含量的减少，相应地硫、镍氧化的转化温度随之增高。吹炼操作必须迅速进行，以保证熔池温度上升的速度永远高于转化温度增高的速度。

硫、镍氧化的转化温度可用一步法按下列反应

$$[S] + 2NiO_{(s)} = 2[Ni] + SO_2$$

进行计算。

热力学分析又指出：

（1）镍锍内含铜全部留在熔体之内，在吹炼过程中不被氧化。

（2）镍锍中的铁最易被氧化，但当降低到 0.8%～1.0% 后即不能被氧化而以残铁留在熔体之内。

（3）镍锍含钴如小于 1%也将留在熔体之内。

通过在卡尔多斜吹旋转炉进行的半工业吹炼实验，在采用上列热力学推论得出的去硫保镍条件下，硫能顺利地降到 1%～2%，充分地证明了理论成功地指导了实践，克服在初期探索性试验中遇到大量镍氧化的困难。在吹炼末期，由于熔体中硫的扩散速度减慢，熔池表面逐渐有 NiO 层累积。采用不吹氧空转还原，可进一步去硫而提高镍的回收率。镍的直接回收率大于 90%，而总回收率大于 95%。镍的主要损失来自高温下镍及其氧化物的挥发。

熔体中残铜、残铁及残钴的存在也通过实验予以证实。

动力学分析指出，熔体中硫的扩散是脱硫反应的控制性环节。硫的传质系数

---

* 原刊于《北京钢铁学院学报》，1981（3）：54～66。
　本文中能量单位原用 cal，1cal=4.184J。
　参与本工作的还有王俭、张千象、张家芝、王鉴等。

β 及扩散系数 D 与温度 T 的关系式分别为:

$$\beta = 8.30\exp\left(\frac{-25000}{RT}\right)$$

$$D = 8.30 \times 10^{-2}\exp\left(\frac{-25000}{RT}\right)$$

镍锍是火法冶金提镍的中间产物。从镍锍提制金属镍通常采用两种方法：(1) 直接电解；(2) 焙烧成为氧化镍再进行还原。为提高镍的回收率及简化冶炼工艺，最近采用镍锍用氧吹炼直接制取金属镍法[1,2]。

用氧吹炼镍锍，其冶炼机理和吹炼铜锍有较大的不同。初次小型探索性试验[3]，证明：在 200kg 氧气斜吹旋转炉吹炼镍锍，发现在吹炼后期大量的 NiO（熔点 1964℃）结成硬壳，逐渐堵塞炉口，有时迫使吹炼中断，而熔体含硫量降到 4%~5% 以下比较困难，同时镍的回收率最高只能达到 75%~80%。从镍锍吹炼直接得金属镍，必须使硫降低到最低值，并同时防止镍被氧化，所以实质上是如何创造条件进行选择性氧化，使硫优先氧化而镍不动，也即做到去硫保镍。因之，在进一步开展小规模半工业试验之前，先对镍、硫的选择性氧化进行热力学的理论分析[4]。

1 热力学分析

1.1 各种硫化物的氧化顺序

镍锍除含 Ni_3S_2 外，尚有 FeS、Cu_2S 及硫化钴等。采用表 1 的热力学数据可以分析它们的氧化顺序。由于钴硫化物在吹炼温度的热力学数据不全，其计算从略。采用纯氧气三种硫化物的标准自由能 ΔF° 数据如下：

$$\frac{2}{3}Cu_2S_{(l)} + O_2 = \frac{2}{3}Cu_2O_{(l)} + \frac{2}{3}SO_2$$

$$\Delta F^\circ = -64100 + 19.40T \tag{1}$$

$$\frac{2}{7}Ni_3S_{2(l)} + O_2 = \frac{6}{7}NiO_{(s)} + \frac{4}{7}SO_2$$

$$\Delta F^\circ = -80600 + 22.48T \tag{2}$$

$$\frac{2}{3}FeS_{(l)} + O_2 = \frac{2}{3}FeO_{(l)} + \frac{2}{3}SO_2$$

$$\Delta F^\circ = -72500 + 12.59T \tag{3}$$

图 1 绘出式 (1)~式 (3) 的 ΔF° 和 T 关系图。可以看出，式 (3) 的自由能负值最大，它的位置最低，而式 (1) 的自由能负值最小，它的位置最高。因此在吹炼过程中，FeS 最容易被氧化，氧化的顺序为 FeS、Ni_3S_2 及 Cu_2S。硫化钴的氧化顺序在 FeS 及 Ni_3S_2 之间。

表1 有关化合物的生成自由能

化 学 反 应	ΔF°/cal	文献
$2Cu_{(l)} + \frac{1}{2}S_{2(g)} = Cu_2S_{(l)}$	$-24830 + 2.54T$	[5]
$Fe_{(l)} + \frac{1}{2}S_{2(g)} = FeS_{(l)}$	$-34200 + 10.20T$	[5]
$\frac{3}{2}Ni_{(s)} + \frac{1}{2}S_{2(g)} = \frac{1}{2}Ni_3S_{2(l)}$	$-28960 + 7.95T$	[6]
$4Cu_{(l)} + O_{2(g)} = 2Cu_2O_{(l)}$	$-69900 + 28.80T$	[7]
$2Fe_{(s)} + O_{2(g)} = 2FeO_{(l)}$	$-109700 + 20.94T$	[7]
$2Fe_{(l)} + O_{2(g)} = 2FeO_{(l)}$	$-113800 + 23.64T$	[7]
$2Ni_{(s)} + O_{2(g)} = 2NiO_{(s)}$	$-111800 + 40.02T$	[7]
$2Ni_{(l)} + O_{2(g)} = 2NiO_{(s)}$	$-120700 + 45.22T$	[7]
$2Co_{(l)} + O_{2(g)} = 2CoO_{(s)}$	$-121000 + 39.06T$	[7]
$\frac{1}{2}S_{2(g)} + O_{2(g)} = SO_{2(g)}$	$-86130 + 17.27T$	[7]
$2C_{(s)} + O_{2(g)} = 2CO_{(g)}$	$-56400 + 40.32T$	[7]
$CO_{(g)} + \frac{1}{2}O_{2(g)} = CO_{2(g)}$	$-66550 + 20.18T$	[7]
$[S]_{Ni} + 2NiO_{(s)} = 2Ni_{(l)} + SO_{2(l)}$	$-70910 - 34.36T$	[7]
$\frac{1}{2}O_{2(g)} = [O]_{Ni}$	$-23200 + 3.93T$	[8]

注：1 cal = 4.184J。

生成的氧化物又可与硫化物相互作用，其反应式如下：

$$Cu_2S_{(l)} + 2CuO_{2(l)} = 4Cu_{(l)} + SO_2$$
$$\Delta F^\circ = 8600 - 14.07T \quad (4)$$

$$\frac{1}{2}Ni_3S_{2(l)} + 2NiO_{(s)} = \frac{7}{2}Ni_{(l)} + SO_2$$
$$\Delta F^\circ = 70230 - 39.80T \quad (5)$$

$$FeS_{(l)} + 2FeO_{(l)} = 3Fe_{(l)} + SO_2$$
$$\Delta F^\circ = 61870 - 16.57T \quad (6)$$

$$2FeS_{(l)} + 2Cu_2O_{(l)} = 2Cu_2S_{(l)} + 2FeO_{(l)}$$
$$\Delta F^\circ = -69060 - 25.64T \quad (7)$$

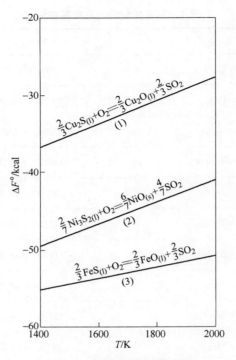

图 1　不同硫化物的氧化

$$2\text{FeS}_{(l)} + 2\text{NiO}_{(s)} = \frac{2}{3}\text{Ni}_3\text{S}_{2(l)} + 2\text{FeO}_{(l)} + \frac{1}{3}\text{S}_{2(g)}$$

$$\Delta F^\circ = 62900 - 58.26T \tag{8}$$

$$2\text{Cu}_2\text{S}_{(l)} + 2\text{NiO}_{(s)} = \frac{2}{3}\text{Ni}_3\text{S}_{2(l)} + 2\text{Cu}_2\text{O}_{(l)} + \frac{1}{3}\text{S}_2(g)$$

$$\Delta F^\circ = 164800 - 45.72T \tag{9}$$

为相互比较，式（4）到式（9）均酌合到 2 个氧原子计算，其 ΔF° 与 T 的关系图见图 2。

从图 2 看出，式（7）的 ΔF° 线远远低于式（4）的 ΔF° 线，说明在有 FeS 存在条件下，FeS 与 Cu_2O 起反应，使之成为 Cu_2S 及 FeO，而让 Cu_2O 没有任何可能进行 Cu_2O 和 Cu_2S 相互作用生成 Cu 的反应。这就是在理论上阐明，为什么吹炼铜锍（冰铜）必须分两个阶段完成：第一阶段去 Fe，第二阶段吹炼成 Cu。

由于式（8）的 ΔF° 线低于式（5）的 ΔF° 线，为吹炼镍锍时，必须先行去 Fe。

在吹炼镍锍时，如果生成 NiO，它是高熔点的固体，根据式（5），只能在一定温度之上才能使 Ni_3S_2 和 NiO 起反应生成金属镍。同时液固反应也不如式（4）的液液反应容易进行。因之，应创造条件避免镍锍吹炼过程中生成 NiO，应使 S

氧化而保留 Ni 在熔体之中。

根据图 2NiO 不能氧化 Cu_2S，说明 Cu_2S 比 Ni_3S_2 更稳定（图2线（9）是反应式（9）的逆式）。

1.2 去硫保镍条件的分析

为了确定去硫保镍的条件，我们提出下列求 S、Ni 氧化的转化温度的方法。

镍锍可认为是含 Cu、Fe、Co 及 S 的镍基熔体。在吹炼过程中随着 S 的氧化，熔体中的含 Ni 量逐渐提高。

文献［1］给出下列反应的平衡常数：

$$[S] + 2NiO_{(s)} = 2Ni_{(l)} + SO_2$$

$$\lg k = \frac{15500}{T} + 7.51$$

因之，该式的自由能为

$$\Delta F^\circ = 70910 - 34.36T \quad (10)$$

由于式（7）

$$2Ni_{(l)} + O_2 = 2NiO_{(s)}$$

$$\Delta F^\circ = -120700 + 45.22T \quad (11)$$

二式相加，得：

$$[S] + O_2 = SO_2$$

$$\Delta F^\circ = -49790 + 10.86T \quad (12)$$

式（10）及式（12）[S] 的活度均按重量1%浓度的标准计算。

以纯物质为标准态镍锍中［Ni］、［Cu］、［Fe］及［Co］的氧化标准自由能见表1。

试采用表2列出的镍锍组成进行计算。

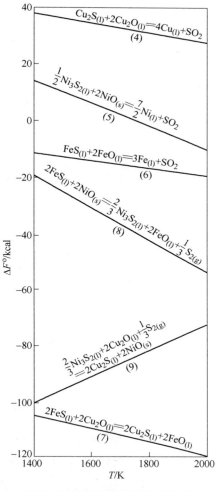

图2 硫化物与氧化物相互作用

表2 镍锍的组成

元素	重量百分数/%	摩尔数 n	摩尔分数 N
Ni	70	0.190	0.595
Cu	5	0.079	0.039
Fe	3	0.054	0.027
Co	0.7	0.012	0.006
S	21.3	0.666	0.333
	100	2.001	1.000

关于硫的活度系数，文献［9］提供镍基熔液中一部分 e_s^j 的值，但其适用范围限于［S］最高达 0.7%。另一文献［10］指出，镍基金属液的 $f_s=1$，其适用范围为 7%≥［%S］>0。假定 $f_s=1$，用纯 O_2 顶吹，$p_{O_2}=1$ atm，炉气根据实验数据含 70%$SO_2^{[3]}$，则：

$$[S] + O_2 = SO_2; \quad \Delta F^\circ = -49790 + 10.86T$$

$$\Delta F = \Delta F^\circ + RS\ln \frac{p_{SO_2}}{F(\%S) \cdot p_{O_2}}$$

$$\Delta F = 19790 + 10.86T + 4.575T\lg \frac{0.7}{21.3} = -49790 + 4.07T \quad (13)$$

在镍基金属液中 Ni 占绝大部分，因之可以认为 $\gamma_{Ni}=1$。其他金属元素的活度系数 γ_i 见表3。

表3　镍基金属液中其他金属元素的 γ_i

元素	γ_i	备注
Ni	1	镍基金属液，$\gamma_i=1$
Cu	2	$N_{Cu}=0\sim0.18^{[11]}$
Fe	0.42（当 $N_{Fe}=0.02\sim0.03$） 0.41（当 $N_{Fe}=0.01\sim0.015$）	内插值[12]
Co	1	缺乏资料，假定 $\gamma_{Co}=1$

对［Ni］、［Fe］、［Cu］及［Co］计算出的氧化反应，其 ΔF 值如下：

$$2[Ni] + O_2 = 2NiO_{(s)}; \quad \Delta F = -120700 + 47.28T \quad (14)$$

$$2[Fe] + O_2 = 2FeO_{(l)}; \quad \Delta F = -113800 + 41.44T \quad (15)$$

$$4[Cu] + O_2 = 2Cu_2O_{(l)}; \quad \Delta F = -69900 + 49.07T \quad (16)$$

$$2[Co] + O_2 = 2CoO_{(s)}; \quad \Delta F = 121000 + 59.39T \quad (17)$$

式（13）到式（17）的 ΔF° 对 T 关系图见图3，相应的线用 $S_{(I)}$、$Ni_{(I)}$、$Fe_{(I)}$、$Cu_{(I)}$ 及 $Co_{(I)}$ 表示。

从图3看出，$Cu_{(I)}$ 及 $Co_{(I)}$ 线均在 $Ni_{(I)}$ 线之上，这说明在大量 Ni 存在条件下，Cu 及 Co 均被 Ni 保护而免于被氧化。$Fe_{(I)}$ 线最低，说明在此吹炼过程中，Fe 首先被氧化。$Fe_{(I)}$ 线和 $S_{(I)}$ 线相交于点 b，其温度为 1440℃，说明超过 1440℃即被氧化，而 Fe 被保护。此 1440℃温度可称为 Fe、S 氧化的转化温度。$Ni_{(I)}$ 线高于 $Fe_{(I)}$ 线不多，实际操作中 Ni 很可能与此低量的 Fe 同时氧化。$Ni_{(I)}$ 与 $S_{(I)}$ 线相交于点 a，其温度为 1368℃，此温度是 Ni、S 氧化的转化温度。所以欲保护 Ni 不被氧化，对我们的镍锍成分来说，开吹温度不能低于 1368℃。

随着吹炼过程的进行，熔体内含 S 量逐渐减少。我们再分析一下当 S 降到

10%而Fe降到1.3%时，熔体温度应提高到多少，才能保证Ni不被氧化？镍锍内因Ni、Cu及Co不被氧化，该三元素保持表2所列的比例。因之，此时镍锍的成分为：Ni82.0%，Cu5.9%，Fe1.3%，Co0.8%，S10%。计算的ΔF的结果为：

$$[S] + O_2 = SO_2; \quad \Delta F = -49790 + 5.58T \tag{13a}$$

$$2[Ni] + O_2 = 2NiO_{(s)}; \quad \Delta F = -120700 + 46.30T \tag{14a}$$

$$2[Fe] + O_2 = 2FeO_{(l)}; \quad \Delta F = -113800 + 44.76T \tag{15a}$$

由于Cu、Co不被氧化，对它们氧化的反应不再进行计算。式（13a）到式（15a）的ΔF°对T关系也绘在图3之内，用$S_{(II)}$、$Ni_{(II)}$及$Fe_{(II)}$三条虚线表示。

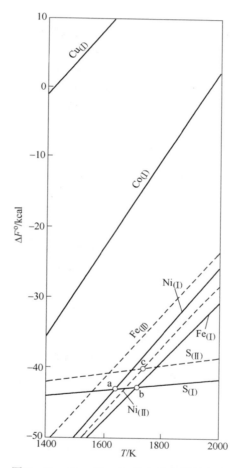

图3 Cu、Co、Ni、Fe及S的选择性氧化
a—1368℃；b—1440℃；c—1468℃

从图3看到，$Fe_{(II)}$线已在$Ni_{(II)}$线之上说明在此成分的镍锍，Fe已被Ni保护而不能被氧化了。因之，在吹炼过程中，一旦镍中的Fe降到一定量（为0.8%~

1.5%）之下，Fe 即不能再被氧化，而以残 Fe 存在于金属镍之中。$Ni_{(II)}$ 和 $S_{(II)}$ 二线相交于点 c，其温度为 1468℃，说明为了保证 Ni 不被氧化，此时熔体的温度必须提高到 1468℃之上。

下面再作熔体内 S 下降到 3.8%、1.0% 及 0.1% 同样的计算。由于 Fe 此时已不能再氧化，故 Fe 的氧化反应也不进行计算。以前两种成分的计算结果和现在三种成分的计算结果合并见表 4。

表 4 镍锍吹炼的热力学分析

例	镍锍成分 %	N	条件	ΔF/cal	转化温度/℃
I	Ni 70 Cu 5 Fe 3 Co 0.7 S 21.3	0.595 0.039 0.027 0.006 0.333	$p_{O_2}=1$ 个大气压 $p_{SO_2}=0.7$ 个大气压 $f_S=1$ $\gamma_{Ni}=1$ $\gamma_{Cu}=2$ $\gamma_{Fe}=0.42$ $\gamma_{Co}=1$	S：$-49790+4.07T$ Ni：$-120700+47.28T$ Fe：$-113800+41.44T$ Co：$-121000+59.39T$ Cu：$-69900+49.07T$	Ni、S：1368 Fe、S：1440
II	Ni 82.0 Co 5.9 Fe 1.3 Co 0.8 S 10.0	0.761 0.050 0.012 0.007 0.170	$\gamma_{Fe}=0.41$ 其他数值同 I	S：$-49790+5.58T$ Ni：$-120700+46.30T$ Fe：$-113800+44.76T$	Ni、S：1468
III	Ni 87.6 Cu 6.3 Fe 1.4 Co 0.9 S 3.8	0.852 0.057 0.014 0.009 0.068	$p_{O_2}=1$ 个大气压 $p_{SO_2}=0.7$ 个大气压 $f_S=1$ $\gamma_N=1$	S：$-49790+7.50T$ Ni：$-120700+45.86T$	Ni、S：1580
IV	Ni 90.3 Cu 6.4 Fe 1.4 Co 0.9 S 1.0	0.900 0.059 0.014 0.009 0.018	同 III	S：$-49790+10.15T$ Ni：$-120700+45.64T$	Ni、S：1725
V	Ni 91.1 Cu 6.5 Fe 1.4 Co 0.9 S 0.1	0.915 0.060 0.014 0.009 0.002	同 III	S：$-49790+14.73T$ Ni：$-120700+45.57T$	Ni、S：2026
VI	同 V	同 V	$p_{SO_2}=0.01$ 个大气压 其他数据同 III	S：$-49790+6.29T$ Ni：$-120700+45.57T$	Ni、S：1532

由于转化温度只决定于熔体的物质及其组成，而后者在吹炼过程中经常地在变化着，所以在某一转化温度下该组成只处于一个极短暂的瞬间平衡。

从表4的理论分析结果，可以得到下列启示：

(1) 欲做到去硫保镍，必须保证：

1) 镍锍熔体开吹温度不能低于1350~1400℃。

2) 吹炼操作必须迅速进行，使熔体温度上升的速度永远高于熔体组成变化带来的转化温度上升的速度。例如当S下降到4%时，炉温必须在1550~1600℃之上。

(2) 例Ⅴ指出，如果使S降低到0.1%之下，则炉温必须大于2026℃。这样的高温炉衬已不能适应。采用低p_{SO_2}或在真空下吹炼，例如使$p_{SO_2} = 0.01$atm，则镍、硫氧化的转化温度可降低到1532℃（例Ⅵ）。

(3) 镍锍内的全部Cu留在金属镍之内。Fe也只能脱掉到0.8%~1.5%。如低于1%也将留在金属镍之内。

再进一步阐述一下选择性氧化的转化温度的意义。转化温度是一个热力学概念。它可以用两步法求出。例如式（13）的ΔF线和式（14）的ΔF线相交，相交点的ΔF相等，也即：

$$-49790 + 4.07T = -120700 + 47.28T$$
$$T = 1368℃$$

合并式（13）及式（14），则：

$$[S] + 2NiO_{(s)} \rightleftharpoons 2[Ni] + SO_2; \quad \Delta F = 70910 - 43.21T \quad (18)$$

令式（18）的$\Delta F = 0$，即

$$70910 - 43.21T = 0$$
$$T = 1368℃$$

这是用一步法求转化温度。利用图4可更好地理解两种求法的关系。从式（18）可看出，氧化的转化温度与氧气的分压p_{O_2}或氧的存在形式（O_2、[O]或FeO）无关。明显地，在转化温度，式（18）是处于平衡状态：

$$[S] + 2NiO_{(s)} \rightleftharpoons 2[Ni] + SO_2$$

1.3 实验结果的分析和讨论

半工业实验是在1.5t卡尔多纯氧斜吹旋转炉进行的❶。镍锍先用喷油副枪熔化，提温到1400℃即开始用纯氧吹炼。氧压6~8kg/cm²，氧流量5~6Nm³/min，炉子转速20r/min。由于[S]的氧化是放热反应，温度迅速上升，[S]迅速下降。在炉衬能允许的温度下，[S]可顺利地去掉到1%~2%。表5说明熔池温度

---

❶ 实验于1973~1974年由金川有色公司负责进行。

一直高于理论计算得出的转化温度。在镍锍熔化过程中，[Fe]即被空气氧化，此时加入石英粉，更有助于[Fe]的氧化，硅酸铁渣在用O_2开吹之前扒净。在镍锍全部熔化后，[Fe]即降到3%~4%，而在吹炼过程中，[Fe]降到约1%即不再下降。镍锍含[Cu]全部留在熔体之内，[Co]也残余在熔体内约1%。以上各点事实都符合上一节理论的分析。

在超过Ni、S氧化的转化温度时S先于Ni氧化。这是从热力学角度，按在熔池表面Ni、S原子同时和O_2相遇的条件时考虑的。从动力学角度出发，由于熔池Ni原子最多，很可能在表面的Ni原子先与O_2结合：

$$2[Ni] + O_2 =\!=\!= 2NiO_{(s)}$$

但由于温度超过Ni、S氧化的转化温度，[S]可将NiO还原成[Ni]而将O_2最后夺去生成SO_2：

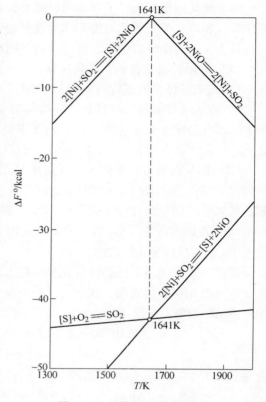

图4　Ni、S氧化的转化温度

$$[S] + 2NiO_{(s)} =\!=\!= 2[Ni] + SO_2;\ \Delta F\text{是负值}$$

最终结果仍是$[S] + O_2 =\!=\!= SO_2$的反应。

但在当熔池含S量降到3%~4%以下时，熔池内部[S]原子扩散到表面的速度减慢，在表面上只可能有[Ni]的氧化反应发生，生成的NiO分解，形成[O]熔于金属镍液中：

$$2[Ni] + O_2 =\!=\!= 2NiO_{(s)N}$$
$$2NiO_s =\!=\!= 2[Ni] - 2[O]_N$$

当$[O]_N$达到饱和后，熔池表面即有不熔解的固态NiO薄膜存在。所以在吹炼末期，随着[S]的大量减少，生成的NiO薄膜特别是当温度不能继续维持超过转化温度时，可能越聚越多，终至炉口被堵塞，如同在最初小型实验所遇到的困难那样。同时，小型实验所用炉子太小，散热损失太大，在吹炼后期，熔体温度上升的速度难以超过转化温度增高的速度，这样便造成困难。因之，在半工业试验中在[S]降到2%~4%之后，加入一个空转还原步骤，其目的即在于消灭生成的NiO而进一步脱S，也即在无O_2的条件下，使式（18）继续进行。由于式

（18）是吸热反应，在此空转还原阶段熔池温度下降，如表 5 所示。最后的实测温度可能高于或低于理论计算的转化温度如图 5 所示。当转化温度低于实测温度时，式（18）的反应仍可进行，其原因是转化温度按 p_{SO_2} = 0.7 个大气压计算的。实际上当空转还原时，冷空气吸入炉内，炉气内 SO_2 含量下降，实测数值指出最低可达 1.5%，也即 p_{SO_2} 可降低到 0.015 大气压。因之实际上转化温度要大为降低，熔池温度仍高于转化温度，反应（18）仍可继续进行（例如表 5 样号 44-8 按 p_{SO_2} = 0.1 大气压计算的转化温度为 1503℃）。为了保证温度不致下降太甚，在空转还原期可在熔池内用副枪燃油提高温度，并可维持还原气氛，以利于 NiO 的还原。

表 5 镍锍吹炼的过程分析

炉次样号	情况	Ni	Cu	Co	Fe	S	熔池温度/℃	按式（18）计算的转化温度/℃	Δt/℃
		%							
44-3	熔化后提温样	69.76	5.27	0.93	1.69	20.27	1420	1373	47
44-5	吹 O_2 10′后	79.04	5.85	0.93	1.74	10.57	1640	1459	181
44-6	吹 O_2 15′后	83.55	6.14	0.81	1.57	5.09	1730	1542	188
44-7	吹 O_2 29′后	84.61	6.82	0.76	0.87	2.48	1760	1622	138
44-8	空转还原 20′后	89.65	6.91	0.79	1.06	1.31	1650	1693	-43
17-1	熔化样	66.25	4.73	0.87	1.41	11.53	1340	—	—
17-2	提温后吹 O_2 15′后	72.00	5.24	0.76	1.33	12.91	1690	1433	257
17-3	吹 O_2 25′后	76.25	5.24	0.78	4.42	8.02	1760	1485	275
17-4	吹 O_2 32′后	78.38	5.95	0.74	3.31	3.05	1800	1593	207
17-5	空转还原 15′后	85.38	5.60	0.89	1.64	2.35	1670	1629	41

熔池在吹炼末期也可能有式（19）发生：

$$[S] + 2[O] = SO_2; \Delta F = -3390 + 3.0T \tag{19}$$

当炉气 SO_2 含量较少时该反应更易发生卡尔多炉的转动引起熔池一定的搅动，更对反应式（18）及式（19）的进行有利。

在吹氧时，维持强度永远高于 Ni、S 氧化的转化温度，这需要提高氧流速度及供氧强度，在一定转速范围内提高炉子转速以加速 [S] 的氧化，使炉壁损失之热量远远小于化学反应发生的热量而保证温度继续上升，使炉温上升的速度超过转化温度增高的速度。在 [S] 下降到一定量之下（2%~4%）以后，为避免 NiO 薄膜继续累积，采用空转还原方法，以消灭不溶解的 NiO 及溶于金属镍液中的 [O]，进一步使 [S] 下降并脱 [O]，而避免在浇铸阳极板时产生 SO_2 气泡。这些步骤使得 Ni 的回收率提高，直接回收率达 90%~91%，而总回收率超过 95%，主要损失是高温下 Ni 及 NiO 的挥发。半工业试验的成功，充足说明选择

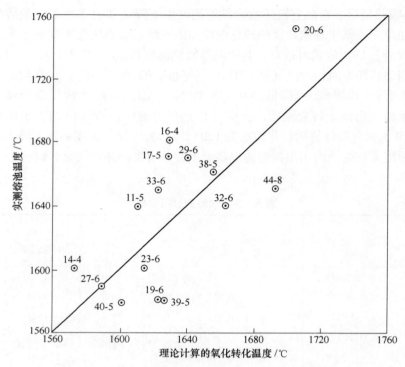

图 5 理论计算的氧化转化温度与实测熔池温度

性氧化和转化温度的理论在指导生产实验上起着显著的作用。

2 动力学分析

[S] 的氧化是界面反应。熔池内 S 原子由内部向表面扩散，在接近表面时有一浓度边界层 δ_N，在熔池表面之上有 O_2 及 SO_2 气体。由于炉气经常处于氧气流吹动之下，气相的浓度边界层很薄，不会是过程的限制性环节，因此可不考虑在气相方面存有浓度边界层。高温时化学反应在界面进行很快达到平衡。图 6 为 S 氧化反应一维模型的示意图。

$$[S] + O_2 \rightleftharpoons SO_2$$

$$-\frac{d[S]}{dt} \cdot \frac{1}{A} = \beta(C_{[S]} - C_{[S]^*}) \qquad (20)$$

式中　$-\dfrac{d[S]}{dt}$——熔池的脱硫速度，mol/s；

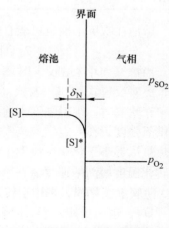

图 6 [S] 氧化的一维动力学模型示意图

A——熔池表面面积，cm^2；

$C_{[S]}$——熔池内部[S]浓度，mol/cm^3；

$C_{[S]}^*$——熔池表面与SO_2及O_2平衡时[S]的浓度，mol/cm^3；

β——传质系数，cm/s。

如[S]的浓度改用重量百分比，则

$$-\frac{d[\%S]}{dt} \cdot \frac{1}{A} = \beta([\%S] - [\%S]^*)/V \tag{21}$$

或改写成：

$$-\frac{d[\%S]}{dt} = \frac{A\beta\rho}{W}([\%S] - [\%S]^*) \tag{22}$$

式中 $-\dfrac{d[\%S]}{dt}$——熔池的脱硫速度，$\%/s$；

[%S]——熔池内部[S]的浓度，%；

$[\%S]^*$——熔池表面与SO_2及O_2平衡时[S]的浓度，%；

V——熔池（也即镍锍）的体积，cm^3；

W——镍锍的重量，g。

A和β的意义同式（20）。

通过计算可以得知$[S]^*$的值很小，可忽略不计（见表6）。

表6 不同温度$[\%S]^*$的计算值

（条件：$p_{O_2}=1$，$p_{SO_2}=0.7$，$f_s=1$）

温度/K	1673	1773	1873	1973
平衡常数k	13532	5814	2734	1387
$[\%S]^*/\%$	5×10^{-5}	1.2×10^{-4}	2.6×10^{-4}	5×10^{-4}

所以有：

$$-\frac{d[\%S]}{dt} = \frac{A\beta\rho}{W}[\%S] \tag{23}$$

从式（23）中可看出，脱硫反应是一级反应。做定积分：

$$\int_{[\%S]_1}^{[\%S]_2} -\frac{d[\%S]}{[\%S]} = \int_{t_1}^{t_2} \frac{A\beta\rho}{W}dt$$

$$\beta = \frac{2.3W}{A\rho} \cdot \frac{1}{t_2-t_1} \lg\frac{[\%S]_1}{[\%S]_2} \tag{24}$$

吹炼分两个炉役进行。第一炉役装入量1t，第二炉役装入量为1.5t。由于炉子倾斜角可以变动（17°~23°），熔池面积基本上不变，A估计为$1.5m^2$。$\rho = 5g/cm^3$。

表7为炉次22的实验数据及计算出的传质系数β值。β值应系恒温下的数值，但吹炼过程中温度一直在升高，故采用平均温度。

表7 炉次22的实验数据及 β 的计算值

阶段	$T/℃$	[%S]	t/min	β	平均温度 T/K	$\dfrac{1}{T}$
I	1400	19.40	0	4.09×10^{-3}	1783	5.61×10^{-4}
II	1580	16.14	10	1.47×10^{-2}	1938	5.16×10^{-4}
III	1750	8.31	20	2.32×10^{-2}	2048	4.88×10^{-4}
IV	1800	3.59	28			

用回归分析求出装入量为1t的13个炉次及装入量为1.5t的8个炉次的 $\lg\beta$ 对 $1/T$ 的关系（图7）为：

$$\lg\beta = 0.919 - \frac{5.46\times10^3}{T} \tag{25}$$

直线式（25）的相关系数 n 为0.887，其置信度>99.9%。

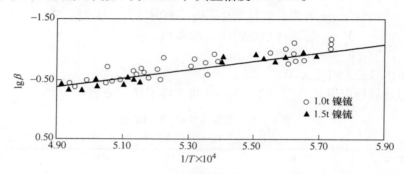

图7 β 与温度 T 的关系

式（25）可换写为

$$\beta = 8.30\times e^{\frac{-25000}{RT}} \tag{26}$$

在1650~1950K的温度范围内，β 值为 $4\times10^{-3} \sim 1.3\times10^{-2}$ cm/s。

由于

$$\beta = \frac{D}{\delta_N} \tag{27}$$

式（27）中，D 为扩散系数，cm²/s。设 $\delta_N = 10^{-2}$ cm，则

$$D = 8.3\times10^{-2}e^{\frac{-25000}{RT}} \tag{28}$$

从式（26）及式（28）得知，传质扩散的活化能为25000cal。此数值较高，可能由于镍锍的黏度较大所致。

结合式（24）及式（26）

$$\Delta t = t_2 - t_1 = 3.69We^{\frac{25000}{RT}}(\lg[\%S_1] - \lg[\%S_2]) \tag{29}$$

式中，W 为熔体重量，t。

由于在空转还原期之前,熔池温度基本上稳定在某个范围(例如 1730~1800℃),可采取温度平均值 T 利用式(29)估算由某原始含 S 量 [%S_1] 降低到 [%S_2] 的时间 Δt,这样对整个吹炼过程可以进行控制。

在空转还原阶段脱硫速度降低,其反应速度较复杂,不再做动力学分析。

3 结束语

利用选择性氧化的转化温度理论作指导可以将镍锍中 [S] 在一般的操作条件下顺利地脱到 1%~2%。动力学分析指出熔池中 [S] 的扩散是决定脱硫反应速度的控制性环节。

参 考 文 献

[1] Queneau P, O'Neill C E, Illis A, Warner J S. J. Mltaes, 1969, 21 (7): 35.
[2] Renzoni L S, Machum G O, McConnell J B, Magee A C. Symposium Nickel, Gesellschaft Deutscher Metllhütten-und Bergleute, Clausthal-Zellerfeld 1970: 49~63.
[3] 转炉氧气斜吹硫化镍探索性试验总结. 内部资料, 1971 年 7 月; 韩其勇. 个人通讯, 1973.
[4] 魏寿昆. 回转炉氧气吹炼冰镍的物理化学. 未发表资料, 1973 年 10 月, 另见: 魏寿昆. 冶金过程热力学. 上海: 上海科学技术出版社, 1980: 99~125.
[5] Ruddle R W. 丁培庸编译. 火法炼铜的物理化学, 1957 年中译本: 141.
[6] Nagamori M, Ingraham T R. Met. Trans., 1970 (1): 1821.
[7] Sims C E. Eletric. Furnace. Steelmaking, 1963, Ⅲ: 134~137, Interscience.
[8] Fischer W A, Ackermann W. Archiv Eisenh. 1966, 37: 43.
[9] Venal W V, Geiger G H. Met. Trans., 1973 (4): 2567.
[10] Alcock C B, Cheng L L. J. Iron Steel Inst, 1960, 195: 169.
[11] Kulkarni A D, et al. Met. Trans., 1973 (4): 1723.
[12] Hultgren R, et al. Selected Values of Thermodynamic Proporties of Metals and Alloys, 1963: 732, Wiley No Sons.

熔锍及熔融金属中元素选择性氧化的热力学*

魏寿昆

摘 要：总结讨论了熔锍及熔融金属中元素选择性氧化的行为，举出镍锍中 Ni 与 S，铁液中 Cr、V、Nb、Mn 或 P 与 C 作为应用的实例。利用热力学分析提出氧化的转化温度的概念，并指出二步及一步计算该温度的方法。在排除新相生成的晶核能的条件下，氧化的转化温度与氧的存在形式（无论是气态 O_2、溶于金属液中的 [O] 或炉渣中的 FeO）以及氧的压力或活度无关，而只决定于参加反应的物质及产物的本质及活度（压力）。同时，转化温度不是一成不变的温度，而是随着熔池组成的改变而不断地在变化。降低气体氧化产物的分压将有助于降低氧化的转化温度。

理论计算的转化温度可提供使熔池中一个元素的优先氧化而使另一元素保留不变的最佳条件。小型实验和工业上实践证明，转化温度的概念可以成功地控制吹炼操作，做到按意图进行选择性氧化。影响熔池内元素氧化顺序的动力学因素也作了简略的分析。

对镍锍脱 S、不锈钢脱 C 以及高碳锰铁降 C 的吹炼，熔池温度永远要高于相应熔池组成的转化温度。而对铁水脱 Cr 和铁水提 V 或 Nb，熔池温度则应保持低于相应熔池组成的转化温度。P、C 在铁水中的氧化顺序，除与转化温度有关外，还取决于熔渣组成以及 CO 承担的压力。

在 Bessemer 法及 Thomas 法炼钢中，Si、Mn 在吹炼初期先于 C 而被氧化，这一事实已早为人所共知（图 1[1] 及图 2[2]），本文利用热力学分析，总结讨论熔锍及熔融金属中元素的选择性氧化行为，采用下列反应，其中一个元素的氧化产物为气体（例如冶金生产中常见的 CO 或 SO_2），对两种元素的氧化顺序进行讨论：

$$x[M] + \frac{y}{2}O_2 = (M_xO_y) \tag{1}$$

$$z[M'] + \frac{y}{2}O_2 = M'_zO_{y(g)} \tag{2}$$

---

* 原刊于《金属学报》，1982，18：115~126。

本文提出了"氧化的转化温度"的概念，并用以分析吹炼过程中某种元素优先氧化的最佳条件。

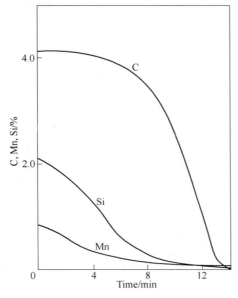

图 1　金属成分随吹炼时间的
　　　变化——Bessemer 法

Fig. 1　Change of metal composition with time of blowing in the Bessemer Process[1]

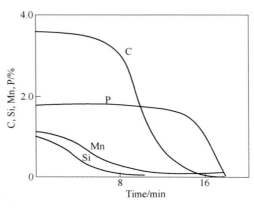

图 2　金属成分随吹炼时间的
　　　变化——Thomas 法

Fig. 2　Change of metal composition with time of blowing in the Thomas Process[2]

1　理论分析

图 3 是根据 Sigworth 及 Elliott[3] 以及 Elliott[4] 的数据绘出的铁液中元素氧化的 $\Delta F°$ 对 T 关系图，也即一种改型的 Ellingham-Richardson 图，其标准态采用 1% 重量溶液。众所周知，铁液中元素氧化的顺序，取决于氧化反应 $\Delta F°$-T 线的位置，有较低 $\Delta F°$-T 线位置的元素能优先于任何有较高 $\Delta F°$-T 线位置的元素而被氧化。例如，当 [Al] 及 [Si] 和 O_2 相遇，[Al] 氧化的 $\Delta F°$-T 线低于 [Si] 氧化的 $\Delta F°$-T 线，所以 [Al] 优先被氧化。此乃因如果 [Si] 先被氧化，生成的 SiO_2 将被 [Al] 还原，因为反应（3）的 $\Delta F°$ 在炼钢温度范围内是负值（ΔF 以 J 为单位，下同）：

$$\frac{4}{3}[Al] + SiO_{2(s)} = [Si] + \frac{2}{3}Al_2O_{3(s)};\quad \Delta F° = -219740 + 35.73T \tag{3}$$

相反地，如果 [Al] 先被氧化，生成的 Al_2O_3 则不可能被 [Si] 还原。

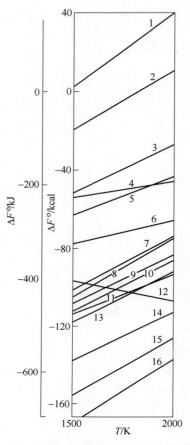

图 3 铁液中元素氧化的 $\Delta F^\circ\text{-}T$ 线

Fig. 3 $\Delta F^\circ\text{-}T$ lines of oxidation of elements in molten iron with oxygen

1—$4[Cu] + O_2 = 2Cu_2O_{(l)}$; 2—$2[Ni] + O_2 = 2NiO_{(s)}$;

3—$\frac{4}{5}[P] + O_2 = \frac{2}{5}P_2O_{5(g)}$; 4—$\frac{2}{3}[Mo] + O_2 = \frac{2}{3}MoO_{3(g)}$;

5—$\frac{2}{3}[W] + O_2 = \frac{2}{3}WO_{3(l)}$; 6—$2[Fe] + O_2 = 2FeO_{(l)}$;

7—$\frac{4}{3}[Cr] + O_2 = \frac{2}{3}Cr_2O_{3(s)}$; 8—$2[Mn] + O_2 = 2MnO_{(s)}$;

9—$\frac{4}{3}[V] + O_2 = \frac{2}{3}V_2O_{3(s)}$; 10—$[Nb] + O_2 = \frac{1}{2}Nb_2O_{4(s)}$;

11—$\frac{4}{3}[B] + O_2 = \frac{2}{3}B_2O_{3(s)}$; 12—$2[C] + O_2 = 2CO_{(g)}$;

13—$[Si] + O_2 = SiO_{2(s)}$; 14—$[Ti] + O_2 = TiO_{2(s)}$;

15—$\frac{4}{3}[Al] + O_2 = \frac{2}{3}Al_2O_{3(s)}$; 16—$\frac{4}{3}[Ce] + O_2 = \frac{2}{3}Ce_2O_{3(s)}$

值得注意的是：[C] 氧化的 $\Delta F^\circ\text{-}T$ 线与不少元素氧化的 $\Delta F^\circ\text{-}T$ 线相交，图 4 的 1a 及 1b 线是 [Cr] 及 [C] 氧化的 $\Delta F^\circ\text{-}T$ 线，二者相交于 1570K，当温度小

于 1570K 时，[Cr] 优先于 [C] 而被氧化；但超过 1570K 时，则 [C] 先被氧化，此相交的温度可称为 [Cr]、[C] 氧化的转化温度。此温度可自图 4 直接读出，或用二步法求出：

$$\frac{4}{3}[Cr] + O_2 = \frac{2}{3}Cr_2O_{3(s)}; \quad \Delta F^\circ = -780310 + 233.59T \quad (4)$$

$$2[C] + O_2 = 2CO; \quad \Delta F^\circ = -281160 - 84.18T \quad (5)$$

在转化温度，反应式（4）及式（5）的 ΔF° 值相等，因之可求出 $T=1570K$。

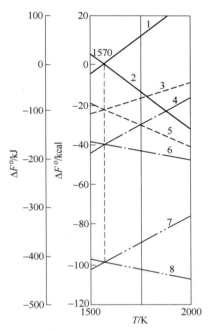

图 4 铁液中 Cr、C 氧化的转化温度
Fig. 4 Transition temperature of oxidation of Cr *vs* C in molten iron

$1-\frac{4}{3}[Cr] + 2CO = 2[C] + \frac{2}{3}Cr_2O_3(4b); \quad 2-2[C] + \frac{2}{3}Cr_2O_3 = \frac{4}{3}[Cr] + 2CO(4a);$

$3-\frac{4}{3}[Cr] + 2(FeO) = 2[Fe] + \frac{2}{3}Cr_2O_3(3b); \quad 4-\frac{4}{3}[Cr] + 2[O] = \frac{2}{3}Cr_2O_3(2b);$

$5-2[C] + 2(FeO) = 2[Fe] + 2CO(3a); \quad 6-2[C] + 2[O] = 2CO(2a);$

$7-\frac{4}{3}[Cr] + O_2 = \frac{2}{3}Cr_2O_3(1b); \quad 8-2[C] + O_2 = 2CO(1a)$

合并式（4）及式（5）得式（6）：

$$\frac{4}{3}[Cr] + 2CO = 2[C] + \frac{2}{3}Cr_2O_{3(s)}; \quad \Delta F^\circ = -499150 + 317.77T \quad (6)$$

式（6）在图 4 以 4b 线表示，此线与 $\Delta F^\circ = 0$ 线的相交点，就是 [Cr]、[C] 氧化的转化温度，因之将式（6）的 $\Delta F^\circ = 0$，得 $T=1570K$。此即是计算转化温度的

一步法。

在图 4 以 4a 线表示的反应式（7）是式（6）的逆式：

$$2[C] + \frac{2}{3}Cr_2O_{3(s)} = \frac{4}{3}[Cr] + 2CO; \quad \Delta F^\circ = 499150 - 317.77T \quad (7)$$

可以看出，在转化温度 1570K 式（6）及式（7）处于瞬间平衡。认为 [C] 是还原剂，从式（7）看出，欲使 [C] 还原 Cr_2O_3，其温度必须超过 1570K，所以 [Cr]、[C] 氧化的转化温度也可以认为是 [C] 还原 Cr_2O_3 最低的还原温度。

图 4 还提供了以铁液内 [O]（2a 及 2b 线）和炉渣中（FeO）（3a 线及 3b 线）作氧化剂的氧化反应的 ΔF°-T 线（每反应均按 2 个氧原子写出）。每对氧化反应的 ΔF°-T 线均相交于 1570K。这样可以得出结论，在排除生成新相（凝聚相或气泡）的晶核能的条件下，无论氧以 O_2 存在，抑或以溶于铁液中的 [O] 或以炉渣中的（FeO）存在，氧化的转化温度与氧存在的形式无关，也与氧的压力（对 O_2）或活度（对 [O] 或 (FeO)）无关，它只决定于参加反应的反应物及产物的本质及活度（压力）。

有一点涉及选择性氧化的动力学问题尚须予以澄清。对上列 [Cr]、[C] 的选择性氧化，如欲去 Cr 保 C，吹炼温度不能高于 Cr、C 氧化的转化温度，但实际上，纵然熔池温度低于转化温度，C 也会氧化一小部分。这里必须考虑动力学因素。吹炼过程中在低于转化温度时 Cr 优先于 C 而氧化，这里有一条件，即 Cr、C 同时和 O_2 相遇。但实际情况是，Cr、C 都分散在铁水之内，当 O_2 和铁水相遇时，如果同时遇到 Cr 及 C 分子，纵然 C 的分子数高于 Cr 的分子数，而 Fe 的分子数则更高，在低于转化温度时，热力学分析指出，Cr 仍优先于 C 或 Fe 而被氧化。但随着铁水表面 Cr 分子的消耗，在内部 Cr 来不及扩散到铁水表面之前，O_2 如遇到 C 分子，当然 C 即被氧化，如果铁水表面连 C 分子也没有，Fe 也会被氧化。这就是在低于转化温度时，C、Fe 都有可能被氧化的动力学原因，由于熔池中 Fe 分子特别多，当 O_2 触及熔池表面时，很可能先生成 FeO，后者以 [O] 溶于 Fe 液向内部扩散，再遇到内部的 Cr、C，再根据熔池温度低于或高于转化温度进行选择性氧化，随着铁液中杂质元素的减少，最终将有一定量自由溶解的 [O] 存于铁液之中。

2 工业实践

2.1 Ni 与 S

镍锍是火法冶金提镍的中间产物，曾设想从它直接提炼金属镍，1971 年在 200kg Kaldo 炉用 O_2 吹炼镍锍的初步试验[5]指出：（1）硫脱到 4%~5% 以下比较困难；（2）吹炼末期生成大量的 NiO 硬壳累积在炉口，有时迫使炉子停吹。

参照工业生产中镍锍的组成，作者[6,7]利用转化温度的概念，进行一系列的热力学计算（表1），得到脱 S 保 Ni 的条件是：

(1) 对已知组成的镍锍，开吹温度必须高于该熔体 Ni、S 氧化的转化温度。

表 1 吹炼镍锍的转化温度
Table 1 Transition temperature during blowing of nickel matte

Item	Ni matte composition/%	Conditions	ΔF/J	Transition temperature/℃
I	Ni 70.0 Cu 5.0 Fe 3.0 Co 0.7 S 21.3	p_{O_2} = 1atm p_{SO_2} = 0.7atm f_S = 1 γ_{Ni} = 1 γ_{Cu} = 2 γ_{Fe} = 0.42 γ_{Co} = 1	[S] + O_2 = SO_2 ΔF = −208150 + 17.24T 2[Ni] + O_2 = 2$NiO_{(s)}$ ΔF = −505010 + 197.82T 2[Fe] + O_2 = 2$FeO_{(l)}$ ΔF = −476140 + 173.34T 4[Cu] + O_2 = 2$Cu_2O_{(l)}$ ΔF = −292460 + 205.22T 2[Co] + O_2 = 2$CoO_{(s)}$ ΔF = −506260 + 248.36T	Ni vs. S 1370 Fe vs. S 1440
II	Ni 82.0 Cu 5.9 Fe 1.3 Co 0.8 S 10.0	p_{O_2} = 1atm p_{SO_2} = 0.7atm f_S = 1 γ_{Ni} = 1 γ_{Fe} = 0.41	[S] + O_2 = SO_2 ΔF = −208150 + 23.18T 2[Ni] + O_2 = 2$NiO_{(s)}$ ΔF = −505010 + 193.76T 2[Fe] + O_2 = 2$FeO_{(l)}$ ΔF = −476140 + 187.40T	Ni vs. S 1470
III	Ni 87.6 Cu 6.3 Fe 1.4 Co 0.9 S 3.8	p_{O_2} = 1atm p_{SO_2} = 0.7atm f_S = 1 γ_{Ni} = 1	[S] + O_2 = SO_2 ΔF = −208150 + 31.63T 2[Ni] + O_2 = 2$NiO_{(s)}$ ΔF = −505010 + 191.88T	Ni vs. S 1580
IV	Ni 90.3 Cu 6.4 Fe 1.4 Co 0.9 S 1.0	ditto	[S] + O_2 = SO_2 ΔF = −208150 + 42.72T 2[Ni] + O_2 = 2$NiO_{(s)}$ ΔF = −505010 + 190.96T	Ni vs. S 1730
V	Ni 91.1 Cu 6.5 Fe 1.4 Co 0.9 S 0.1	ditto	[S] + O_2 = SO_2 ΔF = −208150 + 61.84T 2[Ni] + O_2 = 2$NiO_{(s)}$ ΔF = −505010 + 190.66T	Ni vs. S 2030

（2）吹炼操作必须迅速进行，使熔池温度上升的速度永远高于因熔池组成改变而造成的转化温度上升的速度。

为此，1973~1974 年间在 1.5t Kaldo 炉进行半工业试验，采取相应的吹炼制度，保证上列热力学条件顺利实施，试验结果非常成功[8]，充分说明理论能用以指导实践，金属 Ni 含 S 量顺利地降到 1%~2%，特别在吹炼后期采用不吹 O_2 的空转，以消除由于熔池内部硫扩散速度降低而引起表面上 NiO 的聚集。熔池温度高于理论计算的转化温度（图5）。镍的总回收率超过 95%。

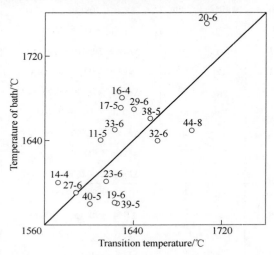

图 5 计算的 Ni、S 氧化的转化温度与实测的熔池温度
Fig. 5 Calculated transition temperature of oxidation of Ni vs S compared with the measured temperature of the bath

如文献［8］指出，硫的扩散是决定脱硫反应速度的控制性环节。传质系数与温度的关系式为：

$$\beta = 8.30 \exp \frac{-104600}{RT}$$

式中　β——传质系数，cm/s；
　　　R——气体常数，8.314J/(K·mol)。

2.2　Cr 与 C

含铬铁水炼钢必须先行脱 Cr，以便能控制钢的最后成分，以脱 Cr 后的半钢在转炉炼钢，半钢又须含有一定的碳以保证热源。1974~1975 年间曾在 10t 摇包用 O_2 吹炼含铬约 4% 的铁水的试验。假定在碳焰开始上升的温度熔池处于瞬间平衡，而此熔池温度认为是 Cr、C 氧化的转化温度，作者[9]根据该试验的一部分数据，利用上面给出的式（6）的 ΔF° 用等温方程式进行反算，得出 $\gamma_{Cr_2O_3}$ 的估算值为 1.8×10^{-6}，$\gamma_{Cr_2O_3}$ 值如此之小，证实铬渣内 Cr_2O_3 和 FeO 结合为稳定的铬尖晶石[10]。

由于种种原因，当时吹炼温度未能控制在 1450~1500℃ 以下，因之铁水的 Cr 含量只能降到 0.8%~1.0%。1977~1978 年间又恢复实验[11]。采用连续测温，摇包空转不吹氧而同时加入冷却剂的措施，使熔池温度控制在预先计算的 Cr、C 氧化的转化温度之下，Cr 含量成功地降到 0.2%~0.5% 或更低。文献[11]指出，由实验室试验的数据计算的 $\gamma_{Cr_2O_3}$ 值，其数量级为 10^{-5}，而利用半工业试验数据作者[12]估计的 $\gamma_{Cr_2O_3}$ 值为 1.4×10^{-6}（表2）。纵然 $\gamma_{Cr_2O_3}$ 值有可非议之处，但根据由它计算出的转化温度控制吹炼操作，可得到降 Cr 到满意的结果（参阅表3[13]）。

冶炼奥氏体 18-9 不锈钢（包括 AOD 法），其目的在于脱 C 保 Cr，认为熔渣以 Cr_3O_4 饱和[14]，作者[15]利用式（8）：

$$\frac{3}{2}[Cr] + 2CO = 2[C] + \frac{1}{2}Cr_3O_{4(s)}; \quad \Delta F^\circ = -465260 + 307.69T \quad (8)$$

进行一系列热力学计算，得出的转化温度见表4，不难看出，采用氩氧混吹，也即降低 p_{CO}，实际上意味着降低了 Cr、C 氧化的转化温度。

表 2 $\gamma_{Cr_2O_3}$ 的估算值
Table 2 Estimated value of $\gamma_{Cr_2O_3}$

Heat No	Metal bath/%			Slag Composition/%			$N_{Cr_2O_3}$	$\lg f_C$	$\lg f_{Cr}$	Temperature /℃	$\gamma_{Cr_2O_3}$
	C	Cr	Si	SiO_2	Cr_2O_3	FeO					
27	3.40	0.14	0.10	24.1	25.8	35.8	0.159	0.481	-0.408	1389	0.8×10^{-6}
28	3.30	0.12	0.10	22.6	27.4	34.0	0.175	0.467	-0.396	1380	0.9×10^{-6}
42	3.15	0.18	0.10	20.4	20.7	34.4	0.143	0.445	-0.378	1399	2.0×10^{-6}
43	3.21	0.29	0.10	21.5	24.0	38.4	0.150	0.451	-0.385	1416	2.5×10^{-6}
55	3.20	0.12	0.10	23.0	21.5	38.9	0.133	0.449	-0.384	1401	0.8×10^{-6}
											av. 1.4×10^{-6}

表 3 Cr 和 C 氧化的转化温度计算值
Table 3 Calculated transition temperature of oxidation for Cr vs. C

%C	4	3.5	3	4	3.5	3	4	3.5	3
%Cr	0.5	0.5	0.5	0.2	0.2	0.2	0.1	0.1	0.1
Transition temp. /℃	1420	1462	1500	1365	1405	1440	1323	1357	1393

表 4 不锈钢冶炼中 Cr 和 C 氧化的转化温度计算值
Table 4 Calculated transition temperature of oxidation for
Cr vs. C during the stainless steelmaking

Cr	Ni	C	p_{CO} /atm	ΔF/J	Trans. temp. /℃	O_2 : Ar : CO
	%					
12	9	0.35	1	$-465260+255.39T$	1555	—
12	9	0.1	1	$-465260+232.46T$	1727	—
12	9	0.05	1	$-465260+220.50T$	1835	—
10	9	0.05	1	$-465260+224.26T$	1800	—
18	9	0.35	1	$-465260+245.06T$	1627	—
18	9	0.1	1	$-465260+222.13T$	1820	—
18	9	0.05	1	$-465260+209.99T$	1945	—
18	9	0.35	2/3	$-465260+251.71T$	1575	1:1:2
18	9	0.05	1/2	$-465260+221.50T$	1830	1:2:2
18	9	0.05	1/5	$-465260+236.86T$	1690	1:8:2
18	9	0.05	1/10	$-465260+248.28T$	1600	1:18:2
18	9	0.02	1/20	$-465260+244.30T$	1630	1:38:2
18	9	1	1	$-465260+268.24T$	1460	—
18	9	4.5	1	$-465260+323.84T$	1165	—

2.3 V 与 C

由含钒铁水提 V 可在转炉（顶吹或底吹）或摇包用 O_2 吹炼，也可在侧吹转炉、槽式炉或用雾化法以空气吹炼，图 6[16] 指出铁水吹炼入渣稳定的氧化物是 V_2O_3，利用式（9）：

$$\frac{2}{3}[V] + CO = [C] + \frac{1}{3}(V_2O_3); \quad \Delta F^\circ = -245890 + 147.82T \quad (9)$$

$\gamma_{V_2O_3}$ 值的数量级估计为 10^{-5}[17]，一般欲从含 0.45%V 的铁水吹炼为 0.03%~0.05%V 的半钢，吹炼温度应低于 1350~1400℃。V、C 氧化的转化温度随半钢含 C 量及含余 V 量而异，如铁水含 Si 较高，则吹炼过程中必须加入冷却剂，钒渣经氧化焙烧后用水法处理以提纯 V_2O_5。

2.4 Nb 与 C

含铌铁水用 O_2 或空气吹炼，Nb 氧化入渣，所用设备和含 V 铁水吹炼的设备相同，含 Nb 铁水又可装入平炉，在初期溢出的熔渣即含有铁水中绝大部分的 Nb，此炉渣再单独处理，图 7[18] 指出，Nb 在铁液内氧化，其氧化物以 Nb_2O_4 最为稳定，欲脱 Nb 保 C，使半钢余 Nb<0.02%~0.03%，吹炼温度一般不宜高于 1400℃。Nb、C 氧化的转化温度在相似组成下略高于 V、C 氧化的转化温度。利用式（10）：

$$[Nb] + 2CO \rightleftharpoons 2[C] + \frac{1}{2}(Nb_2O_4); \quad \Delta F° = -525090 + 305.01T \quad (10)$$

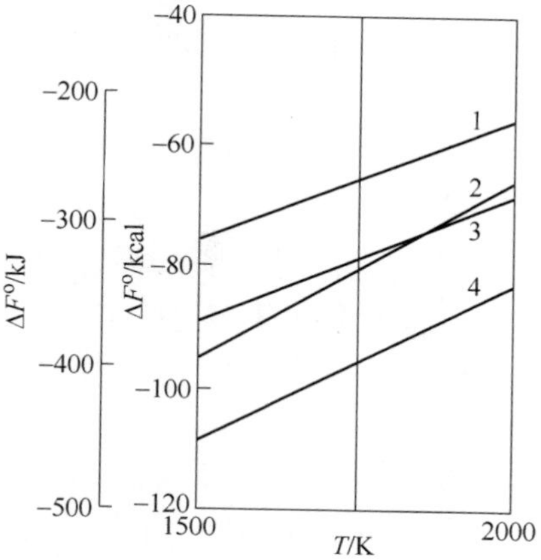

图 6 铁液中 V 氧化生成不同氧化物的 $\Delta F°$-T 线

Fig. 6 $\Delta F°$-T lines of oxidation of V in molten iron with formation of different oxides[16]

$1 - \frac{4}{5}[V] + O_2 = \frac{2}{5}V_2O_5;\ 2 - 2[V] + O_2 = 2VO;$

$3 - [V] + O_2 = \frac{1}{2}V_2O_4;\ 4 - \frac{4}{3}[V] + O_2 = \frac{2}{3}V_2O_3$

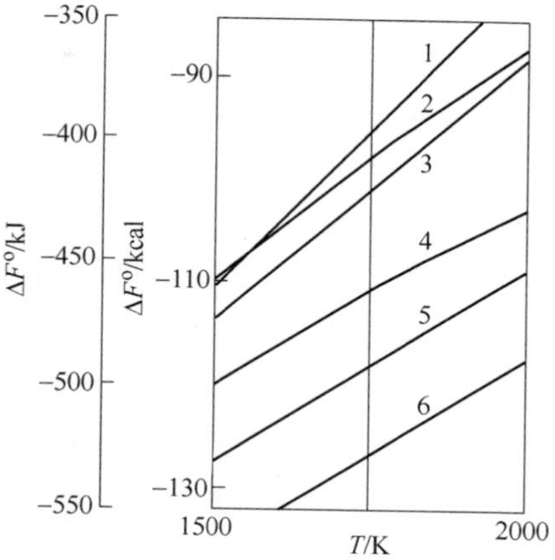

图 7 Nb 氧化生成不同氧化物的 $\Delta F°$-T 线

Fig. 7 $\Delta F°$-T lines of oxidation of Nb with formation of different oxides[18]

$1 - 2[Nb] + O_2 = 2NbO;\ 2 - \frac{4}{5}[Nb] + O_2 = \frac{2}{5}Nb_2O_5;\ 3 - [Nb] + O_2 = \frac{1}{2}Nb_2O_4;$

$4 - \frac{4}{5}Nb + O_2 = \frac{2}{5}Nb_2O_5;\ 5 - Nb + O_2 = \frac{1}{2}Nb_2O_4;\ 6 - 2Nb + O_2 = 2NbO$

作者[19]估计 $\gamma_{Nb_2O_4}$ 的数量级为 $10^{-9} \sim 10^{-10}$。假定氧化产物为 Nb_2O_5,林宗彩

等人[20]根据实验室数据估计 $\gamma_{Nb_2O_5}$ 为 $4.37×10^{-11}$，而张庆宜[21]利用 1966 年半工业试验的数据进行计算，得出 $\gamma_{Nb_2O_5}$ 的估计值为 $0.6×10^{-9}$。

利用底吹转炉吹炼含 Nb 铁水，在不加 CaO 条件下，由于渣内含 FeO 较少，P 可留于半钢之内而不入渣。

2.5　Mn 与 C

对铁液中 Mn、C 的选择性氧化，邵象华[22]曾进行过一系列的热力学计算，并建议一种新方法，在常压下向高碳锰铁熔池吹 O_2，以制备低碳锰铁，欲脱 C 保 Mn，吹炼温度必须在 Mn、C 氧化的转化温度之上，例如，当温度高于 1640℃，可得到 1%C 的中碳锰铁，此时 Mn 的挥发损失约为金属量的 0.5%，如欲达到 0.1%C 的低碳锰铁，则吹炼温度必须高达 1980℃ 以上，此时 Mn 的挥发损失增加到约为金属量的 2.5%，如在低压 0.1atm 时吹炼，则得到 0.1%C 的低碳锰铁，温度只须高于 1680℃，但由于低压，Mn 的挥发损失将达到金属量的 3%，原文献未提供 γ_{MnO} 的数据，对利用氧气转炉在适当温度下吹炼中低碳锰铁的可能性原文进行了讨论。

2.6　P 与 C

P 与 C 及其他元素在吹炼铁水过程中的氧化顺序作者[23]曾进行过分析讨论，根据不同的碱性炼钢方法脱 P 反应可分为三种类型：

（1）脱 P 基本上在冶炼前期较低温度下进行，先脱 P 而后脱 C（例如搅炼炉法及碱性平炉法）。

（2）脱 P 和脱 C 同时进行（LD 或 LD-AC 法）。

（3）脱 P 在脱 C 之后较高温度下进行（Thomas 法）。

在炼钢温度，P_2O_5 虽以气态存在，但从铁液中 P 以气态被氧化掉是不可能的；只有在碱性高氧化渣存在时，P 才能被氧化而入渣。示意图[24]8（a）指出，脱 P 反应从不可能进行的 ΔF 线 a 下移到有碱性高氧化性渣条件下可以进行的 ΔF 线 b，这条线和 C 氧化（p_{CO} =1atm）的 ΔF 线 e 相交于转化温度 t_T，当熔池温度低于 t_T，则 P 优先氧化；高于 t_T，则 P 的氧化被抑制而 C 大量氧化，e 线是 C 在 p_{CO} 为 1atm 的 ΔF 线。但实际上在熔渣下由于缺乏气泡晶核，新生气相必须克服更大的压力才能生成（曾有人[25]估计为 30atm）。这意味着 C 的氧化受到抑制，C 氧化的 ΔF 线上移到线 f，因而在 P 氧化的 ΔF 线之上。这样便保证了脱 P 的优先进行。在平炉的冶炼初期自废钢开始熔化到熔毕铁水含 P 量一般可脱去 95% 以上。

对 Thomas 法炼钢，脱 P 与脱 C 的关系和平炉法完全不同，吹入空气中的 O_2 气很快被消耗掉，剩余 N_2 对 CO 来说类似真空，所以 C 的氧化被促进，其氧化的

ΔF 线自 e 下降到 f（图 8b），由于铁水中 Si、Mn 及 C 的猛烈氧化，渣中 FeO 比较少，加入的 CaO 难以溶解，因而 P 氧化的 ΔF 线不可能下移的很低，由 a 只能下降到 b，而后者高在 C 氧化的 ΔF 线 f 之上，所以在整个 Si、Mn 及 C 氧化期间，P 基本上不能被氧化。只有碳焰下降之后，在后吹期间，炉渣 FeO 增多，CaO 全部溶解后，P 才有机会被氧化进入炉渣。

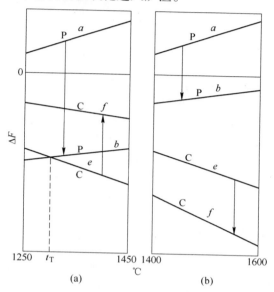

图 8 脱 P 与脱 C 的相互关系示意图

Fig. 8 Relation of dephosphorization with decarburization schematically presented[24]

(a) Dephosphorization before decarburization; (b) Decarburization before dephosphorization

对于氧气顶吹转炉（LD 或 LD-AC 法），P 和 C 同时氧化，这是由于氧气流股造成大量铁球滴和泡沫渣，对脱 P 脱 C 都创造了有利的条件。

3 结束语

上列分析明确地指出，氧化的转化温度的概念对指导氧化冶炼的生产实践有重大意义。

氧化冶炼可以在两方面有利地进行：

（1）在高于转化温度的条件下进行，例如镍锍的脱 S，不锈钢冶炼（包括 AOD 法）的脱 C，以及高碳锰铁的降 C 等等。

（2）在低于转化温度的条件下进行，例如含 Cr 铁水的脱 Cr，含 V 铁水的提 V，以及含 Nb 铁水的提 Nb 等等。

对于 P 和 C，除转化温度的影响外，变更炉渣成分（碱度及 FeO 含量）以及气体产物 CO 的压力可改变二者氧化的先后顺序。

参 考 文 献

[1] Bodsworth C, Bell H B. Physical Chemistry of Iron and Steel Manufacture, 2nd, Longman, London, 1972: 198.
[2] Bodsworth C, Bell H B. ibid: 203.
[3] Sigworth G K, Elliott J F. Met. Sci., 1974 (8): 298; Elliott J F. Electr. Furnace Proc., 1974 (32): 62.
[4] Elliott J F, Gleiser M Q, Ramakrishna V. Thermochemistry for Steelmaking, Vol. II, Addison-Wesley, New York, 1963: 620.
[5] 韩其勇. 个人通讯, 1973; 转炉氧气斜吹硫化镍探索性试验总结 (内部资料), 1971.
[6] 魏寿昆. 回转炉氧气吹炼冰镍的物理化学, 1973 (未发表论文).
[7] 魏寿昆. 冶金过程热力学. 上海: 上海科技出版社, 1980: 103.
[8] 魏寿昆, 洪彦若. 镍锍选择性氧化的热力学及动力学. 1980 年广州冶金过程物理化学学术报告会论文, 有色金属, 1981, 33 (3): 50.
[9] 北京钢铁学院冶金物化专业 72 年级冶炼学习汇报材料汇编, 1975: 5~9.
[10] Schenck H, Wenzel W, Joshyula G K M. Arch. Eisenhuettenwes, 1963, 34: 503.
[11] 林宗彩, 周荣章. 钢铁, 1979, 14 (3): 1.
[12] 文献 [7]: 70; 1.5 吨摇包脱铬补充试验 (内部资料), 1978.
[13] 文献 [7]: 71.
[14] Sims C E. Electric Furnace Steelmaking, Vol. II, Interscience, New York, 1963: 172.
[15] 魏寿昆. 冶炼过程热力学 (内部讲义), 1974: 2~22; 文献 [7]: 76.
[16] 魏寿昆. 炼钢冶金原理 (内部讲义), 1977: 2~88; 文献 [7]: 86.
[17] 魏寿昆. 含钒铁水炼钢的一些物理化学问题. 攀钢技术处 (内部资料), 1976; 文献 [7]: 87.
[18] 文献 [16]: 2~100; 文献 [7]: 29.
[19] 文献 [16]: 2~99; 文献 [7]: 90.
[20] 林宗彩, 周荣章, 等. 北京钢铁学院学报, 1980 (2): 25.
[21] 张庆宜. 钢铁, 1980, 15 (8): 13.
[22] 邵象华. 金属学报, 1977, 13: 182.
[23] 魏寿昆. 金属学报, 1964, 7: 240.
[24] 文献 [16]: 4~95; 文献 [7]: 280.
[25] Rellermeyer H, Knuepple H, Sittard J. Stahl Eisen, 1957, 77: 1296.

选择性氧化——理论与实践[*]

魏寿昆

摘 要: 本文分析讨论选择性氧化过程中转化温度概念的意义及计算方法,并总结近几年来该概念在生产实践中应用的实例。

1 理论分析

火法冶金吹炼过程中选择性氧化的转化温度的概念对控制操作起着显著的作用[1]。两个元素的氧化反应的 ΔG-T 线如相交,其相交点就是转化温度。Ellingham-Richardson 图(图1)中 C 的 ΔG°-T 线与 W、Cr、Mn、Nb、V、B、Si

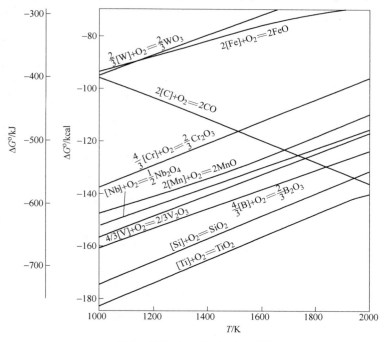

图 1 Ellingham-Richardson 图

---

[*] 原刊于《北京钢铁学院庆祝建校三十周年论文选集》(上册),1982:201~206.

及 Ti 等的相应线均相交。这是由于上列各元素及其纯氧化物的 S^o 几乎相等，氧化物的生成熵 ΔS^o 略等于 O_2 的 S^o，而其值是负值，使得各元素的 ΔG^o 线几乎平行。C 氧化产物 CO 的 S^o 值相当大，使得 C 氧化反应的 ΔS^o 为正值，而它的 ΔG^o 线可以和上列各元素的 ΔG^o 线相交。溶于铁液内元素的氧化反应也有类似现象[2]（图 2）。值得注意的是，镍锍中 Ni、S 的氧化反应的 ΔG^o 线也有相交点[3]，因之可以进行选择性氧化（图 3）。

以铁液内 Cr、C 的选择性氧化为例（为简便计，采用标准态计算），两个氧化反应的 ΔG^o 线相交于 1570K（图 4）。可以看出，在 Cr、C 同时和 O_2 相遇时，在小于 1570K 时，Cr 先于 C 而被氧化，而大于 1570K 时，则 C 先于 Cr 而被氧化。此转化温度可自图中读出，也可用二步法求出：

$$\frac{4}{3}[Cr] + O_2 = \frac{2}{3}Cr_2O_{3(s)}$$

$$\Delta G^o = -780310 + 233.59T(J) \quad (1)$$

$$2[C] + O_2 = 2CO$$

$$\Delta G^o = -281160 - 84.48T(J) \quad (2)$$

在转化温度，二式的 ΔG^o 相等，因之可求出：

$$T = 1570K$$

将式（1）及式（2）合并为式（3）：

$$2[C] + \frac{2}{3}Cr_2O_{3(s)} = \frac{4}{3}[Cr] + 2CO$$

$$\Delta G^o = -499150 + 317.77T(J) \quad (3)$$

式（3）在图 4 中以线 4a 表示，后者与 ΔG^o 的零值线相交于 1570K。使式（3）的 ΔG^o = 0，即可求出转化温度为 1570K。此即求转化温度的一步法。

由式（3）可看出，欲使铁液中的 C 还原 Cr_2O_3，温度必须大于 1570K，所以 Cr、C 氧化的转化温度实质上又是 C 还原 Cr_2O_3 的"最低还原温度"。

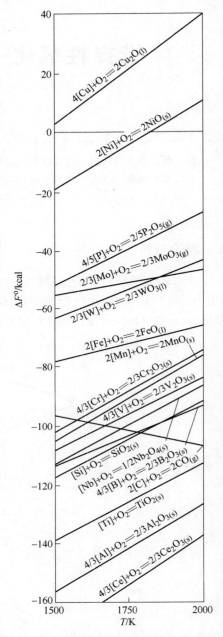

图 2　溶于铁液内元素直接氧化的 ΔG^o

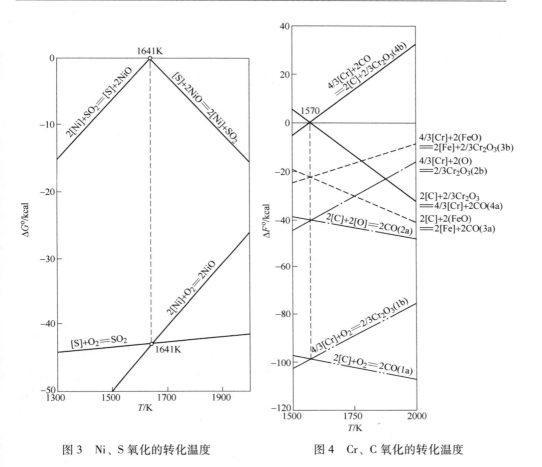

图 3　Ni、S 氧化的转化温度　　　图 4　Cr、C 氧化的转化温度

图 4 中线 4b 是式（3）的逆式。明显地在转化温度该反应式处于瞬间平衡，反应可向正逆两方向进行。

图 4 中线 2a 及线 2b 是用溶解氧氧化的反应，而线 3a 及线 3b 是用炉渣中 FeO 氧化的反应。二者的转化温度同为 1570K。在排除氧化产物晶核自由能的条件下，通过图 4 可以看出，两个元素氧化的转化温度与氧存在的形式，O_2、[O] 或 (FeO)，以及氧气的压力、[O] 或 (FeO) 的活度无关，而只决定于该二元素及氧化产物在熔体中的活度（如是气体，则决定于其压力）。同时，转化温度不是一成不变的温度，而是随着熔池中组成的改变而经常地在变化。

有一点须予以澄清。欲去 Cr 保 C，熔池温度必须低于转化温度。但实际上，纵然熔池温度低于转化温度，C 也要氧化一小部分。这里 C 的氧化是基于熔池的动力学因素。当 O_2 气和铁水相遇时，基于 Fe 分子的大量存在，以及 Cr、C 分子分散于铁水之中，很可能 Fe 先被氧化为 FeO，后者以 [O] 熔于铁水之中。当 [O] 同时遇到 Cr、C，纵然 C 分子数大于 Cr，在低于转化温度时 Cr 将被氧化但如果[O]只遇到C，则CO将生成，在上升过程中如CO遇到Cr，则由于温度低于转化温度，则Cr将还原CO而夺走其氧。但如果CO在上升过程中无机会遇到Cr，特别当熔池中含Cr较少时，CO最终以气体逸出于熔池之外。这就是在低于转化温度时，C有可能被氧化一小部分的动力学原因（参阅图5）。

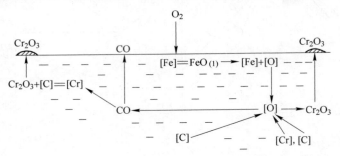

图5 铁液内Cr、C选择性氧化机理（熔池温度小于转化温度）

2 工业上实践

（1）镍锍是火法冶金提镍的中间产物，其组成为Ni 65%~70%，S 20%~25%，其余为Fe、Cu及Co等。1971年在小型Kaldo炉用O_2吹炼镍锍进行去S提Ni的初步试验[4]。当时曾遇到下列困难[5]：1）S脱到4%~5%以下比较困难；2）吹炼末期生成大量NiO硬壳，累积在炉口，有时迫使炉子停止吹炼；3）Ni回收率最高只能达75%~80%。利用Ni、S选择性氧化的转化温度的概念，作者[6]以式（4）进行：

$$[S] + 2NiO_{(s)} = 2[Ni] + SO_2; \quad \Delta G^\circ = 296690 - 143.76T(J) \quad (4)$$

热力学分析，明确了去S保Ni的条件是：

1）对给定组成的镍锍，开吹温度必须高于该熔体组成Ni、S氧化的转化温度，一般不能低于1400℃。

2）吹炼操作必须迅速进行，使得熔池温度上升的速度永远高于因熔池组成改变而造成的转化温度上升的速度。

1973~1974年间在1.5t Kaldo炉由金川有色金属公司组织进一步试验，采取相应的吹炼制度以保证上列的热力学条件，试验结果非常成功[7]。镍锍含S量顺利地降到1%~2%，满足后工序进一步提纯的要求。实测的熔池温度基本上高于理论计算的转化温度。Ni的总回收率超过95%。

（2）利用式（5）：

$$2[C] + \frac{1}{2}Cr_3O_{4(s)} = \frac{3}{2}[Cr] + 2CO; \quad \Delta G^\circ = 465260 - 307.69T(J) \quad (5)$$

作者对奥氏体不锈钢冶炼（包括AOD法）中C、Cr氧化的转化温度进行热力学分析[8]，得出下列结论：

1) 矿石氧化不能做到去 C 保 Cr，其原因在于该氧化反应不能提高熔池温度超过脱 C 所需的转化温度。

2) 返回吹氧法不能将含 Cr 量一次配足，如若配足，则熔池温度必须提高到炉衬远远不能承受的很高的转化温度。

3) 氩氧混吹（AOD 法）实质上大大降低了转化温度。

4) 高 C 配足 Cr、Ni 的铁水有足够低的转化温度能保证脱 C 保 Cr。

(3) 含 Cr 铁水炼钢必须先行脱 Cr，以便能控制钢的最后成分。但在转炉炼钢，必须保 C。根据 1974~1975 年间由上钢一厂组织的 10t 摇包含 Cr 4%铁水脱 Cr 试验的数据，利用式（6）：

$$\frac{4}{3}[Cr] + 2CO = \frac{2}{3}(Cr_2O_3) + 2[C]; \Delta G^\circ = -499150 + 317.77T(J) \quad (6)$$

假定熔池在碳焰开始上升的温度处于瞬间平衡并认为此开始上升温度为 Cr、C 氧化的转化温度，作者进行反算，得出 $\gamma_{Cr_2O_3}$ 的估计值为 1.8×10^{-6}[9]。当时摇包的终点温度经常难以控制在 1450~1500℃ 之下，铁水的 Cr 只能降到 0.8%~1.0%。由 $\gamma_{Cr_2O_3}$ 估计值计算出的转化温度说明，欲使 Cr 继续下降，摇包温度必须小于 1350~1400℃。当时由于十年动乱，降温建议无法被采纳。1977~1978 年由冶金部组织北京钢铁学院等单位恢复铁水脱 Cr 试验[10]。采用连续测温、停吹空转及同时加冷却剂的措施，使熔池温度控制在 Cr、C 氧化的转化温度之下，Cr 成功地降到 0.2%~0.5%，$\gamma_{Cr_2O_3}$ 的估计值其数量级为 10^{-5}。根据上钢一厂 1.5t 摇包脱 Cr 试验的数据[11]，作者对 $\gamma_{Cr_2O_3}$ 又进行计算，得出估计值 $\gamma_{Cr_2O_3} = 1.4\times10^{-6}$[12]，和以前的估计值颇一致。$\gamma_{Cr_2O_3}$ 值如是之小，说明 Cr_2O_3 和 FeO 结为牢固的铬尖晶石，后者可能在渣内呈饱和状态。

(4) 铁水提 V 国内已在工业上进行，一般采取雾化提 V 或侧吹转炉提 V。利用式（7）：

$$\frac{2}{3}[V] + CO = \frac{1}{3}[V_2O_3] + [C]; \Delta G^\circ = -245890 + 147.82T(J) \quad (7)$$

作者[13]根据工业上数据对 $\gamma_{V_2O_3}$ 进行估算，其数量级为 10^{-5}。吹炼温度应低于 V、C 氧化的转化温度。

(5) 含 Nb 铁水用 O_2 气或空气吹炼，可使 Nb 进入炉渣。利用式（8）：

$$[Nb] + 2CO = \frac{1}{2}(Nb_2O_4) + 2[C]; \Delta G^\circ = -525090 + 305.01T(J) \quad (8)$$

作者[14]采用工业上数据曾估计 $\gamma_{Nb_2O_4}$ 值的数量级为 10^{-9}~10^{-10}。假定氧化物为 Nb_2O_5，林宗彩等人[15]根据实验室数据估计 $\gamma_{Nb_2O_5}$ 为 4.37×10^{-11}；而张庆宜[16]由 1966 年半工业试验的数据计算，$\gamma_{Nb_2O_5}$ 估计值为 0.6×10^{-9}。Nb、C 氧化的转化

温度在相似组成下一般略高于 V、C 氧化的转化温度。

（6）自高碳锰铁在转炉吹炼成为中、低碳锰铁，邵象华[17]曾进行一系列计算，其结果见表 1。

表 1

锰 铁	转化温度/℃	锰挥发损失
1%C	1640	占金属量的 0.5%
0.1%C	1980	占金属量的 2.5%
0.1%C	1680（$p_{CO}=0.1atm$）	占金属量的 3.0%

利用 0.8t 顶吹氧气转炉吹炼高碳锰铁，在 1750~1960℃ 的吹炼温度下，已得出 0.39%~1.43%C 的中、低碳锰铁，锰回收达 80%[18]。关于 γ_{MnO} 的数值则未进行报道。

（7）关于炼钢过程中 P、C 的氧化顺序已进行过分析讨论[19,20]。碱性平炉在熔化期中在较低温度以高氧化铁低碱度渣能脱 P90% 以上，这样操作便于高拉碳出钢。侧吹转炉有类似的前期脱 P。$\gamma_{P_2O_5}$ 值[21]对炼钢渣为 10^{-14}~10^{-18}。

（8）我国华南铁矿含 W。含 W 铁水在顶吹转炉进行初步的吹炼试验[22]，由含 W 0.13%~0.20% 的铁水可炼成含 W 0.03%~0.06% 的钢，炼钢渣中 γ_{WO_3} 的估计平均值为 $5×10^{-7}$。W 的行为与 P 相似，有可能在低于 W、C 氧化的转化温度在高碱度下将 W 富集在前期渣内而加以提取。

3 小结

表 2 作为本文的小结。

表 2

项 目	$\gamma_{M_xO_y}$	文献及年限
（1）镍锍提镍去硫	$\gamma_{NiO}=1$	[6], 1973
（2）不锈钢去碳保铬	$\gamma_{Cr_3O_4}=1$	[8], 1974
（3）铁水脱铬	$\gamma_{Cr_2O_3}=1.8×10^{-6}$	[9], 1975
	$\gamma_{Cr_2O_3}=10^{-5}$	[10], 1979
	$\gamma_{Cr_2O_3}=1.4×10^{-6}$	[12], 1980
（4）铁水提钒	$\gamma_{V_2O_3}=10^{-5}$	[13], 1976
（5）铁水提铌	$\gamma_{Nb_2O_4}=10^{-9}$~10^{-10}	[14], 1977
	$\gamma_{Nb_2O_5}=4.37×10^{-11}$	[15], 1980
	$\gamma_{Nb_2O_5}=0.6×10^{-9}$	[16], 1980
（6）高碳锰铁去碳	γ_{MnO} 未提供数值	[17], 1977
（7）炼钢前期去磷	$\gamma_{P_2O_5}=10^{-14}$~10^{-18}（炼钢渣）	[21], 1953
（8）铁水提钨	$\gamma_{WO_3}=5×10^{-7}$（炼钢渣）	[22], 1981

热力学指出,选择性氧化的转化温度的概念对指导生产实践有重大的意义。

氧化吹炼可以在两方面有利地进行:(1)在高于转化温度的条件下进行,例如表2内第(1)、(2)及(6)项;(2)在低于转化温度的条件下进行,例如表2内第(3)~(5)、(7)及(8)项。

参 考 文 献

[1] 魏寿昆. 第一届中美双边冶金学术会议,1981:45~60.
[2] 魏寿昆. 冶金过程热力学. 上海:上海科学技术出版社,1980:21.
[3] 魏寿昆,洪彦若. 有色金属,1981,33(3):50.
[4] 转炉氧气斜吹硫化镍探索性试验总结. 内部资料,1971年7月.
[5] 韩其勇. 个人通讯,1973.
[6] 魏寿昆. 回转炉氧气吹炼冰镍的物理化学. 未发表资料,1973年10月,另阅文献[2]第三章:99~125.
[7] 见文献[3].
[8] 魏寿昆. 冶炼过程热力学. 北京钢铁学院冶金物化教研室讲义,1974:2-1~2-33;另见文献[2]:73~83.
[9] 北京钢铁学院物化专业72年级冶炼学习汇报资料汇编,1975:5-8~5-11.
[10] 林宗彩,周荣章. 钢铁,1979,14(3):1.
[11] 1.5吨摇包脱Cr补充试验. 内部资料,1978.
[12] 见文献[2]:70.
[13] 魏寿昆. 含钒铁水炼钢的一些物理化学问题. 攀钢技术质量处单印本,1976.
[14] 魏寿昆. 钢铁冶金原理. 北京钢铁学院冶金物化教研室讲义,1977,2~99;另见文献[2]:90.
[15] 林宗彩,周荣章,等. 北京钢铁学院学报,1980(2):25.
[16] 张庆宜. 钢铁,1980,15(8):13.
[17] 邵象华. 金属学报,1977,13:182.
[18] 张惠棠,贾宗生,等. 钢铁,1981,16(5):15.
[19] 魏寿昆. 金属学报,1964,7:240.
[20] 见文献[14]:4-90~4-96,另见文献[2]:276~282.
[21] Turkdogan E T,Pearson J. J. Iron Steel Inst.,1953,175:398.
[22] 李名生,尹伯扬. 含Cu、As、W铁水炼钢时的某些特点——热力学分析与实践. 广东省金属学会1981年年会论文.

稀土钢冶炼的物理化学问题[*]

魏寿昆

1 稀土在钢中的行为和作用

从物理化学角度看，稀土在钢中的作用包括以下四个方面：脱氧、脱硫、固溶作用（合金化）、去除有害夹杂（As、Bi、Sb、Sn、Pb等）。以下主要讲稀土的脱氧及脱硫作用。

稀土除了脱钢水中的氧，还要脱耐火材料的氧。稀土和耐火材料能起化学反应，将它们所含的氧化物，特别是 SiO_2 进行还原。判断各种元素的脱氧能力，可用溶解于铁液中各元素与 1mol 氧气化合生成的氧化物的标准自由能进行分析和比较（见图1）。氧化反应一般可写为：

$$\frac{2x}{y}[M] + O_2 = \frac{2}{y}M_xO_y$$

采用 1mol O_2 为标准，是因为氧气是铁液中各元素争夺的对象。

在图1中曲线位置越低，所代表的氧化物越稳定。"下面的"元素可还原"上面的"氧化物。例如，Ca 可以还原 Ce_2O_3、MgO、ZrO_2、Al_2O_3、TiO_2 等。从图1可以看出 Ce 可还原 SiO_2，即易与酸性耐火材料发生反应。稀土氧化物与 ZrO_2、MgO、Al_2O_3 的标准自由能相差不多，故稀土不易与这种材质的耐火材料发生反应。

在钢液中除了氧以外，尚有 S、N、O 等元素时，加入稀土以后，这些元素都能与稀土发生作用，此时加入的稀土是钢液中被夺取的对象，所以应以 1mol 稀土为标准，来比较其标准生成自由能的大小，从而确定反应进行的先后次序（见图2）。

稀土碳化物和稀土氮化物的标准生成自由能数值较大，代表它们的曲线在图2没有画出。一般来讲，稀土加入钢液内，碳化物及氮化物均难以生成。还应该提到，上面两个图都是标准状态的自由能图（即各元素的活度都等于1时的情况）。在一般情况下，各元素在钢液中的活度并非为1，这时要根据具体情况，按照各元素的实际活度进行计算。

---

[*] 1978年9月中国稀土钢冶炼工艺和稀土加入方法会议（柳州会议）发表论文。

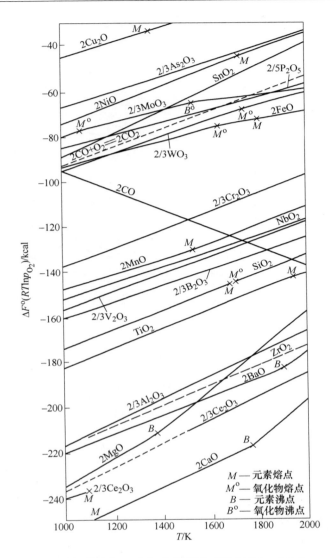

图 1 氧化物的 ΔF° 与 T 关系图（折合到 1mol O_2）

$$\left(\frac{2x}{y}M + O_2 = \frac{2}{y}M_xO_y\right)$$

2 理论计算和分析

2.1 稀土氧化物及硫化物生成顺序的分析

我们以 Ce 为例进行计算。为便于计算，将不同夹杂物在钢液内的溶度积 m 列于表 1。溶度积 m 是该反应平衡常数 k 的倒数。

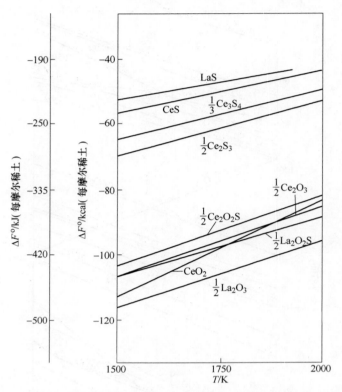

图 2 溶于铁液内 [RE] 和 [S] 及 [O] 生成化合物的 ΔF° 对 T 关系图

表 1 不同夹杂物的溶度积

反 应	溶度积 m（1900K）
$[Ce] + 2[O] = CeO_{2(s)}$	4×10^{-11}
$2[Ce] + 3[O] = Ce_2O_{3(s)}$	3×10^{-21}
$2[Ce] + 2[O] + [S] = Ce_2O_2S_{(s)}$	1.3×10^{-20}
$[Ce] + [S] = CeS_{(s)}$	4.8×10^{-6}
$3[Ce] + 4[S] = Ce_3S_{4(s)}$	6.9×10^{-19}
$2[Ce] + 3[S] = Ce_2S_{3(s)}$	4×10^{-13}
$[Mn] + [S] = MnS_{(s)}$	$0.61(1573K)$
$[Mn] + [S] = MnS_{(l)}$	$4.04(1873K)$

（1）CeO_2 与 Ce_2O_3 生成可能性的比较：

$$2[Ce] + 4[O] = 2CeO_{2(s)}; \quad \Delta F_1^\circ = 2RT\ln m_1 \tag{1}$$

$$2[Ce] + 3[O] = Ce_2O_{3(s)}; \quad \Delta F_2^\circ = RT\ln m_2 \tag{2}$$

式 (1)-式 (2):

$$Ce_2O_3 + [O] = 2CeO_2 \ ; \ \Delta F_3^o = RT\ln \frac{m_1^2}{m_2} \quad (3)$$

$$\Delta F_3 = \Delta F_3^o + RT\ln \frac{1}{a_O}$$

$$= RT\ln \frac{m_1^2}{m_2} + RT\ln \frac{1}{a_O}$$

$$= RT\ln \frac{m_1^2}{m_2 \cdot a_O}$$

若反应 (3) 能够进行,则 ΔF_3 必须为负值,也即 $\frac{m_1^2}{m_2 \cdot a_O} < 1$ 或 $a_O > \frac{m_1^2}{m_2}$。

从表 1 查出 $m_1 = 4 \times 10^{-11}$,$m_2 = 3 \times 10^{-21}$,代入后计算得:a_O 必须大于 0.53。

假定 $a_O \approx [\%O]$,即,$w[O]$ 必须大于 0.53% 时才能生成 CeO_2。

1873K(1600℃)时钢中饱和氧含量为 0.23%,所以在实际情况下,钢水中加入稀土金属 Ce 只能生成 Ce_2O_3,而不能生成 CeO_2。

(2) Ce_2O_3 与 Ce_2O_2S 生成可能性的比较:

$$2[Ce] + 3[O] = Ce_2O_{3(s)} \ ; \ \Delta F_2^o = RT\ln m_2$$

$$2[Ce] + 2[O] + [S] = Ce_2O_2S_{(s)} \ ; \ \Delta F_4^o = RT\ln m_4 \quad (4)$$

式 (4)-式 (2):

$$Ce_2O_3 + [S] = Ce_2O_2S + [O] \ ; \ \Delta F_5^o = RT\ln \frac{m_4}{m_2} \quad (5)$$

$$\Delta F_5 = \Delta F_5^o + RT\ln \frac{a_O}{a_S}$$

$$= RT\ln \frac{m_4}{m_2} + RT\ln \frac{a_O}{a_S}$$

$$= RT\ln \frac{m_4 \cdot a_O}{m_2 \cdot a_S}$$

若反应 (5) 向右进行,则 ΔF_5 必须为负值,即 $\frac{m_4 \cdot a_O}{m_2 \cdot a_S} < 1$ 或 $\frac{a_S}{a_O} > \frac{m_4}{m_2} = \frac{1.3 \times 10^{-20}}{3 \times 10^{-21}} = 4.33$。

假定 [S]、[O] 的活度系数为 1,则:$[\%O] < 0.23[\%S]$。

假如，当钢液中[S] = 0.03%，若$w[O] > 0.0069\%$时生成Ce_2O_3，若$w[O] < 0.0069\%$时，则生成Ce_2O_2S。

（3）用同样的方法可计算CeS与Ce_3S_4以及Ce_3S_4与Ce_2S_3生成的可能性，其结果如下：

$$\frac{1}{3}Ce_3S_4 = CeS + \frac{1}{3}[S]，当 a_0 < 0.006 时 \Delta F 为负值；$$

$$\frac{1}{2}Ce_2S_3 = \frac{1}{3}Ce_3S_4 + \frac{1}{6}[S]，当 a_0 < 0.13 时 \Delta F 为负值。$$

将上列计算的结果汇总为表2。可以看出，在钢液含氧量大于60~90ppm时，加入[Ce]，则生成Ce_2O_3。如果钢液经过脱氧，[O]小于60~90ppm，则生成Ce_2O_2S。此化合物的生成，可以使[O]降低到很小的值。以后[Ce]即进行脱[S]，生成Ce_3S_4（此化合物可以溶解一部分[S]，形成固溶体，所以文献上常写成Ce_3S_{4+x}）。当[S]降到小于0.006%时，再继续脱[S]则生成CeS。所以对脱氧比较好的钢液，加入稀土后，生成夹杂物的顺序一般是：$Ce_2O_2S \rightarrow Ce_3S_4 \rightarrow CeS$。

表2 铈加入钢液后生成的夹杂物（适用于1900K）

反　应	生成物	备　注
[O] > 0.53%	CeO_2	实际上不可能生成
[%O] > 0.23[%S]；在[S] = 0.025% ~ 0.04%、[O]大于60~90ppm时	Ce_2O_3	如有Al_2O_3存在，可能生成$(Ce_2O_3)_x(Al_2O_3)_y$的铝酸物
[%O] < 0.23[%S]；在[S] = 0.025% ~ 0.04%、[O]小于60 ~ 90ppm时	Ce_2O_2S	
0.13% > [S] > 0.006%	Ce_3S_4	
[S] < 0.006%	CeS	
[S] > 0.13%	Ce_2S_3	实际上难以生成

图3是一个实际冶炼操作中，所得到的夹杂物的示意图。可见首先生成的是Ce_2O_2S，然后才是稀土硫化物。这证明了热力学计算的结果与实际情况是吻合的。

由于热力学数据分歧较大，采用不同数据计算出来的结果可能不一样，但总的趋势还是基本一致的。

2.2　稀土加入量的计算

假定钢液脱氧后，[O]小于100ppm，即不再生成RE_2O_3，并假定按以下两

图 3　夹杂物示意图

式进行脱氧、脱硫：

$$2[RE] + 2[O] + [S] = RE_2O_2S$$
$$3[RE] + 4[S] = RE_3S_4$$

根据化学平衡的计算可得（Ce 为例）：

$$稀土加入量 = \frac{自由[RE] + 消耗于RE_2O_2S 及 RE_3S_4 的 RE}{回收率 \times 合金含 RE 的百分数}$$

$$= \frac{自由[RE] + \frac{2 \times 140}{32}[O] + \frac{3 \times 140}{4 \times 32}([S]-[O])}{回收率 \times 合金含 RE 的百分数} \quad (6)$$

分析钢中 [O]、[S] 含量后，可按上式计算稀土加入量。

上式中的自由 [RE] 如果残留过多，将会造成两种恶果：一是水口结瘤，二是钢材产生"热脆"。

水口结瘤问题：水口结瘤的重要原因之一是稀土与耐火材料起作用。钢液中的稀土夹杂物，总的趋势是上浮的。钢液中的 RE 将与水口砖中的 SiO_2 发生如下反应：

$$2[RE] + [S] + SiO_2 = RE_2O_2S + [Si]$$

水口砖一般有空隙，其温度比钢液低，生成的稀土夹杂物在空隙中凝固、生根，成了以后生成夹杂物的核心，特别是当稀土加入时烧损大，钢内稀土夹杂大，在新生的核心上慢慢堆积长大，造成结瘤，最后堵塞水口（见图4）。

从自由能图看出，黏土砖易被 RE 还原。是否可考虑改用更为稳定的水口砖，如镁砖、铝镁砖、熔融刚玉砖、氧化锆砖或含 Al_2O_3 大于 90%的高铝砖。

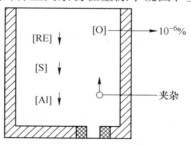

图 4　水口硅结瘤示意图

此处，水口砖的表面质量要考虑。熔融的材质，几乎没有空隙，表面光滑，夹杂物不易附着。

熔融石英水口虽不结瘤，但水口扩大，说明稀土侵蚀生成的夹杂全部进入钢中。

钢中 Al 过多，加剧了水口结瘤，其反应是：

$$4[Al] + 3SiO_2 = 2Al_2O_3 + 3[Si]$$

总之，稀土加入量要适当，即除了生成氧硫化物、硫化物之外，只要有微量自由 RE 起合金化作用就可以了。否则，自由 [RE] 残留过多势必造成结瘤。

关于"热脆"问题：从 La-Fe 相图（图 5）看出，熔体冷却到 785℃时，将有低熔点共晶体在晶界析出。在轧制过程中，升温到 785℃时，晶界上 La-Fe 共晶物开始熔化，从而造成"热脆"。

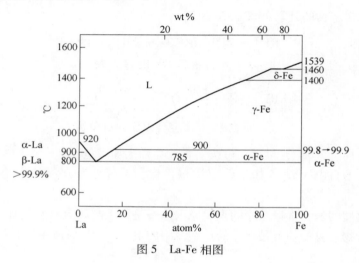

图 5　La-Fe 相图

从 Ce-Fe 相图（图 6）看出，在含少量 Ce 的铁液冷却后其组织为 $CeFe_5$ 及 α-Fe。但这只有在缓慢冷却的平衡条件下，才能有此种组织。但在一般实际冷却条件下，仍有少量熔体在 640℃转变成 $Ce-CeFe_2$ 共晶体，在晶界析出，引起"热脆"。同时 $CeFe_5$ 在 1060℃分解，析出液体，也要促使在低温进行轧制，在高温轧制将要产生裂断现象。

3　稀土的加入方法

国外钢中稀土加入方法有炉内法、包内法和模内法。加入方式包括钢流冲入法、吊挂法、插入法以及金属弹射入法等。采用的稀土合金 1 号合金和混合稀土金属。

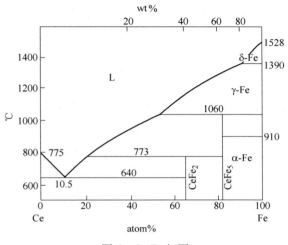

图 6　Ce-Fe 相图

表 3 列出了国外各种稀土加入方法 RE 回收率和成本的比较。

表 3　稀土的回收率及成本

方　法	硅铁稀土（30%RE，密度 5.5g/cm³）		混合稀土（96%RE，密度 6.7g/cm³）	
	回收率/%	成本/美元·t⁻¹	回收率/%	成本/美元·t⁻¹
钢包脱氧，加稀土	3~25	10	—	—
炉内脱氧，钢包加稀土	10~30	8	—	—
钢包脱氧，中性气体搅拌中加稀土	20~40	6	40~65	5
锭模加稀土	20~48	4.5~8	28~58	5.7
换钢包后加稀土	25~50	5~7	—	—
稀土捆包后插入钢包	—	—	55~80	4.3
锭模吊挂	—	—	60~85	3.5
锭模插入	—	—	65~95	3

　　从表中看出模内法有最高的回收率。尽管混合稀土金属较贵，但最后成本却很低（硅铁稀土每公斤价格是 3.3 美元，而混合稀土每公斤价格是 6.4 美元）。

　　硅铁稀土由于密度小（5.5g/cm³），所以易于上浮，因此模中加入硅铁稀土合金易生成皮下夹杂。

　　我们加稀土似以用混合稀土为宜。现在生产的 1 号合金，可考虑改生产 4 号合金用于铸铁中的球化剂，而以 60% 的稀土精矿生产混合稀土金属供炼钢使用。

稀土合金的回收率越低，加入量越大。而稀土回收率低，主要是由于加入过程中稀土的烧损所致。提高稀土的回收率有如下途径：

（1）钢水脱氧后再加稀土。尽管稀土是强的脱氧剂，但是未经铝脱氧时加稀土作脱氧剂是不经济的。钢水中的氧量高要生成 Ce_2O_3，其密度为 $6.86g/cm^3$（见表4），难于上浮。所以加稀土前钢中 [O] 应降到100ppm以下为宜。

表4 不同夹杂物密度和熔点

物质	密度/$g \cdot cm^{-3}$	熔点/℃
铁	7.86	1535
铁水	7.2	—
Ce_2O_3	6.86	1692
CeO_2	7.2	—
La_2O_3	6.51	2305
Ce_2O_2S	6.0	1950
La_2O_2S	5.87	1940
CeS	5.98	2100
LaS	5.86	1970
Ce_2S_3	5.20	2150
Ce_3S_4	—	2050~2075
Al_2O_3	4.0	2030
MnS	4.0	1530

注：密度指该物质密度和水密度（$1g/cm^3$）之比值。

（2）减少稀土与渣的接触。出钢时要尽量少放渣，并且在出钢后静置一段时间再加稀土，使混入钢中的渣子有时间上浮，以减少稀土与渣作用。

（3）尽量避免二次氧化。要真正得到高质量合金钢，采用氩气保护是必要的。这将获得钢质量的改善，减少钢中的夹杂物，同时又提高稀土的回收率。

（4）固定工艺稳定操作。每个车间应根据自己的情况稳定稀土的回收率。回收率不准很容易造成加入过量的稀土，不只产生水口结瘤，甚至导致热脆。在轧制过程中产生裂纹。因此，需要在工艺稳定的条件下找到稀土回收率的值（或找到一个范围），以达到比较准确地计算稀土加入量的目的。

还有一个问题需要加以说明。一些资料中提到稀土置换 MnS 生成稀土硫化铈，这种提法是不科学的。因为在高温下难以有 MnS 生成，MnS 是钢水冷却过程中形成的。根据 $MnS_{(l)}$ 的溶度积在1873K时为4.04（见表1）。到1900K时要高一些，这样大的溶度积实际上根本不可能达到，因此在高温下 MnS 不可能生成。实际上是钢水中 [Ce] 和 [S] 直接反应生成稀土硫化铈。不存在置换 MnS 的问题。

4 物理化学研究中的有关问题

4.1 热力学数据分歧很大

现将 Ce、La 溶于铁水中的标准自由能及 γ^o 数据分别列于表 5 及表 6。

表 5　Ce、La 溶于铁水中的标准自由能及 γ^o（1600℃）

元素	溶解自由能 ΔF^o /cal	γ^o	备注
Ce	—	4×10^{-3}	Richerd　1962
Ce	$-8340-11.0T$	1.05	Teplitskii　1972
Ce	$-13000-11.0T$	0.03	Sigworth　1974
Ce	$-5350-10.76T$	0.268	Kulikov　1975
Ce	$-9.16T$	2.5	Vahed & Kay　1976
La	—	1	Richerd　1962
La	$-5.5T$	15.7	Vahed & Kay　1976
La	$-7000-10.64T$		Kulikov　1975

表 6　Ce 的脱氧常数或 Ce_2O_3 的溶度积 （1600℃）

$2[Ce]+3[O]=Ce_2O_{3(s)}$ $a_{Ce}^2 \cdot a_O^3$	来源	备注
3.5×10^{-11}	Leary，1968	30%SiO_2 黏土坩埚
1.6×10^{-13}	Tucker，1968	MgO 坩埚
1×10^{-20}	Vahed & Kay，1975	—
3×10^{-21}	Vahed & Kay，1976	
9.38×10^{-13}	Janke & Fischer，1978	Ce_2O_3 坩埚

表 5 中元素溶于铁液中 $M_i \rightarrow [M_i]$ 的溶解自由能 ΔF_i^o 与 T 的关系是：

$$\Delta F_i^o = RT\ln\gamma_i^o \frac{55.85}{100M_i}$$

式中，M_i 为元素的原子量。

[O] 的溶解自由能有大家公认的数据，比较准确。但 [Ce] 的溶解自由能差别很大。此外，通过实验直接测定稀土的脱氧常数，表 6 列出分歧的数据。可以看出，此脱氧常数与坩埚材料有关，这说明稀土与坩埚材料发生反应，达不到真正的平衡。数据最大为 10^{-11}，最小为 10^{-21}，最大与最小相差 10^{10} 倍。1978 年最新的数据是 9.38×10^{-18}，是用 Ce_2O_3 坩埚测定的，看来比较可靠。

4.2 分析手段未解决

对于 RE、O、Al 热力学计算上需要的是自由 [RE]、自由 [O]、自由 [Al]。用固体电解质电池直接定氧,可测出自由氧的活度,这已成为当前钢铁研究工作的三大成就之一。其他定氧法如真空熔融法、库仑法、脉冲红外法、中子活化法等等定的都是全氧,即包括了夹杂中的氧。自由 [Al] 可用酸溶法测定,而与全 Al 分开。但是对于自由 [RE] 的分析,目前尚未获得解决。

4.3 高温试验技术上的困难,造成数据上有很大误差

综上所述,由于热力学数据上的分歧,造成计算结果数据不一致。数据的分歧影响图 1 中 Ce_2O_3 自由能线的位置。有的数据将该线大幅度地下移,说明 [Ce] 还原 MgO、ZrO_2 及 Al_2O_3 的能力比较强。在选用数据时,最好不要同时选用多种来源的数据。在计算过程中,如果始终使用同一组数据,所得结论在总的趋势上是比较可信的。

5 小结

(1)稀土钢水口结瘤的重要原因之一是由于钢中自由稀土、自由铝与耐火材料相互作用造成的。所以减少钢中自由稀土和自由铝含量是克服水口结瘤的根本措施。当稀土加入时烧损太大,钢中夹杂太多时,在通过水口时,更容易产生结瘤。

(2)解决稀土钢水口结瘤还可从改变水口砖材料及表面性质的途径加以解决。国外采用 ZrO_2 水口砖或熔融石英水口砖,并通过固体电解质直接测定钢水中氧含量,以控制钢水中的 [O]、[Al] 含量的办法,解决了连铸中水口堵塞问题。稀土钢的结瘤问题也可以考虑用 ZrO_2、镁砖、铝镁砖、刚玉等材质的水口砖,并可做成这些材料的环形或管状的复合水口。熔融石英作水口砖,虽然不结瘤,但水口扩大,侵蚀的水口材质顺流而下,进入钢液增加了钢中杂质,因此不是一个好办法。

(3)稀土加入方法在现阶段应尽量稳定工艺和操作条件,要尽量避免二次氧化。关于稀土加入方法的选择问题,建议对高质量的合金钢考虑采用模内法,而对一般钢可采用包内压入法或插入法,最好采用金属弹射入法。但必须改变目前包内随意投入的加入方式。

此外加入稀土,尽可能使用成分高的混合稀土金属。1 号合金品位低,加入量又太大,在包中加入也不太合适,可制作 4 号合金在生产球墨铸铁中使用。建议有关部门以含稀土 60% 的包头稀土精矿作原料组织混合稀土的生产。

(4)今后加强研究,通过电解、同位素等方法解决测定钢中自由稀土的问

题；加强电子探针定量检测稀土夹杂物等技术；采用固体电解质电池测定并控制钢液含氧量。

（5）钢中应保持少量的溶解态稀土，以达到合金化作用。目前，稀土在固态铁中的溶解度数据存在较大的分歧。国内应通过金属物理、热处理和金属学等研究方法，研究确定在固态铁中稀土的溶解度问题，并进一步研究明确稀土的合金化作用及机理。

在中国国家自然科学基金的资助下，中国的冶金学者在 20 世纪 80~90 年代对稀土在金属溶液中的物理化学进行了系统研究。

含钒铁水炼钢的一些物理化学问题[*]

魏寿昆

来攀钢只有几天,时间很短,学习的很不够。现仅就含钒铁水炼钢的一些物理化学问题提出讨论,错误地方一定很多,请批评指正。

关于提钒问题,拟提出三点来讨论:
(1) 各种提钒方法的初步比较;
(2) 提钒机理及提高钒渣质量的建议;
(3) 从钒渣直接冶炼钒铁的问题。

关于半钢冶炼问题,拟提出三点来讨论:
(1) 热源问题;
(2) 脱硫问题;
(3) 造渣问题。

1 各种提钒方法的初步比较

各厂提钒所用的铁水成分不同。例如铁水含 V 量以承钢为最高,一般为 0.45%,马钢铁水含 V 量最少,一般为 0.2%~0.3%,而攀钢铁水含 V 一般为 0.4%。马钢铁水还有一特点,即 Si 高（0.5%~1.0%）和 P 高（0.6%~1.0%）。由于铁水含 V 量不同,V 渣含 V_2O_5 量也有高低。比较各种提钒方法的优缺点,最好排除所用原料不同的因素而讨论。下面只从设备能力及经济观点进行分析。

1.1 雾化法（攀钢采用）

优点：(1) 设备简单；(2) 基建投资少；(3) 效率高,处理铁水量大,最大可达 280t/h,80~100t 铁水 20min 即可处理完毕。

缺点：(1) 铁水流量难以控制,这包括两方面。第一,铁水从中间罐流向雾化器,开始时铁水面高流的快,快流完时,铁水剩的不多,流的慢；第二,中间罐的水口再用几炉后被局部堵塞,铁水流量和新的水口就不同,铁水流量不能控制,因而影响了钒的氧化。(2) V 渣 FeO 含量很高,一般平均为 54%~56%（也即全 Fe 40%~42%）。(3) 渣中的金属铁含量比较高,一般平均为渣量的 20%~25%（波动范围为 11%~45%）。以上两点都是由本法设备所造成的。铁水用压缩空气雾化,造成 FeO 含量很高。雾化器本身没有熔池,渣和提 V 后的半钢流

---

[*] 攀钢学术报告会讲稿,攀钢技术质量处印,1976,5。

入半钢罐，后者温度比较低，流出的半钢通过黏稠的 V 渣汇集在一起，因之有相当大量的铁珠留存在渣内。（4）中间罐炉衬寿命低，一个雾化器需要五个中间罐。

1.2 槽式法（马钢采用）

优点：（1）设备简单，比雾化法还要简单。（2）上马快，投资少。（3）处理铁水量也比较大，1min 处理铁水 1t。

缺点：（1）铁水流量难以控制。这和雾化法一样，中间罐铁水面和水口断面在变化，使初期的铁水流速大，而后期的流速小。（2）半钢的余 V 较高，一般为 0.1%，这是因为槽内的氧化带较短所造成。（3）V 渣内 FeO 含量高，一般平均为 40%（全铁约 30%）。金属 Fe 为渣量的 18%~20%，也比较高。（4）中间罐炉衬寿命低，槽式炉风眼砖改用铬砖后寿命为 60~70 炉，也不高。

1.3 转炉法（承钢及马钢采用压缩空气侧吹，马钢在试验纯氧底吹）

优点：（1）通过调整枪位的高低（顶吹），或通过面吹、浅吹或深吹（侧吹）可控制 V 渣的 FeO 含量，底吹转炉 V 渣的 FeO 含量最低。V 渣内的金属 Fe 含量也低。

V 渣的 FeO 和金属 Fe 含量如下：

	纯氧顶吹	空气侧吹	纯氧底吹
FeO 含量/%	20~25	30~44	10~15
金属 Fe 占渣量的百分比/%	15	13~20	5~10

（2）炉龄长：顶吹（换用高铝砖）500~600 炉；侧吹 700 炉；底吹 600 炉。

（3）生产能力大，适用于大规模生产。

缺点：占用一个炉子的炼钢设备，这是一个比较重大的缺点。

1.4 摇包法（国外有厂采用，国内用于提 Cr）

优点：（1）提 V 效率高，可大于 90%。（2）用空摇或加碳粉空摇可以降低 V 渣的 FeO 含量。

缺点：（1）设备复杂。（2）摇包不能太大，为 50~60t，太大则使摇动设备复杂化。

我们怎样选择提 V 方法呢？个人意见是根据各厂具体条件，因地制宜，没有必要都用统一的提 V 方法。例如攀钢场址设在山坡之上，场地面积比较小，发展受到限制。腾出一个转炉以炼钢而用雾化法来提 V 是比较合适的。马钢采用多种设备来提 V，用槽式炉在高炉前处理低 V 铁水，而用侧吹处理含 V 较高的铁水，同时又进行底吹纯氧转炉提 V 的试验，为更大规模的生产做准备。承钢 V 渣质量在全国首屈一指，侧吹转炉提 V 的经验丰富，也没有必要改用其他提 V 的方法。

2 提钒机理及提高钒渣质量的建议

2.1 提钒保碳问题

大家都知道，为了将生铁里的 V 提出来，而又同时保住生铁的 C 不大量被氧化，以便在半钢冶炼时能有足够的热源，吹炼生铁时必须在低温进行。为什么要在低温吹炼？低温低到多少？这在理论上涉及元素在铁液内的氧化顺序问题。

对纯氧化物，人们经常采用氧化物的分解压来判断元素氧化的顺序。但对生铁水（或钢水）来说，用分解压这个名词不大适宜，最好采用以 1mol O_2 为标准的各元素氧化反应的自由能来判断元素氧化的顺序。图 1 绘出常见的元素在铁水内直接用 O_2 氧化的 $\Delta F°$ 曲线，$\Delta F°$ 的标准状态是：一个大气压 O_2，溶在铁水内的 1%元素以及纯氧化物。如果采用间接氧化反应，例如任何元素 M 用 [O] 来氧化：

$$[M] + 2[O] = MO_2$$

则各元素氧化的顺序和图 1 完全相同。

图 1 中 V 氧化生成物是 V_2O_3，这是因为自生铁提钒时在酸性炉衬下进行吹炼，不可能生成酸性强的 V_2O_5。V 渣含 V 量通常按 V_2O_5 计算，这是一种习惯，实际上 V 以 V_2O_3 形式在冷却后与 FeO 化合成为钒尖晶石 $FeO \cdot V_2O_3$。

从图 1 可以看出以下几点：

（1）任何反应的 $\Delta F°$ 为正值时即不能自动进行，所以铁水的 [Cu] 在一般冶炼的温度下不能被氧化。Ni 较高的冶炼温度也不能被氧化。

（2）某化学反应的 $\Delta F°$ 值负值越大，则该反应越容易进行，根据这一原则就可以判断元素的氧化顺序。例如在 1400℃时（暂时不考虑 C 的氧化），铁水内所含元素的氧化顺序是：Al、Ti、Si、V、Nb、Mn、Cr、Fe。

（3）$\Delta F°$ 曲线较低的元素可以将 $\Delta F°$ 曲线较高的元素从它的氧化物还原出来，例如：

$$[Si] + \frac{2}{3}V_2O_3 = \frac{4}{3}[V] + SiO_2$$

（4）C 有它的特殊性，其他所有元素的 $\Delta F°$ 线都是向右上方倾斜，也即其他所有元素的 $\Delta F°$ 值都随温度上升而减小，唯独 C 则不然，它的 $\Delta F°$ 曲线是向右下方倾斜，也即 C 的 $\Delta F°$ 值随温度上升而增加。

不难看出 C 氧化的 $\Delta F°$ 线和其他某些元素氧化的 $\Delta F°$ 曲线有一相交点。图 2 画出 V 和 C 氧化的 $\Delta F°$ 曲线，它们相交点的温度是 T。按还原反应来看，当温度

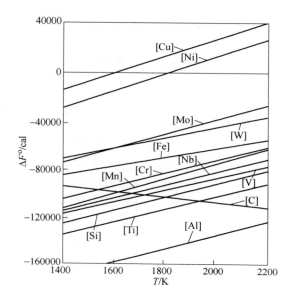

图1 铁液内各元素直接氧化的标准自由能

铁液内各元素直接氧化的反应式：[Cu]：$2[Cu]+O_2=2Cu_2O_{液}$；
[Ni]：$2[Ni]+O_2=2NiO_{固}$；[Mo]：$[Mo]+O_2=MoO_{2固}$；[W]：$\frac{2}{3}[W]+O_2=\frac{2}{3}WO_{3液}$；
[Fe]：$2Fe_{液}+O_2=2FeO_{液}$；[Cr]：$\frac{4}{3}[Cr]+O_2=\frac{2}{3}Cr_2O_{3固}$；[Mn]：$2[Mn]+O_2=2MnO_{固}$；
[Nb]：$[Nb]+O_2=\frac{1}{2}Nb_2O_{4固}$；[V]：$\frac{4}{3}[V]+O_2=\frac{2}{3}V_2O_{3固}$；[Si]：$[Si]+O_2=SiO_{2固}$；
[Ti]：$[Ti]+O_2=TiO_{2固}$；[Al]：$\frac{4}{3}[Al]+O_2=\frac{2}{3}Al_2O_{3固}$；[C]：$2[C]+O_2=2CO$

$>T$时，[C]可以从V_2O_3中把V还原出来：

$$2[C]+\frac{2}{3}V_2O_3=\frac{4}{3}[V]+2CO$$

所以T是C能还原V的氧化物的"最低还原温度"。按氧化反应来讲，当温度$<T$时，由于[V]氧化的ΔF^o曲线低于[C]氧化的ΔF^o曲线，所以[V]先被氧化，而[C]可被保住不动；当温度$>T$时，[C]能先被氧化而[V]不动。所以T是[V]和[C]的"氧化转化温度"。我们

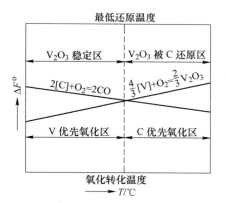

图2 最低还原温度和氧化转化温度

从生铁吹炼提V，就是要去V保C，所以吹炼温度不应高于转化温度T。

2.2 氧化反应的转化温度的计算

氧化反应的转化温度怎样去计算呢？从图 2 可以看出，在转化温度 T，两个氧化反应的 ΔF° 是相等的。

$$\frac{4}{3}[V] + O_2 =\!=\!= \frac{2}{3}V_2O_3 ; \quad \Delta F^\circ = -186400 + 50.53T \tag{1}$$

$$2[C] + O_2 =\!=\!= 2CO ; \quad \Delta F^\circ = -66600 - 20.32T \tag{2}$$

$$-186400 + 50.53T = -66600 - 20.32T$$

$$-119800 + 70.85T = 0$$

所以 $T = 1690K(1417℃)$

这个转化温度是利用标准自由能 ΔF° 计算的；也就是说，它是在 1% [V] 生成纯 V_2O_3 和 1% [C] 生成 1 个大气压 CO 的条件下的转化温度。但是生产实践中我们用的生铁含 [V] 远远高于 1%，而 [V] 经常低于 1%，生成的 V_2O_3 是熔于炉渣之内而非纯 V_2O_3，CO 一般可认为在 1 个大气压左右。对于这些具体情况，我们必须进行具体分析。这样我们不能用 ΔF° 进行计算，而必须用化学反应等温方程式来计算 ΔF，用后者以计算转化温度。

$$\frac{4}{3}[V] + O_2 =\!=\!= \frac{2}{3}V_2O_3$$

$$\Delta F = \Delta F^\circ + RT\ln\frac{(\gamma_{V_2O_3}N_{V_2O_3})^{\frac{2}{3}}}{\{f_V[\%V]\}^{\frac{4}{3}}P_{O_2}}$$

$$= -186400 + 50.53T + RT\ln\frac{(\gamma_{V_2O_3}N_{V_2O_3})^{\frac{2}{3}}}{\{f_V[\%V]\}^{\frac{4}{3}}P_{O_2}}$$

式中　$\gamma_{V_2O_3}$——炉渣中 V_2O_3 的活度系数；

$N_{V_2O_3}$——炉渣中 V_2O_3 的摩尔分数；

f_V——铁水中 V 的活度系数；

P_{O_2}——O_2 的压力。

$$2[C] + O_2 =\!=\!= 2CO$$

$$\Delta F = \Delta F^\circ + RT\ln\frac{P_{CO}^2}{\{f_C[\%C]\}^2 P_{O_2}}$$

$$= -66600 - 20.32T + RT\ln\frac{P_{CO}^2}{\{f_C[\%C]\}^2 P_{O_2}}$$

式中　P_{CO}——生成的 CO 的压力，一般为 1atm；

f_C——铁水中 C 的活度系数。

将两式的 ΔF 列为相等且考虑到两个反应遇到同一的 O_2，因之两式的 P_{O_2} 相消。又考虑到 $P_{CO}=1atm$，简化后得：

$$RT\ln\frac{(\gamma_{V_2O_3}N_{V_2O_3})^{\frac{2}{3}}\{f_C[\%C]\}^2}{\{f_V[\%V]\}^{\frac{4}{3}}} = 119800 - 70.85T \tag{3}$$

式（3）就是具体情况下计算转化温度的公式。

通过式（3）可以看出，转化温度与氧化时所用 O_2 气的压力无关。同样地可以证明，转化温度又与氧存在的形式，无论是 [O] 还是 (FeO) 都无关。所以无论氧化是直接氧化还是间接氧化，转化温度都是一样的。

铁水内元素的活度系数 f 可按通常采用的方法计算。所用的活度相互作用系数的数据是：

$e_C^C = 0.22$; $e_V^V = 0.02$

$e_C^{Si} = 0.107$; $e_V^C = -0.17$

$e_C^V = -0.038$; $e_V^{Si} = 0.27$

$e_C^P = 0.042$; $e_V^P = -0.008$

$e_C^{Ti} = -0.08$; $e_V^{Ti} = 0$

炉渣内 V_2O_3 的活度系数 $\gamma_{V_2O_3}$ 文献上没有资料。V_2O_3 在炉渣内一般认为结合为尖晶石 $FeO\cdot V_2O_3$，所以它的活度系数很小。我们可以估计 $\gamma_{V_2O_3}=10^{-5}$。

根据马钢及攀钢现场数据利用式（3）对转化温度我们做了一些计算，其结果见表 1。

表 1 提 V 的转化温度

马钢 I		马钢 II	
%V = 0.4	$f_V = 0.346$	%V = 0.12	$f_V = 0.243$
%Si = 0.8		%Si = 0.09	
%C = 4.0	$f_C = 9.45$	%C = 3.73	$f_C = 7.09$
%P = 0.6		%P = 0.62	
$N_{V_2O_3} = 0.041$	$\gamma_{V_2O_3} = 10^{-5}$	$N_{V_2O_3} = 0.041$	$\gamma_{V_2O_3} = 10^{-5}$
$T = 1410℃$		$T = 1340℃$	
攀钢 I		攀钢 II	
%V = 0.1	$f_V = 0.255$	%V = 0.059	$f_V = 0.255$
		%Ti = 0.024	
%C = 3.5	$f_C = 5.84$	%C = 3.5	$f_C = 5.83$
$N_{V_2O_3} = 0.07$	$\gamma_{V_2O_3} = 10^{-5}$	$N_{V_2O_3} = 0.0858$	$\gamma_{V_2O_3} = 10^{-5}$
$T = 1350℃$		$T = 1320℃$	

这里对转化温度的意义再进一步明确。以马钢 I 为例：当生铁含 C 4.0%、

含 V 0.4%以及一些其他杂质时，如果让 C 维持在 4.0%而不被氧化，0.4%的 V 能够被氧化，则吹炼时熔池温度绝不能大于 1410℃；如果吹炼温度>1410℃，则 C 即保不住而被氧化。但实际上开始吹炼时，温度一般都低于 1410℃，可是总有一小部分 C 被氧化，只是被氧化的 C 量很少而已。为什么 C 终能被氧化一小部分呢？这不是由于热力学的原因而是由于动力学的原因。计算转化温度的热力学条件是：4%的 C，0.4%的 V，还有大约 94%的 Fe（Si、P 暂不考虑），在熔池内各自形成一个独立体，如果它们遇到了 O_2 气，当温度小于 1410℃时，则只有 V 优先被氧化，C、Fe 维持不动。但实际上熔池内的情况不是这样，熔池内大部分是 Fe 原子，而 V、C 原子均匀地分散在铁原子之中，而且 O_2 气只能和熔池在表面层接触。当熔池表面层含有 C、V 原子（当然还有大量的 Fe 原子）时，一遇 O_2 气，首先被氧化的是 V。当 O_2 气继续喷射时，如果熔池内部的 V 原子来不及扩散到表面层，后者上面如有 C 原子，C 即被氧化；如果 C 原子也没有，则 Fe 将被氧化。所以熔池实际的动力学条件造成一小部分 C 和 Fe 也被氧化。

表 1 中有三个例子是计算半钢的转化温度（马钢Ⅱ，攀钢Ⅰ及攀钢Ⅱ）。计算半钢的转化温度比计算生铁水的转化温度更为有用。我们要求半钢的余 V 低，而考虑到半钢吹炼的热源，半钢含 C 不能过低。规定出半钢的余 V 和 C 的适当成分，按计算出的转化温度以控制吹炼温度更为有利。

由上例转化温度计算的结果可以得到以下结论：

（1）转化温度不是一成不变的温度。铁水及炉渣成分一变，转化温度即随之而变。

（2）余 V 越低，保 C 越难。如果想得到余 V 很低的半钢，在吹炼开始时就应维持较低的吹炼温度。

（3）吹炼温度一般以不高于 1350℃为宜。如生铁含 Si 过高，吹炼过程中必须加冷却剂。

2.3 雾化提 V 的特点

通过雾化器压缩空气将铁水分散为很多铁水珠，造成大量 Fe 的氧化。根据生产实践，铁水中 V 的 60%在雾化炉内氧化，40%在半钢罐中氧化；而 C 则在雾化炉及半钢罐各降半。在雾化炉中直接氧化反应及间接氧化反应都有，在半钢罐中则主要是间接氧化反应：

$$(FeO) \longrightarrow Fe_{液} + [O]$$

$$\frac{4}{3}[V] + 2[O] \longrightarrow \frac{2}{3}(V_2O_3)$$

$$2[C] + 2[O] \longrightarrow 2CO$$

这种间接氧化反应的转化温度和上面直接氧化反应在相同铁水及炉渣成分时

是一样的。

攀钢和马钢两种 V 渣有何区别？取两个现场数据加以计算，结果见表 2。

表 2 炉渣组成物的计算   （%）

项 目	攀 钢	马 钢
V_2O_5	16.60（重量）	8.36（重量）
TFe	42.25	42.93
SiO_2	12.84	19.16
铁橄榄石 $2FeO \cdot SiO_2$	69.4（摩尔）	84.6（摩尔）
钒尖晶石 $FeO \cdot V_2O_3$	22.5	12.2
磁铁矿 $FeO \cdot Fe_2O_3$	8.1	3.2

可以看出，马钢 V 渣含铁橄榄石较多，钒尖晶石较少。这主要是由马钢生铁含 Si 高而 V 低所造成的。攀钢的 V 渣含磁铁矿较高，反映了雾化提 V 生成大量 FeO 的特点。这种粗略的计算只能提供一个估计，更准确的判断有赖于岩相的大量分析工作。

2.4 提高 V 渣质量的建议

欲使 V 渣 V_2O_5 量高，除 SiO_2、CaO 及 P_2O_5 含量应降低外，必须降低 FeO 含量及金属 Fe 含量。SiO_2 及 P_2O_5 取决于生铁的含 Si、P 量，这方面对攀钢问题不大，因攀钢矿石含 P 低，而高炉又采取低温低硅的操作。攀钢铁水含 Ti 也不高，对 V 渣影响也不大。采用含 CaO 少而致密耐冲刷的炉衬（优质镁砖或铬镁砖）可减少 V 渣的 CaO 含量。

攀钢 V 渣最大的问题是 FeO 及金属 Fe 都很高（渣内全 Fe 达 40%~42%，外有金属 Fe 20%~25%，V 渣本身是一个很富的铁矿）。除在翻半钢时尽量将半钢水翻尽外，建议采取下列措施以减低 FeO 及金属 Fe 含量：

（1）半钢罐的半钢尽量翻尽后，V 渣适当用煤气加热，加入一定量的碳粉，使 V 渣中的 FeO 得以还原。如能在半钢罐装入摇动装置则更好。

（2）扩大现有的精整车间，增加细磨、磁选设备。除大块金属 Fe 用人工选出外，含小铁珠的 V 渣经过细磨磁选，分离出的铁质经烧结后进入高炉。这样既提高了 V 渣质量，又节省了运费。

3 由钒渣直接冶炼钒铁问题

攀钢 V 渣运送到东北锦州铁合金厂以冶炼钒铁，需要经过下列几道工序：V 渣→磨细→磁选去铁→回转窑氧化焙烧（得 Na_3VO_4）→水浸→加酸沉淀→过滤→干燥→熔化→电炉冶炼钒铁。

为了节省运费，降低成本，攀钢同有关部门进行从 V 渣直接冶炼钒铁的试验。试验在电炉内分两步进行：（1）预还原去 Fe；（2）冶炼钒铁。试验的结果是很乐观的。不只含 V_2O_3 较高的雾化 V 渣可以直接冶炼钒铁，即使含 V_2O_3 较低（约8%）的在转炉吹炼半钢所得的炉渣也能在电炉用作直接冶炼钒铁的原料，炼出的钒铁含 V 最高可达 30%。

下面讨论两个问题：

（1）预还原温度问题。V 渣在电炉进行预还原，其目的是将混在 V 渣中的大量铁珠（实际是半钢）熔化，同时用 C 粉将 V 渣内大量的 FeO 还原成 Fe，以提高 V 液的 V：Fe 比，使 V 渣的 V_2O_5 含量提高，而 FeO 含量大为降低。但还原温度不宜太高，过高则渣内的 V_2O_5 也被还原，造成 V 的损失。图 3 绘出还原温度的范围在 $T_1 \rightarrow T_2$ 之间，根据 V 渣及半钢的成分 T_2 在 1400℃左右。为了提高 V 渣的流动性，可适当地加入一定量的石灰。

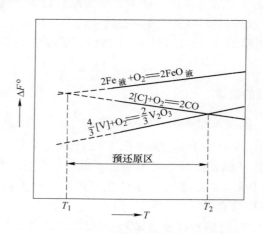

图 3　冶炼钢铁预还原温度区

（2）如何避免或减少钒铁含 Ti 的问题。去 Fe 后的富集钒铁除含 V_2O_3 及 FeO 外，尚含有 P_2O_5、Cr_2O_3、MnO、SiO_2、TiO_2、Al_2O_3、MgO 及 CaO。加入的还原剂是 C 粉、硅铁或 Al。无论用哪一种还原剂，P、Cr 及 Mn（当然 V 也在内）都将被还原成为铁合金，而 Al_2O_3、MgO 及 CaO 不能被还原而进入炉渣。用 C 作还原剂，小部分 SiO_2 被还原，TiO_2 很可能不被还原（参阅图 1），所以铁合金不含 Ti 或含 Ti 很少。用 C 作还原剂的缺点是钒铁含 C 较高不适用于炼低 C 含 V 的合金钢。用 Al 作还原剂肯定的钒铁将含较大量的 Ti，而 Ti 在铁合金内是不受欢迎的元素。用硅铁作还原剂我们能不能阻止 Ti 不被还原进入铁合金呢？

$$[Ti] + O_2 = TiO_{2固}; \quad \Delta F^\circ = -210400 + 52.25T \quad (4)$$

$$[Si] + O_2 = SiO_{2固} \quad \Delta F^\circ = -198000 + 53.59T \quad (5)$$

式（5）- 式（4）　　$[Si] + TiO_{2固} = [Ti] + SiO_{2固}; \quad \Delta F^\circ = 12400 + 1.34T \quad (6)$

从图 1 和式（6）的计算可以看出 Si 是不能还原 TiO_2 的，一则因为 Si 氧化的 ΔF° 线在 Ti 氧化的 ΔF° 线之上，二则根据式（6）Si 还原 TiO_2 反应的 ΔF° 在任何温度下都是正值。但是图 1 画出的和式（6）计算出的自由能 ΔF° 是标准状态下的自由能，它反映的情况是：铁水内 1% 的 Si 还原纯固态 TiO_2 得到含 1%Ti 的铁水和纯固态 SiO_2。但是我们具体的情况则是：向钒渣内加入硅铁，金属熔体含 Si 远大于 1%，同时 SiO_2 及 TiO_2 都是在炉渣之内而不是纯物质。我们试按下列假定作出两种情况的粗略的估计。

（1）采用 75%Si 的硅铁，暂不考虑 Si 的活度系数 $a_{Si}=75$；还原所得的 Ti 为 1%，也不考虑 Ti 的活度系数 f_{Ti}，$a_{Ti}=1$。假定 $a_{SiO_2}=0.15$，$a_{TiO_2}=0.20$。计算式（6）在 1600℃ 时的 ΔF：

$$\Delta F = \Delta F^\circ + RT\ln\frac{a_{Ti}a_{SiO_2}}{a_{Si}a_{TiO_2}}$$

$$= 12400 + 1.34T + 4.575T\lg\frac{1\times 0.15}{75\times 0.20}$$

$$= 12400 + 1.34T - 9.15T$$

$$= 12400 - 7.81T \tag{7}$$

当 $T=1873K$ 时，$\Delta F=-2200\text{cal}$（$-9204.8J$）。

这说明虽然 ΔF° 的推断，[Si] 还原 TiO_2 是不可能的，但在上列给定的条件下，[Si] 是可以从炉渣内把 Ti 还原进入铁合金的。

（2）采用 45% 的硅铁，暂不考虑 Si 的活度系数 f_{Si}，$a_{Si}=[\%Si]=45$。还原所得的 Ti 仍为 1%，也不考虑 Ti 的活度系数 f_{Ti}，$a_{Ti}=[\%Ti]=1$。采用炉渣碱度较小，$a_{SiO_2}=0.30$，a_{TiO_2} 仍不变而为 0.20。

$$\Delta F = \Delta F^\circ + RT\ln\frac{a_{Ti}a_{SiO_2}}{a_{Si}a_{TiO_2}}$$

$$= 12400 + 1.34T + 4.575T\lg\frac{1\times 0.30}{45\times 0.20}$$

$$= 12400 + 1.34T - 4.575T\lg 30$$

$$= 12400 - 5.41T \tag{8}$$

当 $T=1873K$ 时，$\Delta F=+2300\text{cal}$（$+9623.2J$）。

这说明在新的给定条件下，硅铁不能将 V 渣中的 TiO_2 还原。

上面的计算是一个极粗略的估计。由于我们未考虑 f_{Si}，同时把原来适用于稀溶液的 ΔF° 扩展应用到浓溶液，我们可能犯大的错误。虽然如此，我们可以通过上面的计算，说明一个很重要的问题，就是元素的氧化还原顺序不是一成不变的，通过适当改变参加反应的物质的浓度把元素氧化还原顺序予以颠倒，并且向着有利于我们意图的方向而颠倒。上面我们企图在还原富集 V 渣的过程中抑制

TiO_2 的还原，以便钒铁内不含或少含 Ti。这样，上列的计算告诉我们，在采取下列措施时，有可能达到我们的目的：

（1）不用 Al 还原。

（2）用含 Si 少的硅铁还原，例如 45%Si 的硅铁。

（3）提高炉渣 SiO_2 含量，减小炉渣的碱度（当然在不增加炉渣的黏度条件下）以提高 a_{SiO_2}。

（4）有可能时适当采用低 TiO_2 含量，也即低 a_{TiO_2} 的炉渣。

上列措施是否可行，取决于大量试验的结果。热力学的计算是否合理，取决于不考虑 f_{Si} 及 f_{Ti} 造成的误差有多么大。有时热力学上原则上没有问题，而动力学的原因却需要考虑。采用弱还原剂可以抑 TiO_2 的还原，但有可能还原速度太慢，还原时间拖的太长，以致工业上难以采用。

最后，再谈几句火法富集 V 渣冶炼钒铁的方法。西德曾采用小高炉冶炼含 V 生铁，在转炉吹成 V 渣后，将此 V 渣返回高炉，得到含 V 较高的生铁。在转炉吹炼后，将较富的 V 渣再送回高炉，如此循环，最后得到含 V 很富的 V 渣，送电炉还原炼成钒铁。这个方法需要专用小高炉及转炉。我们不妨利用适当设备进行试验。

4 半钢冶炼的热源问题

转炉炼钢的特点是热源自给自足。对半钢来说，由于在提钒吹炼时生铁的 Si（Ti 及 Mn）被氧化，半钢的热源只依靠它的含 C 量。马钢生铁含 P 较高（一般为 0.6%），提 V 过程中 P 不被氧化，所以马钢的半钢冶炼热源比较充分，没有不足的顾虑。攀钢的半钢所含的 C 量究竟以多少为适宜，可以通过简单的计算作出估计。

生铁内各种元素每氧化 1%能提高铁水多少温度可采用表 3 的估计值。

表 3　每氧化 1%元素铁水提高的温度

1%元素	用 O_2 吹炼	用空气吹炼
C	85℃	30℃
Mn	55℃	35℃
Si	150℃	100℃
P	190℃	140℃
Fe	30℃	20℃
Cr	90℃	
V	92℃	

上列计算我们作的假定是：渣及一部分炉衬和铁水同时吸收元素氧化放出的

热量，渣量为铁水量的15%，而被加热的炉衬为铁水量的10%。

这样，如果半钢含C 3.5%，吹炼过程中熔池将升温 3.5×85=298℃，即大约300℃。当出钢温度为1620℃时，则开始吹炼时半钢的温度在1320℃即可，而一般情况下半钢有这样的温度是不难做到的。

对上面的估计，我们没有考虑加入石灰造渣的吸热量以及 Fe 被氧化的放热量。可以认为，这两项的热量，粗略地相等而相互抵消。

在吹炼半钢时，如确实感到热量有不足的可能，在转炉内可以适当地加入一些焦炭。承钢吹炼半钢时配入硅铁，但这种做法不一定在攀钢予以提倡。

5 半钢冶炼的脱硫问题

5.1 生铁含硫问题

国标规定，一、二号生铁含 S 0.05%，而三号生铁含 S 0.07%。攀钢生铁认为 S 0.08%即为合格。这是根据本厂的具体情况而规定的。我厂炼铁时因要控制很低的含 Ti 量，因之必然地也要控制很低的含 Si 量，这样高炉采取的低温操作对脱 S 不利，所以才规定了 S 的合格量为 0.08%。但实际情况是生铁含 S 达到 0.1%时很多，有时甚至高达 0.15%。对此问题应予以足够的重视，因为含 S 高的生铁给炼钢带来很大的困难。应该查明原因，是高炉原料出了问题，还是操作不当引起转炉铁水含 S 反常。例如，可以检查一下，在铁包或混铁炉里是否高炉渣不慎混入铁水，以致一并倒入转炉后而使铁水含 S 量反常。

5.2 脱硫和炉渣的碱度

炉渣和铁液含 S 量的分配比反映炼铁及炼钢操作脱 S 的效率，这个硫分配比与炉渣碱度有关：

$$\frac{(\%S)}{[\%S]} = \Phi(R)$$

最简单的炉渣碱度用 $\frac{(\%CaO)}{(\%SiO_2)}$ 来表示，但对冶炼半钢来说，这种表示方法不一定合适。但为工人用着方便，我们也不能把碱度的表示方法规定得太复杂，以致计算时很麻烦。

人们为了使脱 S 效率和炉渣碱度有一定的规律，经常假定炉渣内含有某些化合物。攀钢对于半钢的炉渣假定有下列的化合物：$CaO \cdot SiO_2$、$3CaO \cdot V_2O_5$ 及 $3CaO \cdot 2TiO_2$。这样：

$$\frac{\%CaO}{\%SiO_2} = \frac{56}{60} = 0.93$$

$$\frac{3(\%CaO)}{\%V_2O_5} = \frac{3 \times 56}{182} = 0.93$$

$$\frac{3(\%CaO)}{2(\%TiO_2)} = \frac{3 \times 56}{2 \times 80} = 1.05$$

因之

$$0.93 : 0.93 : 1.05 = 1 : 1 : 1.13$$

半钢炉渣的碱度采用下列式子进行计算：

$$R = \frac{\%CaO}{\%SiO_2 + \%V_2O_5 + 1.13\%TiO_2}$$

但根据相图，$3CaO \cdot V_2O_5$ 和 $3CaO \cdot 2TiO_2$ 两化合物在固态时即分解，这种不稳定的化合物人们认为一般在液相内不可能存在。为了简便起见，似乎规定半钢炉渣碱度

$$R = \frac{\%CaO}{\%SiO_2 + \%V_2O_5 + \%TiO_2}$$

更好些。

5.3 硫分配比的实践和理论计算

生产实践的大量统计提供了下面的硫分配比数值，见表4。

表4 各种冶炼方法的硫分配比

炉子	$\frac{(\%S)}{[\%S]}$	备注
高炉	20~80	平衡时达 100~400
顶吹转炉	5~10	
电炉氧化期	3~8	
电炉还原期	30~50	
平炉	3~8	

在实际操作中，高炉内铁水和炉渣的脱 S 反应未达到平衡，但由于炉渣含 FeO 很低（1%~2%），而 S 在生铁中的活度系数 f_S 可高达 5~6，所以高炉中尚未达到平衡的硫分配比根据化学分析数据计算即达到 20~80，远远高于炼钢的脱 S 能力。所以钢铁中的脱 S 任务应该主要落在高炉上。但对攀钢来讲，由于高炉操作必须抑制 Ti 的还原，因此不可能采取大焦比、高风温的高温操作，S 不能在高炉内更好地脱除，脱 S 任务不得不由炼钢多负一些责任。

生铁如果含 S 太高，对炼钢很不利。任何提 V 的方法都很难脱 S，半钢太高的 S 就需要采取双渣操作，或进行后吹，二者都延长冶炼时间，影响产量。后吹实际上是大量地吹 Fe，钢水中［O］特别多，造成脱氧后夹杂物多而影响成品钢

的质量。所以必要时（也即当S超过0.1%时），可以进行炉外脱S，在铁水包里加入 Na_2CO_3 或吹入 CaC_2 粉，更可采用摇包处理。对攀钢铁水是否可行，尚须做一些实验。

利用炉渣的离子理论，硫分配比可以用下列公式计算：

$$\frac{(\%S)}{[\%S]} = \frac{32 K_S f_S}{\gamma_{Fe} \gamma_S} \cdot \frac{\sum n_+ \sum n_-}{n_{FeO}} \tag{9}$$

式中 K_S——S 的分配常数，在炼钢温度等于 0.08；

f_S——铁水或钢水中 S 的活度系数；

$\sum n_+$——100g 炉渣中阳离子摩尔数之和；

$\sum n_-$——100g 炉渣中阴离子摩尔数之和；

n_{FeO}——100g 炉渣中 FeO 的摩尔数；

γ_{Fe}，γ_S——炉渣中 Fe、S 的活度系数。

上列计算公式适用于高炉及任何炼钢操作，平炉、电炉或转炉，只不过炉渣内所含的离子，应当根据炼铁或炼钢的具体情况作出相应的假定。

对于攀钢的半钢冶炼，假定炉内生成下列的离子：

$$CaO = Ca^{2+} + O^{2-}$$

$$MgO = Mg^{2+} + O^{2-}$$

$$MnO = Mn^{2+} + O^{2-}$$

$$FeO = Fe^{2+} + O^{2-}$$

$$SiO_2 + 2O^{2-} = SiO_4^{4-}$$

$$Al_2O_3 + 3O^{2-} = 2AlO_3^{3-}$$

$$V_2O_5 + 3O^{2-} = 2VO_4^{3-}$$

$$TiO_2 + O^{2-} = TiO_3^{2-}$$

$$\sum n_+ = n_{Ca^{2+}} + n_{Mg^{2+}} + n_{Mn^{2+}} + n_{Fe^{2+}}$$

所以
$$\sum n_+ = n_{CaO} + n_{MgO} + n_{MnO} + n_{FeO}$$

$$\sum n_- = n_{O^{2-}} + n_{S^{2-}} + n_{SiO_4^{4-}} + n_{AlO_3^{3-}} + n_{VO_4^{3-}} + n_{TiO_3^{2-}}$$

生成的
$$n_{O^{2-}} = n_{CaO} + n_{MgO} + n_{MnO} + n_{FeO} = \sum n_+$$

消耗的
$$n_{O^{2-}} = 2n_{SiO_2} + 3n_{Al_2O_3} + 3n_{V_2O_5} + n_{TiO_2}$$

此外
$$n_{S^{2-}} = n_S，n_{SiO_4^{4-}} = n_{SiO_2}，n_{AlO_3^{3-}} = 2n_{Al_2O_3}$$

$$n_{VO_4^{3-}} = 2n_{V_2O_5}，n_{TiO_3^{2-}} = n_{TiO_2}$$

所以 $\sum n_- = \sum n_+ - n_{SiO_2} - n_{Al_2O_3} - n_{V_2O_5} + n_S$

对于 $\gamma_{Fe^{2+}} \gamma_{S^{2-}}$ 采用对于平炉及实验室平衡实验适用的公式：

$$\gamma_{Fe^{2+}} \gamma_{S^{2-}} = 1.53 \sum N_{SiO_4^{4-}} - 0.17 \tag{10}$$

而
$$\sum N_{SiO_4^{4-}} = \frac{n_{SiO_2} + 2n_{Al_2O_3} + 2n_{V_2O_5} + n_{TiO_2}}{\sum n_-} \quad (11)$$

f_S 在吹炼终期可认为等于 1。

采用炉号 7311574 的半钢冶炼的记录,[%S] = 0.031,终点渣的成分计算结果见表 5。

表 5 炉渣组成物的计算

项 目	炉渣		100g 炉渣中各组成物的摩尔数
	原分析	调整到 100%	
CaO	54.33	56.74	1.011
MgO	6.70	7.01	0.174
MnO	0.72	0.75	0.0106
FeO	13.23	13.82	0.193
SiO_2	9.02	9.44	0.157
Al_2O_3	4.65	4.86	0.0476
V_2O_5	5.68	5.93	0.0326
TiO_2	1.12	1.17	0.0146
S	0.268	0.28	0.00875
合 计	95.718	100.00	

$$\sum n_+ = 1.388 \quad \sum N_{SiO_4^{4-}} = 0.286$$
$$\sum n_- = 1.160 \quad \gamma_{Fe^{2+}}\gamma_{S^{2-}} = 1.85$$

所以
$$\frac{(\%S)}{[\%S]} = \frac{32 \times 0.08 \times 1}{1.85} \times \frac{1.388 \times 1.160}{0.193} = 11.5$$

根据化学分析的结果计算

$$\frac{(\%S)}{[\%S]} = \frac{0.28}{0.031} = 9.1$$

理论计算和实际的由化学分析计算的数值有一定的差异,但在数量级上是一致的。造成差异的原因可能有下列几方面:

(1) 炉渣脱 S 尚未达到平衡;

(2) 假定在炉渣中存在的离子,特别是 VO_4^{3-} 及 TiO_3^{2-} 不正确;

(3) 经验公式 (10) 不适用于攀钢的半钢炉渣;

(4) 炉渣含有不溶解的 CaO,计算时没有减去。

通过炉渣离子理论的计算可以看出,欲提高炉渣的脱 S 能力,应考虑下面几种因素:

(1) 在不影响炉渣黏度时提高炉渣碱度,也就是说,提高碱性氧化物含量,降低酸性氧化物含量。这一方面加大 $\sum n_+$ 值,而另一方面同时减小 $\gamma_{Fe}\gamma_S$ 值,这两方面都能提高硫分配比。

(2) 式 (9) 中分子及分母都有 n_{FeO}。由于分母中 n_{FeO} 的影响比分子中 n_{FeO} 的影响大,尽可能减少其中 FeO 含量。后吹应尽量避免,因它增加渣中 FeO,对脱 S 不利。

(3) 高拉 C 可提高 f_S,对脱 S 有利。

(4) 式 (9) 中 K_S 受温度影响不大,所以炉渣脱 S 的能力与温度的关系不大。当然高温可提高动力学因素,加快反应速度而有利于脱 S。

6 半钢冶炼的造渣问题

炼好钢必须造好渣,这是炼钢工人的宝贵经验。对于半钢冶炼来说,由于在提 V 过程中 Ti、Si、Mn 等元素绝大部分已被氧化掉,缺乏这些酸性氧化物 SiO_2、TiO_2 和碱性氧化物 MnO,对造渣带来一定的困难。为了保证脱 S 脱 P,为传递氧创造条件,快造渣、造好渣,也即快速地造成流动性很好、碱度适当的炉渣对半钢冶炼更是非常必要的。

6.1 造渣的条件

造渣必须满足下列的条件:

(1) 有适当高的温度。没有足够高的温度渣子不能熔化,即使能熔化渣子也没有良好的流动性。但只靠高温度不能造成熔点比较低的渣子,人们必须采用熔剂。

(2) 用多种熔剂。CaO 的熔点是 2570℃。必须加入熔剂以降低 CaO 的熔点,成为低熔点的炉渣。表 6 列出从相图查出来的一些不同组成炉渣的熔点。可以看出,使用一种炉渣虽然能造成较低熔点的炉渣,但很不经济。例如 CaF 能与 CaO 造成熔点为 1400℃ 的炉渣,但 CaF_2 含量至少要达到 65%。FeO 可以降低 CaO 的熔点到 1100℃,但 FeO 含量要高达 73%。Al_2O_3 含量要达到 48%,才能得到熔点为 1400℃ 的 CaO-Al_2O_3 渣。过多量的 CaF_2、FeO 或 Al_2O_3 作造渣材料都不经济,同时大量的 CaF_2 会严重地侵蚀炉衬。单用 SiO_2 以熔化 CaO,在高碱度时例如 CaO 54%、SiO_2 36%,则生成高熔点 2130℃ 的硅酸二钙 $2CaO \cdot SiO_2$,这是在造渣过程中应当尽可能避免的。提高 SiO_2 含量显然地降低 $CaO \cdot SiO_2$ 渣的熔点,但渣的碱度降低变为酸性渣,不能起更好的脱 S 脱 P 作用。从表 6 中可以看出,用两种或两种以上的熔剂造成三元或多元炉渣或引用 MnO 均可大大地降低熔点,得到熔点为 1400℃ 左右的炉渣。在配渣时一般使渣内含有 Al_2O_3 5%~8%、MnO 5%~8%,MgO 5%~9%、在碱度较高时,则能保证炉渣有较低的熔点(约为 1400℃)及较好的流动性。MgO 可以降低炉渣对炉衬的侵蚀,但如果要超过 9%~12%,则炉渣的黏度便增高。

表 6 一些炉渣的成分和熔点

组成/%	熔点/℃	备 注
CaO	2570	
CaO 40，CaF_2 60	1420	
CaO 35，CaF_2 65	1400	
CaO 18，CaF_2 82	1360	共晶体
CaF_2	1420	
CaO 33，FeO 67	1500	
CaO 30，FeO 70	1300	
CaO 27，FeO 73	1100	$CaO \cdot 2FeO$ 与 FeO 共晶体
CaO 41，Fe_2O_3 59	1449	$2CaO \cdot Fe_2O_3$ 化合物
CaO 20，Fe_2O_3 80	1205	$CaO \cdot Fe_2O_3$ 与 $CaO \cdot 2Fe_2O_3$ 共晶体
CaO 52，Al_2O_3 48	1400	
CaO 64，SiO_2 36	2130	$2CaO \cdot SiO_2$ 化合物
CaO 55，SiO_2 45	1460	$CaO \cdot SiO_2$ 与 $3CaO \cdot 2SiO_2$ 共晶体
CaO 47，SiO_2 53	1544	$CaO \cdot SiO_2$ 化合物
CaO 36，SiO_2 64	1436	SiO_2 与 $CaO \cdot SiO_2$ 共晶体
CaO 51，FeO 10，SiO_2 39	1500	
CaO 43，FeO 24，SiO_2 33	1500	
CaO 30，FeO 38，SiO_2 32	1210	$CaO \cdot FeO$，SiO_2 化合物
FeO 47，MnO 20，SiO_2 33	1170	三元共晶体
FeO 47，Al_2O_3 13，SiO_2 40	1073	三元共晶体
FeO 32，Al_2O_3 20，SiO_2 48	1205	三元共晶体
FeO 15，Al_2O_3 3，SiO_2 16，CaO 48，MgO 3.5，MnO 9，P_2O_5 1.5，Fe_2O_3 4	1400	碱性炼钢炉渣

（3）适当高的枪位及合理操作。当采用恒氧压操作时，氧枪适当高的枪位对于加速化渣及避免渣子返干，从而减少喷溅提高金属收得率以及缩短冶炼时间，起着决定性的作用。在吹炼初期，为了生成较大量的 FeO 以快速熔化石灰，防止高熔点的 $2CaO \cdot SiO_2$ 的生成，枪位较高比较合适。在吹炼中末期，为了避免渣子返干造成大量喷溅，氧枪也应及早适当地提高。一般来讲，双高双低的枪位操作比较合适。

6.2 快速造渣的途径

对半钢冶炼，为了加速造渣，除必须适当及时地合理调整氧枪距熔池的高度外，还可以采取下列某一种的措施。

（1）提高石灰质量，包括选用优质石灰石以提高 CaO 的含量，合理焙烧石灰以降低石灰的生烧和死烧的百分率。

（2）使用活性石灰。活性石灰有孔隙度大、表面积大、结晶颗粒小以及被 FeO 溶解的速度快等特点。首钢活性石灰的工艺性试验证明，使用活性石灰炼钢能提高生产率、缩短冶炼时间和改进钢的质量。

（3）使用石灰粉。石灰粉随氧气喷入熔池，能加速化渣，可吹炼高磷生铁，并有利于高拉碳以炼高碳钢。

（4）加入酸性材料。可加入黏土砖块（主要是 Al_2O_3 及 SiO_2）、铁矾土（大部分是 Al_2O_3，杂质有 Fe_2O_3 及 SiO_2 等）或河沙（主要是 SiO_2，杂质有 Al_2O_3 及 Fe_2O_3 等）。这些材料的缺点是熔点较高，一般在 1600℃ 以上。攀钢曾用过辉绿岩和玄武岩，效果良好。它们熔点比较低，在 1200℃ 左右，可促使来渣早，缩短化好渣的时间（参阅表 7 及表 8）。辉绿岩及玄武岩都含有较大量的水分，使用前最好利用石灰窑焙烧把水分赶掉。

表 7　辉绿岩及玄武岩的成分及熔点

项目	成分/%								熔点/℃
	FeO	Fe_2O_3	SiO_2	CaO	MgO	Al_2O_3	TiO_2	灼减	
辉绿岩	3.23	10.61	48.72	8.50	9.57	9.69	2.41	5.98	1215
玄武岩	4.55	8.25	46.44	7.02	14.10	9.13	2.03	6.33	1220

表 8　半钢冶炼时各种造渣材料试验记录（$CaO/SiO_2 = 3$）

项目	渣料/kg·t^{-1}（半钢）	来渣时间	化好渣时间	返干时间	纯吹氧时间
河沙和黏土砖	74	5′46″	9′9″	13′31″	16′6″
辉绿岩	65	3′25″	4′12″	15′44″	17′2″
玄武岩	63.5	5′34″	8′7″	15′23″	20′30″

（5）采用自熔合成渣。合成渣的成分主要是铁酸钙，熔点约为 1450℃，是用由转炉烟尘回收的红泥和石灰粉末两种废品经过处理烧制而成。马钢与上钢三厂已进行试验成功。它的熔点较低，脱 P 效果好，有利于高拉碳操作。

（6）加锰矿作熔剂。MnO 能降低炉渣熔点并提高炉渣流动性。鉴于合理使用我国锰矿资源，我们不提倡使用锰矿作熔剂。可试用铁合金冶炼厂炼制锰铁的炉渣废品代替锰矿充作熔剂。

（7）加铁矿石或烧结矿作熔剂。

（8）适当地配半钢加硅铁或锰铁，提高半钢含 Si 或 Mn 量。这一方法既提供半钢吹炼的热源，又提供造渣需要的 SiO_2 或 MnO 作熔剂。鉴于硅铁和锰铁都是有价值的铁合金，这种方法我们不提倡使用。

含 Nb 及 Mn 的铁液中 Mn 对 Nb 活度系数影响的研究*

张圣弼 佟 亭 王济舫 魏寿昆

摘　要：在前文[1]内，作者曾利用固体电解质定氧电池对含 Nb 铁液中 Nb 的自身活度相互作用系数 e_{Nb}^{Nb} 进行了研究。本文是铁液中 Nb 的热力学行为研究的继续，旨在求出其中 Mn 对 Nb 活度系数的影响。本文利用前文[1]同一实验方法及设备，在 1853K 及 1873K 两个温度进行实验。和前文不同，渣层不用固态 NbO_2 而是用液态 Nb_2O_5。热力学分析证明，铁液中的［Nb］可以将 Nb_2O_5 还原为 NbO_2，而在铁液中有足够量［Nb］的条件下，［Mn］将不参加还原 Nb_2O_5 的反应而无 MnO 生成。因之，电池组装和前文相同，可写为：

$$Mo\,|\,Mo,\,MoO_2\,\|\,ZrO_2(MgO)\,\|\,[Nb],\,NbO_2\,|\,Mo,\,Mo+ZrO_2\ 金属陶瓷$$

同样地根据下式：

$$[Nb]+2[O] = NbO_{2(s)}$$

由测得的 a_O 可计算 a_{Nb}。实验证明，当［Nb］>1%时，［Mn］还原 Nb_2O_5 的反应即可基本上被抑制。对渣层进行 X 射线结构分析证实 Nb_2O_5 已基本上在平衡条件下变成 NbO_2。根据前文的 Fe-Nb 二元系的资料，利用同一浓度法及同一活度法计算出 f_{Nb}^{Mn}。作两个温度的 $\lg f_{Nb}^{Mn}$ 对［%Mn］的直线图，两种方法得出基本上一致的结果：1853K，$e_{Nb}^{Mn}=0.18$；1873K，$e_{Nb}^{Mn}=0.11$。其温度关系式可大体上由下式表达：

$$e_{Nb}^{Mn}=\frac{12100}{T}-6.35$$

当铁液中［%Nb］小于 0.5 时，［Mn］就能还原 Nb_2O_5，所得渣层成分复杂，形成 $MnO\text{-}Nb_2O_5\text{-}NbO_2$ 的熔体，而 NbO_2 及 $MnNb_2O_6$ 可能以饱和相出现，使渣层成为黏滞的二相体。经数据处理估计出 $MnNb_2O_6$ 的生成自由能为：

$$Mn_{(l)}+Nb_{(s)}+3O_2 = MnNb_2O_{6(s)};\quad \Delta G^\circ=-367000+13.3T(\text{cal})\ ❶$$
$$MnO_{(s)}+Nb_2O_{5(l)} = MnNb_2O_{6(s)};\quad \Delta G^\circ=156000-93.2T$$

对渣层进行 X 射线结构分析，发现有 $MnNb_2O_6$、Nb_2O_5、NbO_2 及一些 $FeNb_2O_6$ 存在。今后进行低［Nb］量的 Fe-Mn-Nb 三元系研究时，渣层最好采用固态 NbO_2。

---

* 原刊于《金属学报》，1984，20（5）：A348～A356。参加本工作的有谭赞麟及胡明甫。
❶ 1cal=4.184J。

从含 Nb 铁水提 Nb，需要知道 Nb 在铁液中的热力学行为。在前文中[1]，作者曾对含 Nb 铁液中 Nb 的自身活度相互作用系数 e_{Nb}^{Nb} 进行研究。本文是 Nb 在铁液中热力学行为研究工作的继续，其目的在于求出 Mn 对 Nb 活度系数的影响，也即求出 Mn 对 Nb 的活度相互作用系数 e_{Mn}^{Nb}。

向井楠宏及田上俊男[2]利用封闭室法测定 1570℃ 的 e_{Mn}^{Nb} 为 0.0073。由后者计算出的 e_{Nb}^{Mn} 则为 0.0093。该方法实际上是采用等活度原理以计算 f_{Nb}^{Mn}。本文利用固体电解质定氧电池进行测定，所用温度为 1853K 及 1873K，电池组装如下：

　　Mo | Mo, MoO$_2$ | ZrO$_2$(MgO) | [Nb], NbO$_2$ | Mo, Mo + ZrO$_2$ 金属陶瓷

实验室装置及实验方法和条件详见前文[1]。对固体电解质的电子导电所引起的误差进行过校正，而对它的物理及化学渗透所造成的误差，根据本实验的条件则忽略不计[1]。所用高纯 Fe 及金属 Nb 与前文[1]所述者同。电解 Mn 成分为：C 0.021%，P 0.005%，S 0.02%~0.06%，Mn>99.8%。

和前文[1]不同之处在于，本实验所用覆盖在铁液上的渣料不是固态 NbO$_2$ 而是液态 Nb$_2$O$_5$，但平衡反应仍按反应式

$$[Nb] + 2[O] = NbO_{2(s)} \tag{1}$$

计算。

热力学分析指出（所用自由能数据及其来源见表 1），当铁液中的 [Nb] > 1% 时，能将 Nb$_2$O$_5$ 还原为 NbO$_2$：

$$\frac{1}{2}[Nb] + Nb_2O_{5(l)} = \frac{5}{2}NbO_{2(s)}; \quad \Delta G^\circ = -25830 + 9.43T \tag{2}$$

自由能 ΔG° 的单位为 cal，下同。当铁液中的 [Mn] > 0.5% 时也可与 Nb$_2$O$_5$ 起反应。

表 1　引用热力学数据表
Tablel 1　Thermodynamic data cited

Reaction	ΔG°/cal①	Range/℃	Ref.
Mo$_{(s)}$ + O$_2$ = MoO$_{2(s)}$	-117280+28.28T	1400~1700	[3]
$\frac{1}{2}$O$_2$ = [O]	-32750+1.86T	1400~1700	[3]
[Nb] + 2[O] = NbO$_{2(s)}$	-89710+28.27T	1400~1700	[1]
Nb$_{(s)}$ + O$_2$ = NbO$_{2(s)}$	-187300+39.89T	25~2150	[4, 5]
Nb$_{(s)}$ = [Nb]	-32090+7.90T	1400~1700	[1]
2Nb$_{(s)}$ + $\frac{5}{2}$O$_2$ = Nb$_2$O$_{5(l)}$	-426380+86.35T	1512~2000	[4, 5]
2Nb$_{(s)}$ + O$_2$ = 2NbO$_{(s)}$	-198000+41.40T	25~1937	[4, 5]
Mn$_{(s)}$ + $\frac{1}{2}$O$_2$ = MnO$_{(s)}$	-92940+18.24T	25~1244	[5, 6]
Mn$_{(s)}$ = Mn$_{(l)}$	2900-1.90T	1244	[5, 7]
Mn$_{(l)}$ = [Mn]	976-9.12T	1400~1700	[8]

① 1cal = 4.184J。

$$[Mn] + Nb_2O_{5(l)} = 2NbO_{2(s)} + MnO_{(s)}; \quad \Delta G° = -45040 + 22.69T \quad (3)$$

在大量 [Nb] 存在下, [Nb] 可将 MnO 还原:

$$\frac{1}{2}[Nb] + MnO_{(s)} = \frac{1}{2}NbO_{2(s)} + [Mn]; \quad \Delta G° = 19210 - 13.26T \quad (4)$$

式（3）与式（4）合并即得式（2）。所以可以看出，在大量 [Nb] 存在下，[Mn] 可认为也起促进生成 NbO_2 的作用。此外，NbO_2 是稳定的，它不可能再被 [Nb] 或 [Mn] 还原:

$$NbO_{2(s)} + [Nb] = 2NbO_{(s)}; \quad \Delta G° = 21390 - 6.39T \quad (5)$$

$$[Mn] + \frac{1}{2}NbO_{2(s)} = MnO_{(s)} + \frac{1}{2}[Nb]; \quad \Delta G° = -19210 + 13.26T \quad (6)$$

式（6）按标准状态计算，在高于 1449K 时难以进行。当其他物质的活度为 1 及 Mn 大量存在（$a_{Mn} > 4.5$）时，式（6）在 1873K 才能向右进行。式（6）是式（4）的逆式。式（2）~式（6）的氧化还原反应均按一个氧原子写出，以便相互可以比较。

这样，对含 Nb、Mn 的铁液，在有相当多量 [Nb] 存在的条件下，我们可以用式（1）研究 [Mn] 对 [Nb] 活度系数的影响。

但当 [Nb] 较少时，覆盖的 Nb_2O_5 不可能生成较纯的 NbO_2，其成分变为复杂，此点在后面进行讨论。

1 实验结果及处理

表 2、表 3 分别列出 1853K、1873K 下测出的电动势数据及计算出的 a_O 及 a_{Nb}。表内 p'_e 是固体电解质当离子电导率和电子电导率相等时的氧分压，用以校正由固体电解质存在的电子导电产生的误差。

表 2　Fe-Nb-Mn 系 1853K 时的实验数据
Table 2　Experimental data of Fe-Nb-Mn system at 1853K

No.	[Nb] /wt%	[Mn] /wt%	E/V	$p'_e{}^{\frac{1}{4}} \times 10^4$	a_O	a_{Nb}
9	1.16	0.03	0.3194	1.04	0.0070	0.82
10	1.40	0.28	0.3277	1.04	0.0062	1.02
11	1.07	0.37	0.3196	1.04	0.0070	0.82
12	1.60	0.41	0.3350	1.04	0.0057	1.24
13	1.43	0.44	0.3320	1.04	0.0059	1.14
14	1.64	0.67	0.3386	1.04	0.0054	1.37
15	1.54	0.81	0.3370	1.04	0.0055	1.31

表 3 Fe-Nb-Mn 系 1873K 时的实验数据
Table 3 Experimental data of Fe-Nb-Mn system at 1873K

No.	[Nb]/wt%	[Mn]/wt%	E/V	$p_e'^{\frac{1}{4}} \times 10^4$	a_O	a_{Nb}
10	1.22	0.15	0.3153	1.30	0.0082	0.77
11	1.18	0.32	0.3143	1.30	0.0083	0.75
12	1.49	0.44	0.3235	1.30	0.0074	0.96
13	1.85	0.70	0.3333	1.30	0.0064	1.25
14	1.74	0.81	0.3320	1.30	0.0066	1.21

Fe-Mn-Nb 系中的 f_{Nb} 有下列关系：

$$f_{Nb} = f_{Nb}^{Nb} \cdot f_{Nb}^{Mn} \cdot f_{Nb}^{O}$$

由于熔体中 [O] 含量很小，$f_{Nb}^{O} \approx 1$，故可简略为：

$$f_{Nb} = f_{Nb}^{Nb} \cdot f_{Nb}^{Mn}$$

所以

$$f_{Nb}^{Mn} = \frac{f_{Nb}}{f_{Nb}^{Nb}}$$

求 f_{Nb}^{Mn} 有两种方法：(1) 同一浓度法；(2) 同一活度法。同一浓度法强调三元系 Fe-Mn-Nb 中的 f_{Nb} 和二元系 Fe-Nb 中的 f_{Nb}^{Nb} 应有同一浓度的 [%Nb]，因之 f_{Nb}^{Mn} 又等于有同一浓度的 [%Nb] 的三元系及二元系熔体的各自的活度之比：

$$f_{Nb}^{Mn} = \frac{f_{Nb}}{f_{Nb}^{Nb}} = \frac{a_{Nb(Fe-Mn-Nb)}}{a_{Nb(Fe-Nb)}}$$

利用同一活度法求 f_{Nb}^{Mn}，由于

$$a_{Nb} = f_{Nb} \cdot [\%Nb]_{Fe-Nb-Mn} = f_{Nb}^{Nb} \cdot [\%Nb]_{Fe-Nb}$$

所以

$$f_{Nb}^{Mn} = \frac{f_{Nb}}{f_{Nb}^{Nb}} = \frac{[\%Nb]_{Fe-Nb}}{[\%Nb]_{Fe-Nb-Mn}}$$

关于二元系 Fe-Nb 的资料，由前文[1]图 6 及图 7 可分别求出 1853K 及 1873K 时一定浓度 [Nb] 的 $\lg f_{Nb}^{Nb}$。

为了便于由三元系 Fe-Nb-Mn 中 [Nb] 活度的实验值求出二元系 Fe-Nb 中同一活度的 [Nb] 浓度，根据前文[1]的表 2 及表 3 的数据绘出 1853K 及 1873K 的 a_{Nb} 对 [Nb] 浓度的关系图，见图 1。

表 4、表 5 分别列出 Fe-Nb-Mn 系在 1853K 和 1873K 按同一浓度法及同一活度法计算出的 $\lg f_{Nb}^{Mn}$ 值。

利用最小二乘法求 $\lg f_{Nb}^{Mn}$ 对 [%Mn] 的直线斜率。当 [%Mn]→0 时，$f_{Nb}^{Mn} \to 1$，因之此直线必须通过原点。此斜率即是 e_{Nb}^{Mn}，其计算结果见表 6。

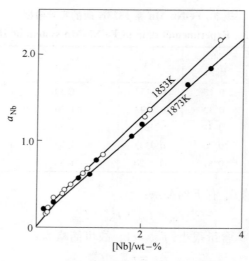

图 1 Fe-Nb 系中 a_{Nb} 对 [Nb] 曲线图

Fig. 1 a_{Nb} vs. [Nb] of Fe-Nb melt

表 4 Fe-Nb-Mn 系 1853K 时按同一浓度法及同一活度法计算出的 $\lg f_{Nb}^{Mn}$ 值

Table 4 Calculated value of $\lg f_{Nb}^{Mn}$ of Fe-Nb-Mn system at 1853K by methods of same concentration and activity

No.	[%Nb]	[%Mn]	a_{Nb}	Same concentration method			Same activity method	
				$\lg f_{Nb}$	$\lg f_{Nb}^{Nb}$	$\lg f_{Nb}^{Mn}$	[%Nb]$_{Fe-Nb}$	$\lg f_{Nb}^{Mn}$
9	1.16	0.03	0.82	-0.151	-0.185	0.034	1.24	0.029
10	1.40	0.28	1.02	-0.137	-0.197	0.060	1.60	0.058
11	1.07	0.37	0.82	-0.116	-0.180	0.064	1.24	0.064
12	1.60	0.41	1.24	-0.111	-0.201	0.090	2.00	0.097
13	1.43	0.44	1.14	-0.098	-0.198	0.100	1.80	0.100
14	1.64	0.67	1.37	-0.078	-0.202	0.124	2.20	0.128
15	1.54	0.81	1.31	-0.070	-0.200	0.130	2.10	0.135

表 5 Fe-Nb-Mn 系 1873K 时按同一浓度法及同一活度法计算出的 $\lg f_{Nb}^{Mn}$ 值

Table 5 Calculated value of $\lg f_{Nb}^{Mn}$ of Fe-Nb-Mn system at 1873 K by methods of same concentration and activity

No.	[%Nb]	[%Mn]	a_{Nb}	Same concentration method			Same activity method	
				$\lg f_{Nb}$	$\lg f_{Nb}^{Nb}$	$\lg f_{Nb}^{Mn}$	[%Nb]$_{Fe-Nb}$	$\lg f_{Nb}^{Mn}$
10	1.22	0.15	0.77	-0.200	-0.210	0.010	1.30	0.028
11	1.18	0.32	0.75	-0.197	-0.205	0.008	1.26	0.028
12	1.49	0.44	0.96	-0.191	-0.230	0.039	1.68	0.052
13	1.85	0.70	1.25	-0.170	-0.250	0.080	2.24	0.083
14	1.74	0.81	1.21	-0.158	-0.245	0.087	2.18	0.098

表6 同一浓度法及同一活度法求出的 e_{Nb}^{Mn} 值和直线关系的相关系数 r

Table 6 Value e_{Nb}^{Mn} obtained by methods of same concentration and activity as well as corresponding value of correlation coefficient, r, for straight-line relation

Temperature /K	Same concentration method		Same activity method	
	e_{Nb}^{Mn}	r	e_{Nb}^{Mn}	r
1853	0.18	0.963	0.19	0.967
1873	0.10	0.961	0.12	0.985

可以看出，用同一浓度法与同一活度法求出的 e_{Nb}^{Mn} 基本上是相同的。理论分析[9,10]指出，只有当 Fe-Nb 及 Fe-Nb-Mn 系服从亨利定律也即 f_{Nb}^{Nb} 及 f_{Nb} 均等于1时，两个方法求出的 e_{Nb}^{Mn} 值才相等。本实验两方法所得的 e_{Nb}^{Mn} 基本上相同，只能是一种巧合。

如选用 e_{Nb}^{Mn} 在1853K时为0.18；1873K时为0.11，则 e_{Nb}^{Mn} 与温度的关系式可粗略地表达为：

$$e_{Nb}^{Mn} = \frac{12100}{T} - 6.35$$

2 讨论

实验结果指出，当 [%Nb]<0.5 时，用式（1）处理的 $\lg f_{Nb}^{Mn}$ 数据无论采用同一浓度法或同一活度法，均属反常，表现在 $\lg f_{Nb}$ 为正值。在 [Nb] 量较小时，渣层不可能是较纯的 NbO_2，而是 NbO_2 与 Nb_2O_5 的溶液或混合体。同时由式（3）生成的 MnO 可与 Nb_2O_5 化合为 $MnNb_2O_6$。因之，渣层是 $MnO-NbO_2-Nb_2O_5$ 的熔体，其中 NbO_2 及 $MnNb_2O_6$ 均有可能呈饱和态。渣层黏而流动性差，也说明它是液固两相体。

采用前文[1]的 e_{Nb}^{Nb} 及本研究的 e_{Nb}^{Mn} 数据，计算低 [Nb] 实验的 a_{Nb} 及 a_{Mn}，并计算 $a_{Mn} \cdot a_{Nb}^2 \cdot a_O^6$ 值，其结果见表7及表8。可以看出，在一定温度下脱氧常数 $K' = a_{Mn} \cdot a_{Nb}^2 \cdot a_O^6$ 基本上是一常数，1853K 时 K' 的平均值为 2.42×10^{-14}，而1873K 时 K' 的平均值为 3.30×10^{-14}。经处理后得下列反应的估计值：

$$[Mn] + 2[Nb] + 6[O] = MnNb_2O_{6(s)}; \lg K = \frac{23400}{T} + 1.0$$

$$\Delta G° = -107000 - 4.57T$$

表7　1853K 下低 [Nb] 浓度时的数据分析
Table 7　Analysis of low [Nb] concentration data at 1853K

($e_{Nb}^{Nb} = -0.21$　$e_{Nb}^{Mn} = 0.18$　$e_{Mn}^{Mn} = 0$　$e_{Mn}^{Nb} = 0.11$)

No.	[%Nb]	[%Mn]	a_O	a_{Nb}	a_{Mn}	$a_{Mn} \cdot a_{Nb}^2 \cdot a_O^6$
1	0.19	0.029	0.0152	0.18	0.030	1.20×10^{-14}
2	0.20	0.056	0.0145	0.19	0.059	2.00×10^{-14}
3	0.20	0.089	0.0137	0.19	0.094	2.24×10^{-14}
4	0.21	0.090	0.0133	0.20	0.095	2.10×10^{-14}
5	0.28	0.12	0.0118	0.26	0.13	2.37×10^{-14}
6	0.23	0.14	0.0125	0.22	0.15	2.77×10^{-14}
7	0.32	0.19	0.0112	0.30	0.21	3.73×10^{-14}
8	0.48	0.27	0.0090	0.43	0.30	2.95×10^{-14}
						av. 2.42×10^{-14}

表8　1873K 下低 [Nb] 浓度时的数据分析
Table 8　Analysis of low [Nb] concentration data at 1873K

($e_{Nb}^{Nb} = -0.22$　$e_{Nb}^{Mn} = 0.11$　$e_{Mn}^{Mn} = 0$　$e_{Mn}^{Nb} = 0.07$)

No.	[%Nb]	[%Mn]	a_O	a_{Nb}	a_{Mn}	$a_{Mn} \cdot a_{Nb}^2 \cdot a_O^6$
1	0.36	0.040	0.0148	0.30	0.042	3.97×10^{-14}
2	0.31	0.048	0.0156	0.27	0.050	5.25×10^{-14}
3	0.33	0.056	0.0139	0.28	0.059	3.34×10^{-14}
4	0.29	0.060	0.0140	0.25	0.063	2.96×10^{-14}
5	0.37	0.070	0.0123	0.31	0.074	2.46×10^{-14}
6	0.37	0.073	0.0126	0.31	0.077	2.96×10^{-14}
7	0.32	0.076	0.0127	0.28	0.080	2.63×10^{-14}
8	0.35	0.12	0.0122	0.30	0.13	3.86×10^{-14}
9	0.43	0.13	0.0104	0.36	0.14	2.30×10^{-14}
						av. 3.30×10^{-14}

故由其元素生成的标准生成自由能估计为：

$$Mn_{(1)} + 2Nb_{(s)} + 3O_2 = MnNb_2O_{6(s)}; \quad \Delta G° = -367000 + 13.3T$$

由其氧化物生成的标准生成自由能估计为：

$$MnO_{(s)} + Nb_2O_{5(1)} = MnNb_2O_{6(s)}; \quad \Delta G° = 156000 - 93.2T$$

由于 $MnNb_2O_6$ 的生成，类似于式（3）的反应在高温更易于进行：

$$[Mn] + 2Nb_2O_{5(1)} = 2NbO_{2(s)} + MnNb_2O_{6(s)}; \quad \Delta G° = 111000 - 70.5T$$

对铁液上的渣进行 X 射线结构分析（表9）证明，当 [%Nb]>1 时，和铁液平衡的渣层基本是 NbO_2（有少量的 Nb_2O_5 及 $MnNb_2O_6$）；而当 [%Nb]<1 时，该渣则含有 NbO_2、Nb_2O_5、$MnNb_2O_6$ 及一些 $FeNb_2O_6$。

表9 铁液上渣层的 X 射线结构分析

Table 9 X-ray diffraction analysis of slag layar over iron melt

Measured				ASTM Standard									
%Nb=1.40 %Mn=0.28		%Nb=0.35 %Mn=0.12		4-0672 FeNb$_2$O$_6$		25-543 MnNb$_2$O$_6$		19-864 γ'-Nb$_2$O$_5$		19-862 β-Nb$_2$O$_5$		9-235 NbO$_2$	
$d/\text{Å}$[①]	I/I_1	$d/\text{Å}$	I/I_1	$d/\text{Å}$	I/I_1	$d/\text{Å}$	I/I_1	$d/\text{Å}$	I/I_1	$d/\text{Å}$	I/I_1	$d/\text{Å}$	I/I_1
		3.74	32	3.74	40					3.75	100		
3.675	6	3.67	26			3.68	100	3.65	100			3.63	30
				3.61	60	3.60	35			3.59	80		
										3.51	40		
3.403	100	3.42	16									3.42	100
								3.32	80			3.21	30
		3.15	23										
		3.04	100	3.01	40			3.08	100				
2.999	15	2.96	34	2.92	100	2.98	100	2.981	100	2.982	60	2.91	30
		2.88	27			2.87	35						
		2.85	25	2.81	40								
2.585	49	2.58	21	2.55	20	2.54	15			2.544	60	2.54	80
		2.52	21	2.50	60	2.50	20	2.503	60	2.498	40	2.491	30
		2.47	24					2.441	40	2.445	40		
2.406	14					2.402	45					2.422	50
2.227	7	2.23	16	2.17	80	2.220	10	2.220	40			2.253	30
		2.13	16					2.198	20			2.166	20
				2.05	80	2.092	10	2.051	20	2.045	60	2.014	10
		1.92	18	1.91	40	1.904	10	1.912	40	1.911	80	1.932	20
		1.87	19	1.87	50			1.885	20			1.897	20
		1.84	18	1.85	40	1.845	25	1.828	40			1.829	20
		1.79	39	1.80	60	1.782	35	1.805	60	1.789	40	1.766	30
1.763	54	1.76	33	1.74	80	1.745	35	1.724	40	1.740	40	1.754	80
1.704	42			1.70	80			1.703	60	1.722	40	1.712	50
1.532	8	1.57	29	1.57	40					1.581	60		
1.523	13	1.53	20			1.539	15	1.540	20	1.529	40		
		1.513	27	1.51	80								
		1.47	27			1.492	10	1.492	20				
1.422	15	1.443	18	1.44	90	1.447	30	1.453	20	1.454	40		
1.398	2	1.386	16	1.37	20			1.366	20	1.398	40		
1.225	10	1.221	16							1.268	40		
		1.104	16										

① 1Å=0.1nm。

为避免复杂反应的发生，本研究工作最好采用纯 NbO_2 与 Fe-Mn-Nb 熔体进行平衡。纯 NbO_2 可由化学纯 Nb_2O_5 在 1100℃ 恒温 2h 用 H_2 气还原制得。

3 结论

对含 Nb>1% 的含 Mn 铁液，当采用 Nb_2O_5 为渣料时，利用固体电解质定氧电池的 $[Nb] + 2[O] = NbO_{2(s)}$ 反应式，通过 a_O 的测定可以研究 Mn 对 Nb 活度系数的影响。用同一浓度法和同一活度法求出基本上一致的数值：

$$1853K,\ e_{Nb}^{Mn} = 0.18;\quad 1873K,\ e_{Nb}^{Mn} = 0.11$$

其温度关系式估计为：

$$e_{Nb}^{Mn} = \frac{12100}{T} - 6.35$$

当铁液中含 Nb 较少时（<0.5%），渣层是 NbO_2-Nb_2O_5-MnO 的熔体，可能有饱和的 $MnNb_2O_6$ 及 NbO_2 的固相存在。经数据加工处理，估计出 $MnNb_2O_6$ 的标准生成自由能为：

$$Mn_{(l)} + 2Nb_{(s)} + 3O_2 = MnNb_2O_{6(s)};\quad \Delta G^\circ = -1535530 + 55.84T(J)$$

$$MnO_{(s)} + Nb_2O_{5(l)} = MnNb_2O_{6(s)};\quad \Delta G^\circ = 652700 - 389.54T(J)$$

致谢：在实验过程中得到冶金工业部钢铁研究总院 14 室及 9 室有关同志以及北京钢铁学院冶金物化教研室有关教师大力协助，谨致谢意。

参 考 文 献

[1] 魏寿昆, 张圣弼, 佟亭, 谭赞麟. 稀有金属, (CSM), 1983, 2 (1): 10.
[2] 向井楠宏, 田上俊男. 鉄と鋼, 1975, 61: 2328.
[3] Janke D, Fischer W A. Arch. Eisenhuettenwes, 1975, 46: 775.
[4] Chase M W, Curnutt J L, Prophet H, Mcdonald R A, Syverud A N. J. Phys. Chem. Ref. Data, 1975, 4: 1.
[5] Turkdogan. E T. Physical Chemistry of High Temperature Technology. Academic Press, New York, 1980: 15~17.
[6] Alcock C B, Zador S. Electrochim. Acta, 1967, 12: 673.
[7] Hultgren R, Desai P D, Hawkins D T, Gleiser M, Kelley K K, Wagman D D. Selected Values of the Thermodynamic Properties of the Elements, ASM, Ohio, 1973: 305.
[8] Sigworth G K, Elliott J F. Met. Sci., 1974, 8: 298.
[9] Turkdogan E T. J. Iron Steel Inst., 1956, 182: 66.
[10] 魏寿昆. 活度在冶金物理化学中的应用. 北京: 中国工业出版社, 1964: 86.

A Study of the Effect of Silicon upon the Activity Coefficienf of Niobium in Liquid Iron with the Solid Electrolyte Oxygen Cell Technique[*]

Wei Shoukun Tung Ting Zhang Shengbi Hu Mingfu

Solid electrolyte oxygen cell technique was used to study the effect of Si upon the activity coefficient of Nb in liquid iron at three temperatures 1823K, 1853K and 1883K. To prevent the reduction of NbO_2 by Si, liquid iron containing Nb and Si was equilibrated in quartz crucible with SiO_2 under purified Ar atmosphere. With the a_O values experimentally determined, the interaction coefficients of Nb upon Si at the three temperatures were evaluated both on the same concentration as well as on the same activity basis. The corresponding values of Si upon Nb was then calculated by the conventional method.

1 Introduction

In the previous papers[1,2] investigation on the thermodynamic behavour of Nb in liquid iron with the evaluation of the interaction coefficients e_{Nb}^{Nb} and e_{Nb}^{Mn} using the solid electrolyte oxygen cell technique was reported. This present paper deals with the study of the effect of Si upon the activity coefficient of Nb is liquid iron with evaluation of e_{Nb}^{Si}, the same oxygen cell technique being used. To prevent the reduction of NbO_2 by Si:

$$[Si] + NbO_2 = [Nb] + SiO_2 \qquad (1)$$

in contrast to the previous papers, liquid iron containing Nb and Si was equilibrated in quartz crucible with pure SiO_2 under purified Ar atmosphere, the following cell assembly being adopted:

$$Mo \mid Mo, MoO_2 \parallel ZrO_2(MgO) \parallel [Si], SiO_2 \mid Mo + ZrO_2 \text{ cerment, } Mo$$

From the values experimentally ascertained, interaction coefficients of e_{Si}^{Nb} were evaluated both on the same concentration as well as on the same activity basis. With the conventional method, the corresponding values of e_{Nb}^{Si} were then calculated.

---

[*] 原刊于《Solid State Irons》,1986,18&19:1232~1236。

2 Experimentals and Results

The experimental apparatus and the composition of high purity iron and metallic Nb were given as in Reference [1]. Silicon used was single crystal Si with a purity of 99.999%. The high-purity quartz crucible used was 40 mm high, with an I. D. of 27 mm and an O. D. of 30mm. The metallic melt weighted about 110g. No slag powder was put above the melt, SiO_2 being formed automatically from deoxidation of Si. Experiments were carried out at three temperatures, 1823K, 1853K and 1883K, details of experimental procedures were similar to those as reported elsewhere[1,2].

The error caused by the electronic conductivity of the MgO-dosed ZrO_2 was corrected with the Schmalzfried formula[3], P'_e being measured as 0.12×10^{-16}, 1×10^{-16} and 8.55×10^{-16} for 1823K, 1853K and 1883K respectively.

Thermodymic data used are as follow:

$$Mo_{(s)} + O_2 = MoO_{2(s)}$$
$$\Delta G^\circ = -490700 + 118.32T \text{ (J)}^{[4]}$$
$$O_2 = 2[O]$$
$$\Delta G^\circ = -274050 + 15.56T \text{ (J)}^{[4]}$$
$$Si_{(s)} + O_2 = SiO_{2(s)}$$
$$\Delta G^\circ = -905840 + 175.52T \text{ (J)}^{[5]}$$
$$Si_{(s)} = Si_{(l)}$$
$$\Delta G^\circ = -50540 - 30.00T \text{ (J)}^{[6]}$$
$$Si_{(l)} = [Si]$$
$$\Delta G^\circ = -131500 - 17.24T \text{ (J)}^{[7]}$$
$$Nb_{(s)} = [Nb]$$
$$\Delta G^\circ = -134260 + 33.05T \text{ (J)}^{[8]}$$
$$Nb_{(s)} + O_2 = NbO_{2(s)}$$
$$\Delta G^\circ = -783600 + 166.90T \text{ (J)}^{[9]}$$

From the values a_O experimentally ascertained, a_{Si} for the ternary system Fe-Nb-Si can be calculated from the following:

$$[Si] + 2[O] = SiO_{2(s)}$$
$$\Delta G^\circ = -550830 + 207.20T \text{ (J)} \tag{2}$$

Calculating f_{Si}^{Nb} on the same concentration basis:

$$f_{Si}^{Nb} = f_{Si}/f_{Si}^{Si} = a_{Si}/a'_{Si} \tag{3}$$

where a'_{Si} denotes the activity of Si in the binary Fe-Si system.
Knowing that:

$$e_{Si}^{Si} = 34.5/T + 0.089 \tag{4}$$

a'_{Si} can be found with equation (5):
$$\log a'_{Si} = e^{Si}_{Si}[\%Si] + \log[\%Si] \qquad (5)$$
Calculating f^{Nb}_{Si} on the same activity basis:
$$f^{Nb}_{Si} = f_{Si}/f^{Si}_{Si} = [\%Si]'/[\%Si] \qquad (6)$$
Where [%Si]' denotes the content of Si for the binary Fe-Si system and can be found from equation (7) by iteration:
$$\log a_{Si} = e^{Si}_{Si}[\%Si]' + \log[\%Si]' \qquad (7)$$

Figs. 1~3 represent the plots of f^{Nb}_{Si} vs. [%Nb] basing on the same concentration at the temperatures 1823, 1853 and 1883 K respectively.

Figs. 4~6 represent the same on the same activity basis.

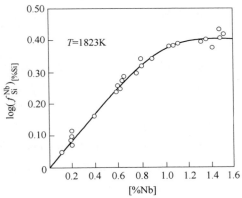

Fig. 1 $\log f^{Nb}_{Si}$ vs. [%Nb] at 1823K
(same concentration basis)

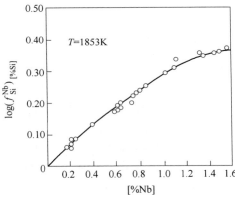

Fig. 2 $\log f^{Nb}_{Si}$ vs. [%Nb] at 1853K
(same concentration basis)

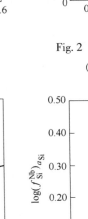

Fig. 3 $\log f^{Nb}_{Si}$ vs. [%Nb] at 1883K
(same concentration basis)

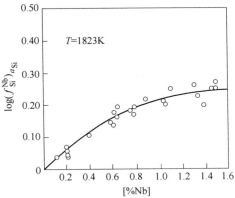

Fig. 4 $\log f^{Nb}_{Si}$ vs. [%Nb] at 1823K
(same activity basis)

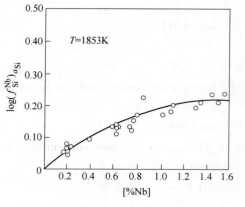
Fig. 5 $\log f_{Si}^{Nb}$ vs. [%Nb] at 1853K
(same activity basis)

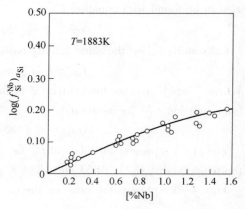
Fig. 6 $\log f_{Si}^{Nb}$ vs. [%Nb] at 1883K
(same activity basis)

By drawing tangent lines at the origin to the curves of $\log f_{Si}^{Nb}$ vs. [%Nb], e_{Si}^{Nb} values at the corresponding temperatures can be calculated. The corresponding values of e_{Nb}^{Si} can then be calculated with the following formula[10]:

$$e_{Nb}^{Si} = 1/230[(230 e_{Si}^{Nb} - 1) M_{Nb}/M_{Si} + 1]$$

where M_{Nb} and M_{Si} represent the atomic weight of Nb and Si respectively. Both values of e_{Si}^{Nb} and e_{Nb}^{Si} at the three temperatures with their relations with temperature as obtained by regression analysis are given below:

Basing on the same concentration:

	e_{Si}^{Nb}	e_{Nb}^{Si}
1823K	0.42	1.38
1853K	0.30	0.98
1883K	0.22	0.72
	11454/T−5.87	37763/T−19.36
	(r=0.994)	(r=0.994)

Basing on the same activity:

	e_{Si}^{Nb}	e_{Nb}^{Si}
1823K	0.31	1.02
1853K	0.27	0.88
1883K	0.16	0.52
	8559/T−4.37	28535/T−14.60
	(r=0.963)	(r=0.967)

3 Discussion

The key to the success of this investigation is to keep the reaction (2) always at equilibrium and to prevent the reaction (9) to occur:

$$[Nb] + 2[O] =\!=\!= NbO_{2(s)}$$
$$\Delta G^\circ = -375350 + 118.29T \text{ (J)} \qquad (8)$$

The judging criterion for this condition is that the actual deoxidation product $a_{Nb} \cdot a_O^2$ should be always less than the equilibrium deoxidation constant $(a_{Nb} \cdot a_O^2)_{equil}$. The latter values are:

        1823K 2.65×10^{-5}

        1853K 3.95×10^{-5}

        1883K 5.83×10^{-5}

This necessitates the calculation of a_{Nb} for every experimental data, and henceforth the calculation of f_{Nb}^{Si} and f_{Nb}^{Nb} from the corresponding values of e_{Nb}^{Si} and e_{Nb}^{Nb}. From Figs. 1~6 it can be seen that the curves of $\log f_{Si}^{Nb}$ vs. [%Nb] tend to flatten out as Nb reaches 0.5%~0.6%. This means that, for those experiments with Nb over 0.5%~0.6%, use of e_{Si}^{Nb} would give too high values of f_{Si}^{Nb}. The same thing holds true for f_{Nb}^{Nb}. It would be difficult to predict the content of Si at which the curve of $\log f_{Nb}^{Si}$ against [%Si] would depart from the straight-line relationship. Yet assumption of Si up to 0.5% for maintaining the straight-line relationship might be reasonable. In view of the fact that many experiments were conducted with Si content much higher than 1% and with Nb content much higher than 0.5%, so these values of $a_{Nb} \cdot a_O^2$ calculated on the straight-line relationship would be to high and not be feasible for comparison. But, however, as the value of e_{Nb}^{Si} drops with increase in temperature, from the 27 experiments at 1883K only 6 experiments with high Si (up to 1.99%) and high Nb (up to 1.32%) give values of $a_{Nb} \cdot a_O^2$ higher than the equilibrium deoxidation constant. These high values would not represent the true values of the deoxidation product because of the high values of f_{Nb}^{Si} and f_{Nb}^{Nb} calculated on the straight-line relationship. In view of the fact that all experiments with comparatively low Si and low Nb satisfy the criterion given above, so it might be inferred that the condition for the non-occurrence of the reaction (9) was realized in most of our experiments.

4 Conclusion

Values of e_{Si}^{Nb} and e_{Nb}^{Si} calculated on the concentration basis are higher than those evaluated on the same activity basis, as shown below:

Basing on the same concentration:

$$e_{Si}^{Nb} = 11454/T - 5.87 \; ; \; e_{Si}^{Nb} \text{ at } 1873K = 0.25$$
$$e_{Nb}^{Si} = 37763/T - 19.36 \; ; \; e_{Nb}^{Si} \text{ at } 1873K = 0.80$$

Basing on the same activity:

$$e_{Si}^{Nb} = 8559/T - 4.37 \; ; \; e_{Si}^{Nb} \text{ at } 1873K = 0.20$$
$$e_{Nb}^{Si} = 28535/T - 14.60 \; ; \; e_{Nb}^{Si} \text{ at } 1873K = 0.63$$

References

[1] Shoukun Wei, Shengbi Zhang, Ting Tung, Zanlin Tan. Rare Metals(CSM), 1983,2(1):10.
[2] Shengbi Zhang, Ting Tung, Jifang Wang, Shoukun Wei. Acta Met. Sinica,1984,20:A348.
[3] Schmalzfried H, Z. Phy. Chem. N. F. ,1963,38:87.
[4] Janke D, Fischer W A. Arch. Eisenh. ,1975,46:755.
[5] Stull D R, Prophet H. Janaf Thermochem. Tables (U. S. Commer., Washington D. C., 1971).
[6] Hultgren R, et al. Selected Values Thermody. Prop. Elem. ASM, Park, Ohio,1973:467.
[7] Elliott J F. Elec. Fur. Proc. ,1974,32: 62.
[8] See Ref. [1].
[9] Chase M W, et al. J. Phy. Chem. Ref. Data, 1978(7): 793.
[10] Schenck H, Frohberg M G, Steinmetz E. Arch. Eisenh. ,1960,31:671.

Nb、Si 在铁液中活度相互作用的研究[*]

佟 亭　魏寿昆　张圣弼　胡明甫

摘　要：利用前文[1,2]的实验方法，本文研究铁液中 Si、Nb 活度的相互作用。为了使脱氧产物稳定，本文和前文不同，采用下列组装的定氧电池：

Mo｜Mo, MoO_2 ‖ $ZrO_2(MgO)$ ‖ [Si]、SiO_2｜Mo + ZrO_2 金属陶瓷, Mo

在石英坩埚内使含 Nb、Si 的铁液与 SiO_2 取得平衡。由半稳定的 ZrO_2 管的电子导电造成的误差，利用 Schmalzried 公式进行校正。由测定的 a_0 值，由反应 [Si] + 2[O] = $SiO_{2(s)}$ 求出 a_{Si}，再根据同一浓度及同一活度法计算三个温度 1823K、1853K 及 1883K 的 e_{Si}^{Nb} 值，然后利用惯用的方法计算出相应的 e_{Nb}^{Si}，同时用回归分析求出其温度关系式。

对每次实验的 $a^{Nb} \cdot a_O^2$ 值进行了计算，并对如何与 Nb 的脱氧常数相比较进行了讨论，可以推论，绝大部分的实验中 NbO_2 不能生成，因此，所得实验结果是可以信赖的。

最后，对如何选择最佳的电动势测量值，以保持可逆的电化学反应，进行了简要的讨论。

从含 Nb 铁水提 Nb，需要知道 Nb 在铁液中的热力学行为，在前文中[1,2]，作者已对含 Nb 铁液中 Nb 本身的活度相互作用系数 e_{Nb}^{Nb} 及 Mn 对 Nb 的活度相互作用系数 e_{Nb}^{Mn} 进行了研究。本文是上述研究工作的继续，其目的在于利用同样电化学方法对 Fe-Si-Nb 熔体进行研究，先求出 Nb 对 Si 的活度相互作用系数 e_{Si}^{Nb}，然后再计算出 e_{Nb}^{Si}。

1　实验原理、方法及结果

如采用前文[1,2]的电池装置，则不可避免地会有下列反应发生：

$$[Si] + NbO_{2(s)} \longrightarrow (SiO_2) + [Nb]$$

这样，一则反应难以达到平衡，二则生成的 SiO_2 可能与 NbO_2 互溶，使计算复杂

---

[*] 原刊于《金属学报》1987，23（2）：B47~B54。参加本工作的尚有谭赞麟、王济舫、王晓英及陈小林。

化。因此，本实验采用下列的电池装置：

Mo | Mo, MoO$_2$ ‖ ZrO$_2$(MgO) ‖ [Si], SiO$_2$ | Mo + ZrO$_2$ 金属陶瓷, Mo

在净化的 Ar 气气氛下，使含 Si 及 Nb 的铁液在石英坩埚内与 SiO$_2$ 取得平衡，使 NbO$_2$ 不能生成，Si 是单晶硅，纯度为 99.999%。金属 Nb 及纯 Fe 的成分见前文[1]。采用外径 30mm、内径 27mm、高 40mm 的纯石英坩埚。每次试验约用试样 110g。熔体表面未覆盖渣料，SiO$_2$ 由脱氧反应自动生成。每次试验依次在 1823K、1853K 和 1883K 下进行。每个温度下测定完毕后均用净化的 Ar 气保护的石英管抽取试样，淬冷后用以作化学分析，然后升温至下一个温度，保温 30min 后重复上述步骤。总的实验装置及实验方法和前文[1]相同。所用热力学自由能数据及其来源见表 1。

表1 引用的热力学数据
Table 1 Thermodynamic data cited

Reaction	$\Delta G^{\ominus}$/J	Ref.
Mo(s) + O$_2$ = MoO$_{2(s)}$	-490700+118.32T	[3]
O$_2$ = 2[O]	-274050+15.56T	[3]
[Nb] + 2[O] = NbO$_{2(s)}$	-375350+118.28T	[1]
SiO$_{2(s)}$ = Si$_{(s)}$ + O$_2$	905840-175.52T	[4]
Si$_{(s)}$ = Si$_{(l)}$	50540-30.00T	[5]
Si$_{(l)}$ = [Si]	-131500-17.24T	[6]

由掺杂 MgO 的半稳定的 ZrO$_2$ 管的电子导电所引起的误差，用 Schmalzfried 公式[7]校正，其特征氧分压 p'_e 用抽氧法测定，测得的 p'_e 值为：

$$1823K \quad p'_e = 0.116 \times 10^{-16}$$
$$1853K \quad p'_e = 0.996 \times 10^{-16}$$
$$1883K \quad p'_e = 8.550 \times 10^{-16}$$

由测量的电动势值计算得出的 a_O 值可以根据式（1）计算 a_{Si} 值：

$$[Si] + 2[O] = SiO_{2(s)}; \quad \Delta G^{\ominus} = -550830 + 207.20T \quad (1)$$

三个温度下的实验测量值 E 及计算出的 a_O 及 a_{Si} 值均列于表 2～表 4。

表2 1823K 时 Fe-Si-Nb 系的实验数据
Table 2 Experimental data of Fe-Si-Nb System at 1823K

No.	[%Nb]	[%Si]	E/V	a_O	a_{Si}	Same concentration basis		Same activity basis	
						a_{Si}^{Fe-Si}	lg f_{Si}^{Nb}	a_{Si}^{Fe-Si}	lg f_{Si}^{Nb}
1	0.11	0.34	0.3318	0.00513	0.416	0.370	0.0508	0.377	0.0445
2	0.20	0.45	0.3448	0.00431	0.589	0.503	0.0680	0.513	0.0572
3	0.20	1.21	0.3890	0.00235	1.99	1.63	0.0849	1.39	0.0608

续表 2

No.	[%Nb]	[%Si]	E/V	a_O	a_{Si}	Same concentration basis		Same activity basis	
						$a_{Si}^{Fe\text{-}Si}$	$\lg f_{Si}^{Nb}$	$a_{Si}^{Fe\text{-}Si}$	$\lg f_{Si}^{Nb}$
4	0.20	1.44	0.4002	0.00204	2.63	2.06	0.107	1.70	0.0721
5	0.20	1.98	0.4148	0.00167	3.93	3.24	0.0841	2.24	0.0530
6	0.39	1.16	0.3944	0.00221	2.25	1.55	0.162	1.50	0.110
7	0.59	1.16	0.4004	0.00203	2.65	1.55	0.233	1.64	0.151
8	0.61	1.96	0.4284	0.00138	5.72	3.19	0.254	2.73	0.144
9	0.62	0.50	0.3645	0.00331	0.999	0.566	0.247	0.753	0.178
10	0.64	1.44	0.4140	0.00169	3.84	2.06	0.271	2.10	0.164
11	0.65	0.39	0.3574	0.00364	0.825	0.430	0.283	0.622	0.203
12	0.76	1.16	0.4055	0.00190	3.04	1.55	0.294	1.77	0.183
13	0.80	1.46	0.4202	0.00155	4.56	2.10	0.337	2.29	0.196
14	0.80	1.99	0.4343	0.00127	6.75	3.26	0.315	2.96	0.172
15	0.89	0.65	0.3834	0.00256	1.67	0.764	0.338	1.09	0.223
16	1.05	1.44	0.4228	0.00149	4.90	2.06	0.377	2.36	0.214
17	1.08	1.90	0.4368	0.00123	7.23	3.05	0.375	3.02	0.201
18	1.11	0.66	0.3883	0.00239	1.90	0.778	0.389	1.17	0.248
19	1.32	0.56	0.3819	0.00262	1.60	0.644	0.395	1.01	0.255
20	1.36	1.31	0.4202	0.00155	4.56	1.81	0.400	2.22	0.229
21	1.41	1.74	0.4314	0.00133	6.22	2.68	0.366	2.76	0.201
22	1.48	1.18	0.4183	0.00159	4.33	1.58	0.437	2.10	0.250
23	1.49	0.73	0.3942	0.00221	2.23	0.875	0.407	1.31	0.254
24	1.51	0.44	0.3742	0.00290	1.30	0.491	0.422	0.826	0.273

表 3 1853K 时 Fe-Si-Nb 系的实验数据

Table 3 Experimental data of Fe-Si-Nb System at 1853K

No	[%Nb]	[%Si]	E/V	a_O	a_{Si}	Same concentration basis		Same activity basis	
						$a_{Si}^{Fe\text{-}Si}$	$\lg f_{Si}^{Nb}$	$a_{Si}^{Fe\text{-}Si}$	$\lg f_{Si}^{Nb}$
1	0.17	0.37	0.3254	0.00648	0.470	0.406	0.0642	0.420	0.0552
2	0.20	0.44	0.3322	0.00591	0.564	0.491	0.0604	0.495	0.0513
3	0.20	1.16	0.3755	0.00329	1.82	1.55	0.0710	1.31	0.0519
4	0.20	1.43	0.3852	0.00288	2.38	2.04	0.0667	1.59	0.0467
5	0.20	1.98	0.4028	0.00226	3.87	3.23	0.0780	2.22	0.0494
6	0.22	0.52	0.3413	0.00523	0.720	0.592	0.0854	0.611	0.0697

续表 3

No	[%Nb]	[%Si]	E/V	a_O	a_{Si}	Same concentration basis		Same activity basis	
						$a_{Si}^{Fe\text{-}Si}$	$\lg f_{Si}^{Nb}$	$a_{Si}^{Fe\text{-}Si}$	$\lg f_{Si}^{Nb}$
7	0.39	1.16	0.3808	0.00306	2.11	1.55	0.134	1.43	0.0934
8	0.60	0.47	0.3448	0.00499	0.791	0.528	0.176	0.640	0.134
9	0.62	0.36	0.3342	0.00576	0.595	0.394	0.179	0.497	0.140
10	0.63	1.43	0.3962	0.00247	3.22	2.04	0.199	1.91	0.126
11	0.63	1.93	0.4100	0.00204	4.73	3.11	0.182	2.48	0.108
12	0.64	1.11	0.3844	0.00291	2.32	1.46	0.202	1.51	0.135
13	0.74	1.28	0.3908	0.00267	2.77	1.76	0.198	1.72	0.129
14	0.75	1.97	0.4140	0.00193	5.29	3.21	0.217	2.63	0.126
15	0.77	1.20	0.3902	0.00269	2.73	1.62	0.228	1.68	0.147
16	0.79	0.46	0.3488	0.00473	0.881	0.516	0.233	0.681	0.171
17	0.86	0.62	0.3710	0.00350	1.61	0.723	0.348	1.05	0.229
18	1.03	1.93	0.4188	0.00180	6.06	3.11	0.289	2.80	0.162
19	1.10	1.43	0.4055	0.00217	4.17	2.04	0.311	2.19	0.184
20	1.12	1.17	0.3980	0.00241	3.39	1.56	0.336	1.87	0.204
21	1.33	1.73	0.4188	0.00180	6.06	2.66	0.358	2.73	0.198
22	1.36	1.30	0.4042	0.00221	4.02	1.79	0.351	2.09	0.207
23	1.46	0.44	0.3570	0.00423	1.10	0.491	0.351	0.760	0.238
24	1.51	1.18	0.4002	0.00234	3.60	1.58	0.357	1.93	0.214
25	1.57	0.70	0.3784	0.00316	1.97	0.833	0.374	1.21	0.239

表 4 1883K 时 Fe-Si-Nb 系的实验数据

Table 4 Experimental data of Fe-Si-Nb System at 1883K

No.	[%Nb]	[%Si]	E/V	a_O	a_{Si}	Same concentration basis		Same activity basis	
						$a_{Si}^{Fe\text{-}Si}$	$\lg f_{Si}^{Nb}$	$a_{Si}^{Fe\text{-}Si}$	$\lg f_{Si}^{Nb}$
1	0.17	0.34	0.3084	0.00922	0.410	0.370	0.0447	0.372	0.0394
2	0.20	1.12	0.3602	0.00455	1.68	1.48	0.0567	1.23	0.0423
3	0.20	1.42	0.3714	0.00389	2.30	2.02	0.0571	1.56	0.0403
4	0.20	1.95	0.3874	0.00311	3.61	3.16	0.0585	2.13	0.0378
5	0.21	0.42	0.3188	0.00802	0.542	0.466	0.0659	0.478	0.0560
6	0.26	0.50	0.3257	0.00730	0.654	0.566	0.0637	0.564	0.0525

续表 4

No.	[%Nb]	[%Si]	E/V	a_O	a_{Si}	Same concentration basis		Same activity basis	
						a_{Si}^{Fe-Si}	$\lg f_{Si}^{Nb}$	a_{Si}^{Fe-Si}	$\lg f_{Si}^{Nb}$
7	0.39	1.14	0.3644	0.00429	1.89	1.51	0.0976	1.34	0.0702
8	0.59	1.94	0.3942	0.00282	4.39	3.13	0.146	2.38	0.0889
9	0.61	0.36	0.3188	0.00802	0.542	0.394	0.139	0.466	0.112
10	0.61	0.41	0.3244	0.00743	0.631	0.454	0.143	0.533	0.114
11	0.64	1.40	0.3786	0.00352	2.82	1.98	0.153	1.77	0.101
12	0.75	1.28	0.3748	0.00371	2.53	1.76	0.159	1.64	0.106
13	0.75	1.99	0.3969	0.00271	4.74	3.25	0.163	2.49	0.0975
14	0.77	1.17	0.3702	0.00396	2.22	1.56	0.153	1.49	0.105
15	0.78	0.44	0.3288	0.00700	0.711	0.491	0.161	0.587	0.125
16	0.87	0.61	0.3449	0.00562	1.10	0.709	0.192	0.844	0.141
17	1.00	0.49	0.3385	0.00622	0.926	0.553	0.212	0.715	0.164
18	1.05	1.16	0.3758	0.00366	2.60	1.55	0.226	1.63	0.148
19	1.06	1.42	0.3843	0.00325	3.31	2.02	0.215	1.94	0.136
20	1.07	1.94	0.4012	0.00255	5.36	3.13	0.233	2.64	0.134
21	1.11	0.61	0.3498	0.00525	1.26	0.709	0.250	0.912	0.176
22	1.30	1.25	0.3808	0.00341	2.99	1.70	0.246	1.79	0.156
23	1.31	0.51	0.3437	0.00571	1.07	0.579	0.266	0.788	0.189
24	1.32	1.72	0.3962	0.00274	4.64	2.63	0.247	2.41	0.146
25	1.41	0.44	0.3369	0.00627	0.886	0.491	0.257	0.675	0.186
26	1.46	1.15	0.3801	0.00344	2.94	1.53	0.284	1.74	0.179
27	1.57	0.67	0.3582	0.00468	1.59	0.791	0.304	1.07	0.204

对于 Fe-Si-Nb 系有：

$$f_{Si} = f_{Si}^{Nb} f_{Si}^{Si} f_{Si}^{O}$$

由于 [%O] 很小，$f_{Si}^{O} \approx 1$，故有：

$$f_{Si} = f_{Si}^{Nb} f_{Si}^{Si}$$

$$f_{Si}^{Nb} = \frac{f_{Si}}{f_{Si}^{Si}} \tag{2}$$

（1）利用同一浓度法求 f_{Si}^{Nb}：

$$[\%Si]^{Fe-Si-Nb} = [\%Si]^{Fe-Si}$$

则：

$$f_{Si}^{Nb} = \frac{f_{Si}}{f_{Si}^{Si}} = \frac{a_{Si}^{Fe-Si-Nb}}{a_{Si}^{Fe-Si}} \tag{3}$$

其中 a_{Si}^{Fe-Si} 由以下公式求得：

$$e_{Si}^{Si} = \frac{34.5}{T} + 0.089^{[6]} \tag{4}$$

$$a_{Si}^{Fe-Si} = 10^{e_{Si}^{Si}[\%Si]^{Fe-Si}}[\%Si]^{Fe-Si} \tag{5}$$

(2) 利用同一活度法求 f_{Si}^{Nb}:

$$a_{Si}^{Fe-Si-Nb} = a_{Si}^{Fe-Si}$$

$$f_{Si}[\%Si]^{Fe-Si-Nb} = f_{Si}^{Si}[\%Si]^{Fe-Si}$$

则:
$$f_{Si}^{Nb} = \frac{f_{Si}}{f_{Si}^{Si}} = \frac{[\%Si]^{Fe-Si}}{[\%Si]^{Fe-Si-Nb}} \tag{6}$$

式 (6) 中 $[\%Si]^{Fe-Si-Nb}$ 已由试验取样用化学分析得出，根据公式 (5):

$$a_{Si}^{Fe-Si-Nb} = a_{Si}^{Fe-Si} = 10^{e_{Si}^{Si}[\%Si]^{Fe-Si}}[\%Si]^{Fe-Si}$$

也即
$$\lg a_{Si}^{Fe-Si-Nb} = e_{Si}^{Si}[\%Si]^{Fe-Si} + \lg[\%Si]^{Fe-Si} \tag{7}$$

因之，由公式 (7) 可用迭代法求出 $[\%Si]^{Fe-Si}$。

计算出的 $\lg f_{Si}^{Nb}$ 值见表 2～表 4。

分别绘出三个温度下 $\lg f_{Si}^{Nb}$ 对 [%Nb] 的关系图 (图 1～图 3)，则曲线在原点处切线的斜率即 e_{Si}^{Nb}。

利用:
$$e_i^j = \frac{1}{230}\left[(230e_j^i - 1)\frac{M_i}{M_j} + 1\right]^{[8]} \tag{8}$$

其中，M_i 为 i 组元的原子量，M_j 为 j 组元的原子量，可将 e_{Si}^{Nb} 换算成 e_{Nb}^{Si}。活度相互作用系数的总结果见表 5。

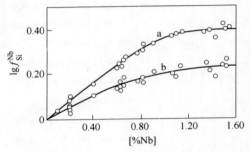

图 1　$\lg f_{Si}^{Nb}$ 对 [%Nb] 图 (1823K)
Fig. 1　$\lg f_{Si}^{Nb}$ vs. [%Nb] at 1823K
a—same concentration basis; b—same activity basis

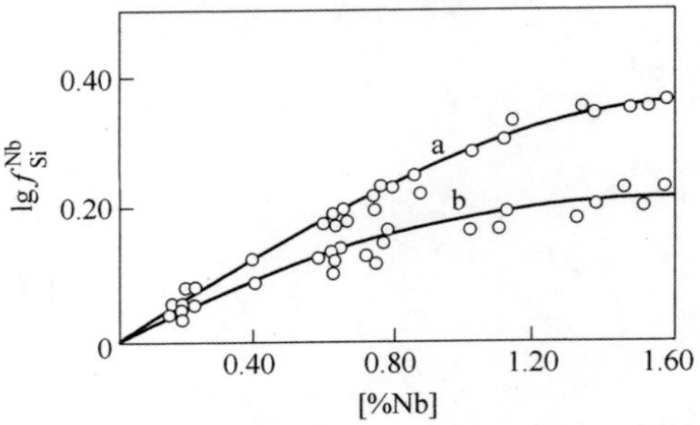

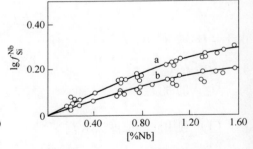

图 2　$\lg f_{Si}^{Nb}$ 对 [%Nb] 图 (1853K)
Fig. 2　$\lg f_{Si}^{Nb}$ vs. [%Nb] at 1853K
a—same concentration basis; b—same activity basis

图 3　$\lg f_{Si}^{Nb}$ 对 [%Nb] 图 (1883K)
Fig. 3　$\lg f_{Si}^{Nb}$ vs. [%Nb] at 1883K
a—same concentration basis; b—same activity basis

表5 不同温度下的 e_{Si}^{Nb} 和 e_{Nb}^{Si}

Table 5 Values of e_{Si}^{Nb} and e_{Nb}^{Si} at different temperatures

Temp. /K	Based on same conc'n		Based on same activity	
	e_{Si}^{Nb}	e_{Nb}^{Si}	e_{Si}^{Nb}	e_{Nb}^{Si}
1823	0.42	1.38	0.31	1.02
1853	0.30	0.98	0.27	0.88
1883	0.22	0.72	0.16	0.52
T	$\dfrac{11454}{T}-5.87$ ($\tau=0.994$)	$\dfrac{37763}{T}-19.36$ ($\tau=0.994$)	$\dfrac{8559}{T}-4.37$ ($\tau=0.963$)	$\dfrac{28535}{T}-14.60$ ($\tau=0.967$)

2 讨论

2.1 NbO_2 能否生成的问题

在测定 e_{Si}^{Nb} 时，本研究工作成功的关键在于避免了 NbO_2 的生成，而使 Si 脱氧反应长久地保持平衡，也即：

$$[Nb] + 2[O] \rightleftharpoons NbO_{2(s)}$$
$$[Si] + 2[O] \rightleftharpoons SiO_{2(s)}$$

达成上列条件的判据是实际的 Nb 脱氧活度积 $a_{Nb}a_O^2$ 应小于 Nb 的平衡脱氧常数 $(a_{Nb}a_O^2)_{平衡}$：

$$a_{Nb}a_O^2 < (a_{Nb}a_O^2)_{平衡}$$

Nb 的脱氧常数为：1823K 2.65×10^{-5}、1853K 3.95×10^{-5}，1883K 5.83×10^{-5}，这就需要对每一项实验计算出 a_{Nb}。

通常采用直线关系的 f_{Nb} 进行计算，即：

$$a_{Nb} = f_{Nb}[\%Nb]$$
$$lga_{Nb} = lg[\%Nb] + e_{Nb}^{Nb}[\%Nb] + e_{Nb}^{Si}[\%Si] \tag{9}$$

人们习惯于用同一浓度法的 e_{Nb}^{Si} 进行计算。显然当采用同活度法的 e_{Nb}^{Si} 计算时，所得的 a_{Nb} 值则较低，但必须注意，e_i^j 只有在浓度 j 较低时才适用，因为 lgf_i^j 对 [%j] 的曲线在高浓度 j 时趋向平坦，其斜率逐渐变小。对高浓度时如仍采用式（9）计算 a_{Nb}，其值将会太高，因而活度积 $a_{Nb}a_O^2$ 值也太高，此不正确的太高值不能用以和平衡脱氧常数进行比较，从而作出 NbO_2 是否生成的论断。按式（9）以同一浓度法的 e_{Nb}^{Si} 计算 $a_{Nb}a_O^2$，在上列三个温度 76 项实验中，有 49% 的 $a_{Nb}a_O^2$ 值小于脱氧平衡常数，但我们不能断言其他 51% 的值是不成功的实验。我们应当求出各浓度真实的 lgf_i^j 值，这可以由 lgf_i^j 对 [%j] 的实验曲线读出。文献 [1] 有 lgf_{Nb}^{Nb} 对 [%Nb] 的曲线，因之可读出相应的 lgf_{Nb}^{Nb} 值。本文未求出 lgf_{Nb}^{Si} 对 [%Si]

的曲线,但按下列推理可以从 $\lg f_{\text{Si}}^{\text{Nb}}$ 对 [%Nb] 曲线求出前曲线的相应值。

式(8)可换写为:

$$e_i^j = \frac{M_i}{M_j}e_j^i + \frac{1}{230}\left(1 - \frac{M_i}{M_j}\right) \tag{10}$$

也即:

$$e_{\text{Nb}}^{\text{Si}} = 3.31 e_{\text{Si}}^{\text{Nb}} + \frac{1}{230}(-2.31) \tag{11}$$

式(11)右方第二项等于 -0.01 可忽略不计,因之得出:

$$e_{\text{Nb}}^{\text{Si}} = 3.31 e_{\text{Si}}^{\text{Nb}} \tag{12}$$

在同一浓度下

$$\lg e_{\text{Nb}}^{\text{Si}} = 3.31 \lg f_{\text{Si}}^{\text{Nb}} \tag{13}$$

式(13)在 $\lg f_i^j$ 曲线的直线范围内适用,但可以认为它也适用于曲线的非直线部分。这样,我们可以按曲线由某浓度的 $\lg f_{\text{Si}}^{\text{Nb}}$ 求出同浓度的 $\lg f_{\text{Si}}^{\text{Nb}}$ 值,从而对 a_{Nb} 进行更合理更正确的计算。由此计算出的 $a_{\text{Nb}}a_{\text{O}}^2$ 值小于平衡脱氧常数的有 54 项,占全部实验 76 项的 71%。鉴于式(1)的 $\Delta G^{\ominus}$ 值有实验误差,假定其相对误差为 2.5%,则 $a_{\text{Nb}}a_{\text{O}}^2$ 值大于平衡脱氧常数的只有 5 项(1823K 4 项,1853K 1 项),因此在该误差范围内成功的实验占 93.5%。

由上列计算可以得出结论,正如作者在文献[9]中所推论的,本研究的绝大多数的实验是符合 NbO_2 不能生成的条件,因而所得结果是可以信赖的。

再次指出,对高浓度范围用直线关系计算 $\lg f_i^j$ 足以导致错误的结论。我们应当尽可能地由实验曲线求出真实的 $\lg f_i^j$ 值。

文献[10]给出 1873K 的 e_i^j 值:$e_{\text{Si}}^{\text{Nb}}$ 为 0、0.04;$e_{\text{Nb}}^{\text{Si}}$ 为 -0.01、0.13。由于未看到两种数据的原始文献,不能作进一步的比较和分析。

2.2 电动势 E 取值的合理性问题

实验测出的电动势随时间变化的曲线一般如图 4 所示,AB 段为定氧测头的温度逐渐达到平衡建立稳定氧分压的过程;由于测头插入熔体前在液面上预热 2min,因此这段很短或不出现。BC 段所对应的 E 值为我们读取的电池的可逆电动势,此时由电子导电引起的极化现象[11]还未出现或不明显。其后

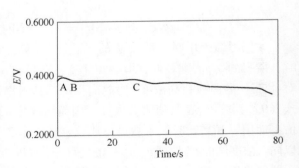

图 4　EMF 随时间变化

Fig. 4　Variation of measured EMF with time

极化现象发生，E 随时间延长而不断下降，氧化锆管内壁由于氧的传递出现一层极薄的金属 Mo，与此同时，氧化锆管的外壁接近铁液处出现一富氧层。将 1823K 下测量 135s 的氧化锆管纵向剖开作电子探针定量分析，在管内壁不同地方得到结果如下：

	Al	Zr	Mo	Ca	Si
1号	22.13	11.38	63.37	1.57	1.55
2号	21.43	7.06	71.02	0.49	—

分析结果表明，E 随时间变化不断下降是由管的内壁生成金属 Mo 薄膜的极化现象引起的，时间越长，极化现象越明显，因此选取最初稳定的电动势值是合理的。

3 结论

采用石英坩埚，利用电化学法测定 Fe-Si-Nb 熔体中的 a_O，用反应式 [Si] + 2[O] = SiO$_{2(s)}$ 研究 Nb 对 Si 活度的影响，用同一浓度法及同一活度法求出 e_{Si}^{Nb}，最后用惯用方法求出 e_{Nb}^{Si}。同一浓度法计算出的值较用同一活度法求出的值为高。

同一浓度法：$\quad e_{Si}^{Nb} = \dfrac{11454}{T} - 5.875;\ 1873K \quad e_{Si}^{Nb} = 0.25$

$\quad\quad\quad\quad\quad e_{Nb}^{Si} = \dfrac{37763}{T} - 19.36;\ 1873K \quad e_{Nb}^{Si} = 0.80$

同一活度法：$\quad e_{Si}^{Nb} = \dfrac{8559}{T} - 4.37;\ 1873K \quad e_{Si}^{Nb} = 0.20$

$\quad\quad\quad\quad\quad e_{Nb}^{Si} = \dfrac{28535}{T} - 14.60;\ 1873K \quad e_{Nb}^{Si} = 0.63$

致谢：实验过程中得到北京钢铁研究总院九室、北京理化测试中心有关同志及北京钢铁学院冶金物化教研室有关教师大力帮助，谨致谢意。

参 考 文 献

[1] 魏寿昆，张圣弼，佟亭，谭赞麟. 稀有金属，1983，2（1）：10.
[2] 张圣弼，佟亭，王济舫，魏寿昆. 金属学报，1984，20：A348.
[3] Janke D, Fischer W A. Arch Eisenbuettenwes, 1975, 46: 755.
[4] Turkdogan E T. Physical Chemistry High Temperature Technology. NewYork: Academic Press, 1980: 20.

[5] Hultgren R, Desai P D, Hawkins D T, et al. eds. Selected Values of the Thermodynamic Properties of the Elements. Ohio: ASM, 1973: 467.
[6] Sigworth G K, Elliott J F. Met Sci, 1974, 8: 298.
[7] Schmalzried H. Z Phys Chem, Neue Folge, 1963, 38: 87.
[8] Schenck H, Frohberg M G, Steinmetz E. Arch Eisenhuettenwes, 1960, 31: 671.
[9] Shoukun Wei, Ting Tung, et al. Solid State Ionics, 1986, 18 & 19: 1232.
[10] 日本铁锅協会. 鉄鋼便覽, I. 基礎. 九善株式会社, 第三版, 1981: 19.
[11] Janke D. Arch Eisenhuettenwes, 1983, S4: 259.

Selective Oxidation of Elements in Metal Melt and Their Multireaction Equilibria*

Wei Shoukun

Abstract: Oxidation of elements in metal melt falls into two categories, as shown from the modified Ellingham-Richardson diagram. For the first category, at which the ΔG-T lines of two reactions intersect, a concept of transition temperature has been introduced. With control of this transition temperature, blowing operation of the metal melt could be so conducted that a certain element is selectively oxidized while the other remains intact in the melt, so that separation of these elements could be accomplished. Practical application of this transition temperature is discussed with respect to Ni vs. S, Cr, V, Nb, Mn or P vs. C. For the second category with ΔG-T lines nearly parallel side by side, thermodynamic analysis has shown that selective oxidation of a certain element can also occur, but as oxidation continues, its free energy change becomes equal to that of another element, so that simultaneous oxidation of both elements takes place. This is followed by simultaneous oxidation of other elements, until the final equilibrium composition of the metal melt is attained. Examples of selective oxidation followed by simultaneous oxidation are set forth in respect to deoxidation of liquid iron with Al and Si, and the oxidizing blowing of Nb-bearing hot metal. Computer calculation of the equilibrium composition of the melt with respect to the referred elements has been made. Lastly, the role in the oxidation played by Fe, the main ingredient of the melt, is briefly discussed.

When metal melts containing more than one element are subjected to oxidation, metallurgists are faced with the problem of investigating the sequence of oxidation of different elements and their residual composition at equilibrium of the multireactions. The selective oxidation of certain elements in preference to others has been wellknown in steelmaking since a long time ago. In the Bessemer process[1], see Fig. 1, Si, Mn are oxidized at the beginning stage of blowing in preference to C, while in the Thomas process[2], see Fig. 2, the same sequence of oxidation of Si, Mn in preference to C occurs, and it is only after the removal of C to minimum that P begins to be oxidized in the afterblow. It is

---

* 原刊于《Steel Research》,1988,59:381~393。

sometimes desirable to take advantage of the selective oxidation for extraction of a certain element or elimination of impurities in industrial metallurgical practices, whereby certain elements are eliminated by being oxidized into slag or escaping as a gas, while the other element remains intact as an ingredient of the metal bath. Principles and governing factors for accomplishing the elimination of certain elements from the metal bath and intensional reservation of some other elements within the bath have to be studied.

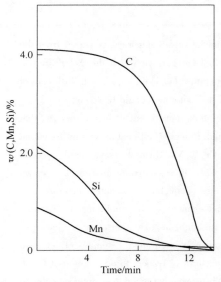

Fig. 1   Change of metal composition with time of blowing in the Bessemer process

Fig. 2   Change of metal composition with time of blowing in the Thomas process

The sequence of oxidation of elements is governed by the principle of minimum free energy of reaction. This reaction which accompanies the smallest change in free energy will take place first, and the sequence of multireactions follows in general the order of magnitude of the free energy of reaction. A modified Ellingham-Richardson diagram[3] showing the relation of the free energy of oxidation of elements in liquid iron (based on 1wt. % standard solution) vs. temperature is given in Fig. 3. For comparison the diagram is drawn on a 1 mol O_2 basis. Obviously, it can be shown that the diagram really consists of three regions with 2 categories. The upper and lower regions with series of ΔG lines nearly parallel one below the other, form one category, while the middle region where, in addition to the series of nearly parallel ΔG lines, also contains a ΔG line of C with a negative slope, forms the other calegory. The temperature at which ΔG lines of C and another element intersect and the sequence of oxidation of the two elements changes in direction, has been called the transition temperature of the two reactions. In the first part of this paper theoretical considerations concerning the concept of transition temperature and its applications in metallurgical industries are dealt with, while in the second part of the paper multireaction equilibria are to be discussed.

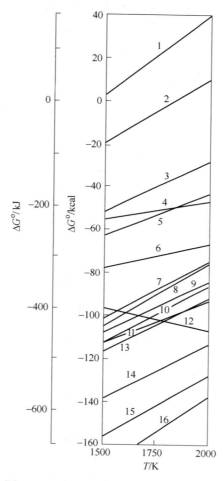

Fig. 3 ΔG°-T lines of oxidation of elements in molten iron with oxygen

1 Concept of transition temperature

Thermodynamic considerations of the sequence of ΔG-T lines of oxidation of elements infers that any elements with its ΔG-T line lower than that of another element will be oxidized preferably to that element. Thus, with Al and Si in iron melt, when they meet O_2 simultaneously, Al with the ΔG-T line lower than that of Si will be oxidized in preference to Si. This is due to the fact that, should Si be oxidized first, then Al will react with the

SiO$_2$, formed and reduce it back to Si with the formation of Al$_2$O$_3$, due to the free energy of reaction (1):

$$\frac{4}{3}[\text{Al}] + \text{SiO}_{2(s)} = [\text{Si}] + \frac{2}{3}\text{Al}_2\text{O}_{3(s)}$$

$$\Delta G° = -219740 + 35.73T \tag{1}$$

In the range of steelmaking temperatures $\Delta G°$ has a negative value; while, on the other hand, if Al$_2$O$_3$ has been formed, it would be impossible for Si to scavenge the oxygen away from the Al$_2$O$_3$. The above calculation is based on the standard state of the reactants and products of the reaction. Should the latter differ from the standard state, the same principle holds true, but conclousious must be drawn according to the calculation of ΔG based on the Van't Hoff's isotherm.

$$4[\text{Cu}] + \text{O}_2 = 2\text{Cu}_2\text{O}_{(l)} \tag{2}$$

$$2[\text{Ni}] + \text{O}_2 = 2\text{NiO}_{(s)} \tag{3}$$

$$\frac{4}{5}[\text{P}] + \text{O}_2 = \frac{2}{5}\text{P}_2\text{O}_{5\ (g)} \tag{4}$$

$$\frac{2}{3}[\text{Mo}] + \text{O}_2 = \frac{2}{3}\text{MoO}_{3(g)} \tag{5}$$

$$\frac{2}{3}[\text{W}] + \text{O}_2 = \frac{2}{3}\text{WO}_{3(l)} \tag{6}$$

$$2[\text{Fe}] + \text{O}_2 = 2\text{FeO}_{(l)} \tag{7}$$

$$\frac{4}{3}[\text{Cr}] + \text{O}_2 = \frac{2}{3}\text{Cr}_2\text{O}_{3(s)} \tag{8}$$

$$2[\text{Mn}] + \text{O}_2 = 2\text{MnO}_{(s)} \tag{9}$$

$$\frac{4}{3}[\text{V}] + \text{O}_2 = \frac{2}{3}\text{V}_2\text{O}_{3(s)} \tag{10}$$

$$[\text{Nb}] + \text{O}_2 = \frac{1}{2}\text{Nb}_2\text{O}_{4(s)} \tag{11}$$

$$\frac{4}{3}[\text{B}] + \text{O}_2 = \frac{2}{3}\text{B}_2\text{O}_{3(s)} \tag{12}$$

$$2[\text{C}] + \text{O}_2 = 2\text{CO}_{(g)} \tag{13}$$

$$[\text{Si}] + \text{O}_2 = \text{SiO}_{2(s)} \tag{14}$$

$$[\text{Ti}] + \text{O}_2 = \text{TiO}_{2(s)} \tag{15}$$

$$\frac{4}{3}[\text{Al}] + \text{O}_2 = \frac{2}{3}\text{Al}_2\text{O}_{3(s)} \tag{16}$$

$$\frac{4}{3}[\text{Ce}] + \text{O}_2 = \frac{2}{3}\text{Ce}_2\text{O}_{3(s)} \tag{17}$$

Under the category where the ΔG-T line of C intersects with that of other elements, such as Cr, Mn, V, Nb, B, Si etc, examples of Cr vs. C with the standard state calculation is illustrated in Fig. 4. In this figure, lines 1a and 1b show the ΔG°-T lines of oxidation of Cr and C, intersecting at 1570K. Below 1570K Cr is oxidized first in preference to C, while above 1570K C will be oxidized selectively. This transition temperature of oxidation of Cr vs. C may be read directly from the figure, or found by calculation with the the two-step method:

$$\frac{4}{3}[Cr] + O_2 = \frac{2}{3}Cr_2O_{3(s)}$$

$$\Delta G^\circ = -780310 + 233.59T \quad (18)$$

$$2[C] + O_2 = 2CO_{(g)}$$

$$\Delta G^\circ = -780310 + 233.59T \quad (19)$$

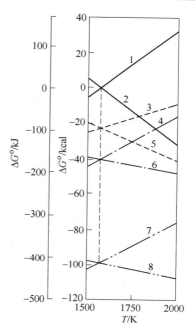

Fig. 4 Transition temperature of oxidation of Cr vs. C in molten iron

At the transition temperature, the ΔG° of the two reactions (18) and (19) are equal, from which one can get T = 1570K. This transition temperature can also be calculated with the one-step method by combining reaction (18) with reaction (19):

$$\frac{4}{3}[Cr] + 2CO = 2[C] + \frac{2}{3}Cr_2O_{3(s)}$$

$$\Delta G^\circ = -499150 + 317.77T \quad (20)$$

which is shown as line 4b in Fig. 4. This line intersects the zero-value line of ΔG° at a temperature which is the transition temperature of oxidation of Cr vs. C. By putting the ΔG° value of reaction (20) equal to zero, the transition temperature comes to be 1570K. The following reaction (21)

$$2[C] + \frac{2}{3}Cr_2O_{3(s)} = \frac{4}{3}[Cr] + 2CO$$

$$\Delta G^\circ = 499150 - 317.77T \quad (21)$$

denoted by line 4a is a reversal of reaction (20). Obviously at the transition temperature the two reactions (20) and (21) are at equilibrium. Taking C as a reducing agent, one can see that to reduce Cr_2O_3, with C, a minimum temperature of 1570K is required, so that this transition temperature of oxidation might also be called the minimum temperature of reduction of Cr_2O_3 by C. In Fig. 4 reactions of oxidation by [O] (lines 2a and 2b) and by (FeO) (lines 3a and 3b) are also shown (drawn with respect to 2 atoms of oxygen).

Each pair of lines intersect with each other at 1570K. Thus, it can be concluded that on the assumption that no nucleation energy is required for the formation of new phases (condensed particles or gaseous bubbles), the transition temperature of oxidation is independent of the form of oxidizing agent, be it gaseous oxygen, [O] dissolved in the melt, or (FeO) from the slag, and independent of the pressure (for gaseous oxygen) or activity (for [O] and (FeO)) of the oxidizing agent, but it depends solely upon the nature and activity (or pressure) of the reactants and products taking part in the reaction. Further, the transition temperature is not a fixed temperature, it varies continuously with the composition of the bath as the oxidation process continues. One point might be clarified as regards the kinetic aspects of the selective oxidation of elements. In the previous example of the selective oxidation of Cr vs. C, the statement has been made that below the transition temperature Cr is preferably oxidized to C, which remains intact or unchanged in the bath, but in actual practice of blowing, part of C is still being oxidized, although the temperature of the bath lies far below the transition temperature corresponding to the composition of the bath. The selective oxidation occurs only on the condition that Cr and C meet O_2 simultaneously. But in the iron bath, as the blowing continues, the elimination of Cr might have gone so far that at the point of oxygenblow, no more Cr atoms could diffuse up to the surface, then C with a concentration much higher than that of Cr would be able to diffuse and meet the oxygen at the surface and thus be oxidized.

(1) $\frac{4}{3}[Cr] + 2CO = 2[C] + \frac{2}{3}Cr_2O_3;$

(2) $2[C] + \frac{2}{3}Cr_2O_3 = \frac{4}{3}[Cr] + 2CO;$

(3) $\frac{3}{4}[Cr] + 2(FeO) = 2[Fe] + \frac{2}{3}Cr_2O_3;$

(4) $\frac{4}{3}[Cr] + 2[O] = \frac{2}{3}Cr_2O_3;$

(5) $2[C] + 2(FeO) = 2[Fe] + 2CO;$

(6) $2[C] + 2[O] = 2CO;$

(7) $\frac{4}{3}[Cr] + O_2 = \frac{2}{3}Cr_2O_3;$

(8) $2[C] + O_2 = 2CO.$

In fact, if no C atoms are present at the surface, iron atoms would surely be oxidized. It would be most likely that, when O_2 attacks the iron bath, the iron atoms present in great abundance would be oxidized to FeO first, which dissolves as O in the bath (refer to Fig. 5). Where O meets C in absence of Cr, CO would form. Should the CO bubble on its way up meet Cr, the latter would reduce CO back to C and Cr_2O_3, formed if the temperature of the bath lies below the transition temperature. Under absence of Cr, CO would escape reduction and float up to the surface. In this way C is eventually oxidized to a certain extent.

```
—Cr₂O₃——————CO       O₂
    ↑        ↑       +
Cr₂O₃+[C]←[Cr]      Fe→FeO₍ₗ₎→Fe+[O]————Cr₂O₃
           +                        ↑
       CO←CO←——[O]           [O]→Cr₂O₃
                  +          +
                 [C]        [C]   [Cr]
```

(Temperature of bath below the transition temperature of Cr vs.C)

Fig. 5 Kinetic aspects of oxidation of Cr vs. C in molten iron

Intersection of ΔG-T lines of oxidation of elements is also encountered in pyrometallurgy of non-ferrous metals. in which S takes the role of C. see Fig. 6, as discussed below. In general, reactions with transition temperature are of the following type:

$$x[M] + \frac{y}{2}O_2 = M_xO_y$$

$$z[M'] + \frac{y}{2}O_2 = M'_zO_{y(g)}$$

where M' is either C or S with a gaseous product.

2 Applications of transition temperature[6]

Ni vs. S(Desulfurization of Ni matte). It has been intended to win metallic nickel directly from nickel matte, an intermediate product from pyrometallurgy of nickel. This matte contains 20%~22% S, about 70% Ni, the rest being Cu, Fe and Co. Preliminary experiments in 1971 carried out in a 200kg Kaldo converter with O_2-blowing have shown that it was very difficult to reduce the S-content to below 4%~5%; and a large amount of solid crust, presumedly NiO, was accumulated especially in the latter period of blowing at the mouth of the Kaldoconverter, resulting sometimes in compulsory stopping-down of the furnace. With

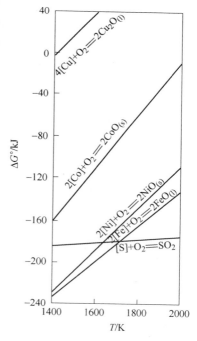

Fig. 6 $\Delta G°$-T lines of oxidation of Cu, Co, Ni, Fe and S in Ni-matte

introduction of the concept of transition temperature, thermodynamic analysis was resorted to with the aim of eliminating S and preserving Ni. Calculations of the transition temperatures using different compositions of the matte as met in practice and with introduction of activity coefficient for the following two reactions : $[S]+O_2 = SO_2$, $2[Ni] + O_2 = 2NiO_{(s)}$ is shown in Table 1[7]. The ΔG-T lines for the initial composition of the matte is shown in Fig. 6. Conclusions as to the guide points for the blowing operation could be drawn as follow:

(1) To avoid the oxidation of Ni, blowing should begin at a bath temperature always higher that the transition temperature of oxidation of Ni vs. S at the specific composition of the melt.

(2) The blowing operation as governed by the flow rate of oxygen should be so conducted that the rate of rise of bath temperature (due to the exothermic oxidation of S) should be always higher than the rate of increase in the transition temperature brought about by the change in composition of the melt.

Table 1 Transition temperature of Ni vs. S in Ni-matte

| Composition/% | | | | | ΔG/J | | Transition |
Ni	Cu	Fe	Co	S	$[S]+O_2 = SO_2$	$2[Ni]+O_2 = 2NiO_{(s)}$	temperature/℃
70.0	5.0	3.0	0.7	21.3	$-208150 + 17.24T$	$-505010 + 197.82T$	1370
82.0	5.9	1.3	0.8	10.0	$-208150 + 23.18T$	$-505010 + 197.82T$	1470
87.6	6.3	1.4	0.9	3.8	$-208150 + 31.63T$	$-505010 + 197.82T$	1580
90.3	6.4	1.4	0.9	1.0	$-208150 + 42.72T$	$-505010 + 197.82T$	1730
91.1	6.5	1.4	0.9	0.1	$-208150 + 61.84T$	$-505010 + 197.82T$	2030

With these deductions in mind, experiments in a 1.5t Kaldo converter were resumed in 1973~1974, measures being taken to maintain the two optimum conditions deduced above. Excellent results were achieved, S in the melt being smoothly reduced down to below 1%~2%, especially when at the last stage of blowing an empty running of the converter without O_2-blowing was adopted to eliminate the layer of NiO formed at the surface of the molten bath due to the slowing-down in diffusion rate of S to the surface. One example of blowing operation is shown in Table 2. A comparison between bath temperature at the end of blowing of some melts and the transition temperatures calculated is shown in Fig. 7. The calculation was made on the assumption of $p_{SO_2} = 0.7$atm, but in reality, particularly in the later period of blowing, p_{SO_2} was much less than this value, so that the maintenance of the bath temperature higher than the transition temperature was assured.

Table 2 Example of the blowing operation of Ni-matte

Condition	Composition/%					Bath temperature /℃	Transition temperature calculated/℃
	Ni	Cu	Co	Fe	S		
begin of blowing	69.76	5.27	0.93	1.69	20.27	1420	1373
blowing after 10min	79.04	5.85	0.93	1.74	10.57	1640	1459
blowing after 15min	83.55	6.14	0.81	1.57	5.09	1730	1542
blowing after 29min	84.61	6.82	0.76	0.87	2.48	1760	1622
empty running after 20min	89.65	6.91	0.79	1.06	1.31	1650	1693

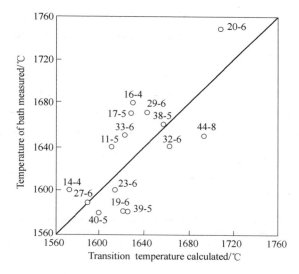

Fig. 7 Calculated transition temperature of oxidation of Ni vs. C compared with the measured temperature of the bath

2.1 Cr vs. C

Two cases need to be considered. In the first case it is aimed to eliminate Cr with reservation of C, blowing being conducted at a temperature lower than the transition temperature of oxidation of Cr vs. C, as is met with in the dechromizing of hot metal. The second case aims at the elimination of C with reservation of Cr, blowing being conducted at a temperature higher than the corresponding transition temperature, as is illustrated by the making of 18-8 stainless steel.

Dechromization of hot metal[8]. Cr-bearing hot metal must first be dechromized before it can be blown into steel of proper composition. It was intended to dechromize hot metal containing about 4% Cr, and experiments were carried out in 1974~1975 in a shaking ladle of 10 t capacity with O_2-blowing. Owing to the inefficient lowering of the temperature of blowing prevailing at that time (mostly 1450~1500℃), it was possible to reduce Cr down to only about 1%. Assuming that at the temperature of the emergence of the carbon flame the bath was at a transient equilibrium between slag and metal, and that this temperature of the bath was supposed to be the transition temperature of oxidation of Cr vs. C for the specific composition of the bath, the writer first proposed to estimate $\gamma_{Cr_2O_3}$, from the reaction (20) mentioned before with the result that the calculated value of $\gamma_{Cr_2O_3}$ was equal to 1.8×10^{-6}. This low value of $\gamma_{Cr_2O_3}$ indicates that Cr_2O_3 is combined with FeO to form the stable chrome-spinel. With the concept of transition temperature in mind, experiments were resumed in 1977~1978, with adaptation of continuous temperature recording, empty running of the furnace without blowing and addition of ore as cooling agent. These measures insured the keeping-down of the bath temperature below that of the transition temperature, with the result that Cr was successfully reduced down to below 0.2%. Table 3 shows one example of the heat records. From the data of a 1.5t shaking ladle the average value of $\gamma_{Cr_2O_3}$ was calculated to be 1.4×10^{-6}, as shown in Table 4. Although the value of $\gamma_{Cr_2O_3}$ needs to be confirmed by more elaborate experiments, use of the transition temperature of oxidation calculated with the estimated value of $\gamma_{Cr_2O_3}$ has brought about very successful results of blowing operation in reducing down the Cr content, refer to Table 5.

Table 3 Example of blowing operation of Cr-bearing hot metal

Condition	Composition/%				Continuous temperature recording/℃
	C	Cr	Si	Mn	
begin of blowing	4.45	4.14	2.12	0.28	1289
O_2-blowing, 6min	4.35	3.08	1.32	0.17	1400
empty running with ore, 6min	4.08	1.24	0.18	0.03	1370
O_2-blowing, 3.5min	3.52	0.45	0.10	—	1390
empty running, 3min	3.15	0.18	0.10	—	1399

Table 4 Estimated value of $\gamma_{Cr_2O_3}$

Heat No	Metal bath/%			Slag composition/%			$N_{Cr_2O_3}$	$\lg f_C$	$\lg f_{Cr}$	Tempetature /℃	$\gamma_{Cr_2O_3}$
	C	Cr	Si	SiO_2	Cr_2O_3	FeO					
27	3.40	0.14	0.10	24.1	25.8	35.8	0.159	0.481	-0.408	1389	0.8×10^{-6}
28	3.30	0.12	0.10	22.6	27.4	34.0	0.175	0.476	-0.396	1380	0.9×10^{-6}

Continued Table 4

Heat No	Metal bath/%			Slag composition/%			$N_{Cr_2O_3}$	$\lg f_C$	$\lg f_{Cr}$	Tempetature /℃	$\gamma_{Cr_2O_3}$
	C	Cr	Si	SiO_2	Cr_2O_3	FeO					
42	3.15	0.18	0.10	20.4	20.7	34.4	0.143	0.445	−0.378	1399	$2.0×10^{-6}$
43	3.21	0.29	0.10	21.5	24.0	38.4	0.150	0.451	−0.385	1416	$2.5×10^{-6}$
55	3.20	0.12	0.10	23.0	21.5	38.9	0.133	0.449	−0.384	1401	$0.8×10^{-6}$
											av. $1.4×10^{-6}$

Table 5 Calculated transition temperature of oxidation for Cr vs. C

C/%	4	3.5	3	4	3.5	3	4	3.5	3
Cr/%	0.5	0.5	0.5	0.2	0.2	0.2	0.1	0.1	0.1
Transition temperature/℃	1420	1462	1500	1365	1405	1440	1323	1357	1393

Decarburization in 18-8 stainless steelmaking[9]. It is wellknown that the lower the C content in 18-8 stainless steel, the better would be the resistance against corrosion. But the elimination of C with reservation of Cr in stainless steelmaking could only be achieved at a bath temperature higher than the transition temperature of oxidation for C vs. Cr. Tracing back the history of the development of 18-8 stainless steelmaking, the important role played by the transition temperature of oxidation for C vs. Cr became evident which not only gives the key explanation of the mechanism of removing C, but also forms a linkage between the three stages of the development of stainless steelmaking. In the twenties, when stainless steel was first invented, no chemical reactions but only melting operation predominated in the electrical furnace. Reutilization of stainless scraps was made impossible by the increase of C content during remelting in the electric furnace. Iron ore, the only adequate oxidizer known at that time, could not bring down the C content of the melt because of the endothermic nature of the oxidizing reaction as a whole (solid Fe_2O_3 as an oxidizer; temperature drop of 20℃ for burning down every 0.1% C, and temperature rise of 8℃ for oxidizing every 1% Cr), so that the bath temperature could never be raised higher than the transition temperature for a specific bath composition. It was only in the late thirties, when the Americans invented the injection of O_2 gas into the bath, that the bath temperature could be raised so that burning down of C and hence use of stainless steel scraps became possible (O_2 as the oxidizer: temperature increase of 12℃ for burning down every 0.1% C, and temperature rise of 110℃ for oxidizing every 1% Cr). But this second development stage was accompanied by a great demerit, in that full charge of Cr up to 18%~20% could not be adopted. Only a partial feed of Cr, approximately 12%~13% could be charged into the furnace, so that after O_2 blowing, a temperature rise of about 200℃ with a residual Cr content of about 10%~10.5% and with the necessary burning-down of C to the desired level could be achieved. As a result, to make 18% Cr stainless steel. extralow-C ferrochrome must be added to fill up the Cr deficiency. which naturally would result in an appreciable increase in cost of production. But why is it impossible to give a full charge of Cr to the electric furnace before O_2 blowing? This can only be answered from results of calculation of the transition temperature.

Taking the slag as saturated with Cr_3O_4 and using the following reaction:

$$\frac{3}{2}[Cr] + 2CO = 2[C] + \frac{1}{2}Cr_3O_{4(s)}$$

$$\Delta G° = -465260 + 307.69T \tag{22}$$

the author has made a series of calculations of the transition temperature of oxidation for C vs. Cr. From Table 6, it can be deduced that the higher the Cr content and the lower the C content in the bath, the higher would be the transition temperature to be surpassed in order to bring about a successful operation of the electric furnace for stainless steelmaking. By comparing row 1 with row 5, it is shown that an initial charge of 12% Cr after melting down can have its C content of 0.35% started to burn down at a bath temperature of 1555℃, whereas an initial charge of 18% Cr would require a bath temperature of 1627℃ to begin C-elimination. This would mean that the temperature would be rather too high to be carried out in the melting-down practice. Furthermore, by comparing row 4 and row 7, it can be shown that a bath of 10% Cr could have its C content brought down to 0.05% at a tolerable temperature of 1800℃, but it would be entirely out of question to accomplish this Celimination if the bath should contain 18% Cr because of the extrahigh temperature of 1945℃. For this reason a charge of only 12% Cr could be given to the bath to bring down the C content to 0.05% with a loss of about 2% Cr during the O_2-blowing. Oxidation of these 2% Cr would help to raise the bath temperature to 1800℃. It is shown also from row 8 to row 12 that lowering the partial pressure of CO would decrease the transition temperature appreciably. The latest development of stainless steelmaking, the AOD process, is founded, from a thermodynamic viewpoint, really on lowering of the transition temperature caused by mixed blowing of Ar and O_2, and as a result stainless steel with ultralow C content can be produced with the AOD process without any diffculty. Besides, as shown by the last two rows in the table, the AOD process could use in regard to the Cr and C content any material for blowing, which, of course, would greatly reduce the cost of production.

Table 6 Calculated transition temperature of oxidation for Cr vs C during the stainless steelmaking

Cr/%	Ni/%	C/%	p/atm	ΔG/J	Transition temperature/°C	O_2 : Ar : CO
12	9	0.35	1	$-465260+255.39T$	1555	—
12	9	0.1	1	$-465260+232.46T$	1727	—
12	9	0.05	1	$-465260+220.50T$	1835	—
10	9	0.05	1	$-465260+224.26T$	1800	—
18	9	0.35	1	$-465260+245.06T$	1627	—
18	9	0.1	1	$-465260+222.13T$	1820	—
18	9	0.05	1	$-465260+209.99T$	1945	—
18	9	0.35	2/3	$-465260+251.71T$	1575	1 : 1 : 2
18	9	0.05	1/2	$-465260+221.50T$	1830	1 : 2 : 2
18	9	0.05	1/5	$-465260+236.86T$	1690	1 : 8 : 2
18	9	0.05	1/10	$-465260+248.28T$	1600	1 : 18 : 2
18	9	0.02	1/20	$-465260+244.30T$	1630	1 : 38 : 2
18	9	1	1	$-465260+268.24T$	1460	—
18	9	4.5	1	$-465260+323.84T$	1165	—

2.2 V vs. C (Extraction of V from V-bearing hot metal)

Vanadium-bearing hot metal can be blown either with O_2 in a top-blown or bottom-blown converter or in a shakingladle : or with air in a side-blown converter, in a runninglaunder or with the spray-blowing method. During blowing, V is selectively oxidized into the slag, from which it can beextracted by roasting and leaching. With the reaction

$$\frac{2}{3}[V] + CO \rightleftharpoons [C] + \frac{1}{3}(V_2O_3)$$

$$\Delta G^\circ = -245890 + 147.82T \tag{23}$$

$\gamma_{Cr_2O_3}$ was estimated to be of the order of 10^{-5}, To blow hotmetal containing 0.3% ~ 0.45% V to obtain a residual V content of 0.03% ~ 0.05%, the blowing temperature should be below 1350 ~ 1400°C. In case that the hot metal contains much Si, cooling reagents should be added to keep down the temperature of the bath.

2.3 Nb vs. C (Extraction of Nb from Nb-bearing hot metal)

Nb can be selectively oxidized into slag by blowing withniobium-bearing hot metal with the same equipment as that for V. The hot metal can also be charged into an openhearth furnace, from which the first flush of slag, which contains most of the Nb in the hot metal, is collected and treated for Nb extraction. To preserve C with removal of Nb to a residual content of 0.02% ~ 0.03%, the blowing temperature should in general not be higher than 1400℃, the transition temperature of oxidation for Nb vs. C is somewhat higher than that for V vs. C. The latest investigation undertaken by the author and his associates[10] has verified the coexistence of Nb_2O_5 and Nb_2O_4 in the slag. From industrial data $\gamma_{Nb_2O_5}$ or $\gamma_{Nb_2O_4}$ has been estimated to be of the order of 10^{-9} to 10^{-10}. In case that the hot metal contains also P, bottomblowing can reserve P in the hot metal owing to the small FeO content of the slag.

2.4 Mn vs. C (Decarburization of high-C ferromanganese)

Thermodynamic calculation of the selective oxidation of C vs. Mn has been reported by Shao[11], who has proposed aprocess of making low-C ferromanganese by O_2-blowing of high-C ferromanganese at a temperature higher than the transition temperature of oxidation of C vs. Mn. For examplem, at 1640℃ or higher, a medium-C ferromanganese of 1% C could be obtained with a volatilization loss of Mn equal to 0.5% of the metal bath. To make a low-C ferromanganese of 0.1% C, the blowing temperature should be higher than 1980℃, but the volatilization loss of Mn would be increased to about 2.5% of the metal bath. Under a reduced pressure of 0.1 atm, the same content of 0.1% C could be obtained at 1680℃, but with a volatilization loss of Mn of 3%, based on the weight of the metal bath. No data of γ_{MnO} has been reported.

2.5 P vs. C (Dephosphorization of hot metal and steel)

The sequence of oxidation of P in respect to C and other elements in liquid iron has been discussed by the author elsewhere[12]. In accordance with the steelmaking processes, dephosphorization can be classified into 3 categories : (1) P being oxidized before C, preferably at a relatively low temperature (the pretreatment of hot metal, the puddling process and the basic open hearth process) ; (2) P being oxidized simultaneously with C (the LD-process) ; and (3) P being oxidized after C at a rather high temperature (the Thomas process). Although P_2O_5 is a gas at steelmaking temperature, P can never be removed as a gas thermodynamically. It is only in presence of a basic ferruginous slag that P can be oxidized and removed into the slag. Fig. 8[13] shows schematically the sequence of oxidation of P and C under different conditions. In Fig. 8(a), the ΔG line of oxidation of P is shifted from position a, at which dephosphorization in never possible, to position b in presence of basic ferruginous slag. This line bintersects the ΔG line e of oxidation of C

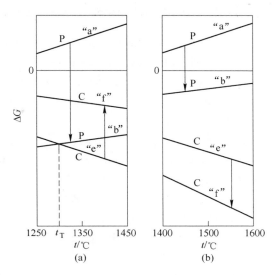

Fig. 8 Relation of dephosphorization with decarburization schematically presented
(a) dephosphorization before decarburization; (b) decarburization before dephosphorization

(p_{CO} supposed to be 1 atm) at t_T, the transition temperature of oxidation for P vs. C. At temperatures below t_T, P is oxidized preferably. But in reality, in absence of gas nuclei, the new phase CO formed must sustain a pressure much higher, 30atm as reported for basic open hearth furnace[14], so that actually the ΔG line of oxidation of C should lie much higher, presumedly shifted to line f, which lies for above line b, with the result that oxidation of P can go on smoothly in preference to C, a case which is met with in the pretreatment of hot metal with basic oxidizing slag and in the basic openhearth furnace, where 95% or more P can be removed during the melting-down stage.

In the Thomas process, O_2 in the blown-in air is consumed very quickly. The gas bubble contains N_2 and forms a vacuum in respect to CO, so that the oxidation of C is enhanced, its ΔG line of oxidation being shifted from position e to f(see Fig. 8(b)). Owing to low FeO and low CaO content of the slag, the ΔG line of oxidation of P cannot be shifted down very much, presumedly only to position b, which lies far above the ΔG line of oxidation of C, with the result that C is oxidized preferably to P. As shown in the Thomas process. P can be oxidized only after the elimination of C at the after-blow stage. For the LD process, P and C are oxidized simultaneously. This is due to the fact that the oxygen lance furnishes a strong oxygen flow into the bath with formation of highly oxidized iron droplets and foamed slag, both of which facilitate the simultaneous oxidation of P and C.

3 Equilibrium study of multireactions of oxidation

As shown above from the Richardson-Ellingham diagram. Al is oxidized in preference to Si. This sequence of oxidation holds true only for standard state conditions. In presence of a relatively excessive amount of Si, or moderate amounts of both Al and Si, more complicated cases can occur. To make a detailed study. elaboration of equilibrium compositions of multireactions would be of great interest to metallurgists.

Solutions of the final compositions of the chemical species taking part in multireactions at equilibrium can be made by writing down equations from the equlibrium constants of the respective equations. But this method usually involves difficulties in calculation. Using the principle of minimum free energy, Smith[15] has mentioned a method of Lagrange's undetermined multipliers to solve the ultimate equilibrium compositions of multireactions. This method is summarized with the following equations:

$$\sum_i n_i a_{ik} = A_k$$

$$\Delta G_{fi}^o + RT\ln a_i + \sum_k \lambda_k a_{ik} = 0 \text{ (for solutions)}$$

$$\Delta G_{fi}^o + RT\ln N_i p + \sum_k \lambda_k a_{ik} = 0 \text{ (for gases)}$$

where i the specific chemical species taking part in the reactions of the system; n_i the no. of moles of the chemical species i; k the specific element present in the chemical species, a_{ik} the no. of atoms of the element k in 1 molecule of the chemical species i; A_k the total no. of atoms of the element k present in the system, as determined by the initial constitution of the system; ΔG_{fi}^o the standard free energy of formation of the chemical species i; a_i activity of chemical species i, N_i mole fraction of chemical species i, λ_k Langrange's undetermined multiplier for element k; p total pressure of the gaseous system. Examples for application of this method of calculation are given below.

4 Deoxidation of liquid iron with Al and Si

Using both Al and Si as deoxidizing agents, the deoxidation of liquid iron could be achieved either by Si alone, by Al alone or by Al and Si simultaneously, as governed by the relative amounts of Al and Si added. With the thermodynamic data cited in Table 7, based on an initial content of O = 0.05%, Si = 1.2%, and Al in various amounts in the iron melt, a series of computer calculations was made, the equations shown in Table 8 being formulated according to Smith's method. From the calculation results in Table 9, it can be shown that under item 1 and 2, Si serves as the scavenger for O, while Al remains intact in the bath. Under item 3 and 4, Al and Si both take part in the deoxidation. Under item 5 and 6, Al plays the role of deoxidizing agent, while Si remains unchanged. Deeper insight into the deoxidation mechanism might be revealed by calculation of the free energy of oxidation at different stages of deoxidation. For item 1, see Table 10, Si as a deoxidizer reaches equilibrium with O, while Al remains unchanged in the bath. For item 2, Si behaves as the scavenger for O, till equilibrium is being reached, while Al, although it remains unchanged in content, reaches equilibrium finally with O similarly as Si. For item 4, Al first reacts with O as deoxidizer until O is reduced from 500 to 336 ppm, when the free energy of oxidation of Si for that specific content of O equals that of Al and henceforward Si and Al will be oxidized simultaneously until final equilibrium both with O is reached. Here we have an example of selective oxidation, of Al with subsequent simultaneous oxidation of Al and Si.

Table 7 Thermodynamic data used for deoxidation of liquidiron with Al and Si at 1873K

$e_{Si}^{Si} = 0.11 e_{Al}^{Al} = 0.045 e_{O}^{O} = -0.20$

$e_{Si}^{Al} = 0.058 e_{Al}^{Si} = 0.0056 e_{O}^{Si} = -0.131$

$e_{Si}^{O} = -0.23 e_{Al}^{O} = -6.6 e_{O}^{Al} = -3.9$

$Si_{(l)} \rightarrow [Si], \Delta G^{\circ} = -163.791 J$

$Al_{(l)} \rightarrow [Al], \Delta G^{\circ} = -115.449 J$

$\frac{1}{2} O_2 \rightarrow [O], \Delta G^{\circ} = -122.558 J$

$\Delta G_f^{\circ}(SiO_2) = -575.849 J$

$\Delta G_f^{\circ}(Al_2O_3) = -1077.794 J$

$a_{Si} a_O^2 = 2.2 \times 10^{-5}$

$a_{Al}^2 a_O^3 = 4.3 \times 10^{-14}$

Table 8 Equations for solving the equilibrium composition for deoxidation of liquidiron with Al and Si

$X_{Si} + \frac{7}{15} X_{SiO_2} = 1.2$

$X_{Al} + \frac{9}{17} X_{Al_2O_3} = A$

$X_O + \frac{8}{15} X_{SiO_2} + \frac{8}{17} X_{Al_2O_3} = 0.05$

$-575894 + \lambda_{Si} + 2\lambda_O = 0$

$-1077794 + 2\lambda_{Al} + 3\lambda_O = 0$

$-163791 + 35853 \lg X_{Si} + 35853(0.11 X_{Si} - 0.23 X_O + 0.058 X_{Al}) + \lambda_{Si} = 0$

$-115449 + 35853 \lg X_{Al} + 35853(0.045 X_{Al} - 6.6 X_O + 0.0056 X_{Si}) + \lambda_{Al} = 0$

$-122558 + 35853 \lg X_O + 35853(-0.20 X_O - 0.131 X_{Si} - 3.9 X_{Al}) + \lambda_O = 0$

$X_i = [\%i]; A = 0.0005, 0.000961, 0.01, 0.02, 0.051, 0.1$

Table 9 Equilibrium composition for deoxidation of liquid iron with Al and Si

Item	Element	Composition/% initial	Composition/% equilibrium	Activity product
I	Si	1.2	1.1610	$a_{Si}a_O^2 = 2.2\times10^{-5}$
	Al	0.0005	0.0005	$a_{Al}^2 a_O^3 = 1.2\times10^{-14}$
	O	0.05	0.005378	
II	Si	1.2	1.1610	$a_{Si}a_O^2 = 2.2\times10^{-5}$
	Al	0.000961	0.000961	$a_{Al}^2 a_O^3 = 4.3\times10^{-14}$
	O	0.05	0.005400	
III	Si	1.2	1.1680	$a_{Si}a_O^2 = 2.2\times10^{-5}$
	Al	0.01	0.000966	$a_{Al}^2 a_O^3 = 4.3\times10^{-14}$
	O	0.05	0.005391	
IV	Si	1.2	1.1758	$a_{Si}a_O^2 = 2.2\times10^{-5}$
	Al	0.02	0.000972	$a_{Al}^2 a_O^3 = 4.3\times10^{-14}$
	O	0.05	0.005381	
V	Si	1.2	1.2	$a_{Si}a_O^2 = 2.2\times10^{-5}$
	Al	0.051	0.000991	$a_{Al}^2 a_O^3 = 4.3\times10^{-14}$
	O	0.05	0.005350	
VI	Si	1.2	1.2	$a_{Si}a_O^2 = 1.3\times10^{-7}$
	Al	0.1	0.044418	$a_{Al}^2 a_O^3 = 1.2\times10^{-14}$
	O	0.05	0.000594	

Table 10 Change in free energy during deoxidation

Item	Composition/% Si	Al	O	$\Delta G/\text{kJ}$ $[Si] + 2[O] = SiO_{2(s)}$	$\frac{4}{3}[Al] + 2[O] = \frac{2}{3}Al_2O_{3(s)}$
I	1.2	0.0005	0.05	−68.72	−40.81
	1.17	0.0005	0.0157	−33.19	−16.31
	1.161	0.0005	0.005378	0(+0.001)	+13.55
II	1.2	0.000961	0.05	−68.60	−54.25
	1.17	0.000961	0.0157	−33.08	−29.77
	1.161	0.000961	0.005400	0(+0.007)	0(−0.007)
IV	1.2	0.02	0.05	−63.34	−112.01
	1.2	0.005	0.0367	−58.17	−82.15
	1.2	0.0015	0.0336	−56.40	−56.35
	1.1758	0.000972	0.005381	0(+0.003)	0(−0.010)

Accordingly, a diagram on the behaviour of Al and Si during deoxidation of liquid iron is schematically summarized in Fig. 9. It might be concluded that with different amounts

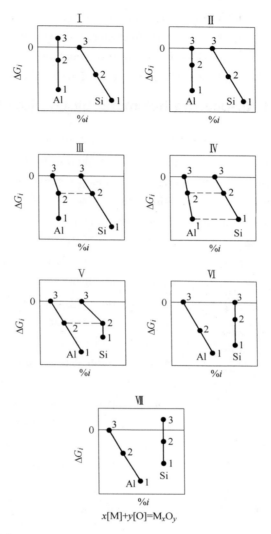

Fig. 9 Schematic representation of deoxidation of liquidiron with Al and Si

of the deoxidizing agents 2 categories for oxidation processes could be determined. First, either Si or Al can allone be. oxidized selectively till equilibrium with O is reached with the other element remaining intact in the bath with or without reaching equilibrium with O. Secondly, either Al or Si can accomplish preliminarily a selective oxidation and reduce the O down to a certain level, from then on both elements are oxidized simultaneously till equilibrium between Al, Si and O is finally reached. This furnishes an example of selective oxidation of one element followed by simultaneous oxidation of both elements. In between these 2 categories, there might be possible theoretically a 3rd category, in which the two elements at the very beginning are of such an appropriate proportion in amount that they havethe same free energy of oxidation at the specific initial Ocontent, so that they canbe oxidized side by side simultaneously till equilibrium both with O is finally reached without any selective oxidation of either element.

5 Oxidation of Nb-bearing hot metal by O_2 blowing

Blowing Nb-bearing hot metal with O_2 yields a slag which is used as a raw material for making ferroniobium. It is highly desirable during blowing to reduce the Nb-content as low as possible and, at the same time, to retain the C-content as high as possible, with the aim that the blown hotmetal(sometimes called "semi-steel") should contain enough C as a heat source to be able to be blown subsequently into steel. The hot metal used contains besides Nb also Si and much Mn. Study of the oxidation of Si, Mn, C and Nb in the melt has been made to elucidate the sequence of oxidation and the effect of different factors governing the final equilibrium composition of the melt. In contrast to the deoxidation of iron with Al and Si, where limited amount of O_2 is present, blowing of hot metal to oxidize the different elements is carried out with abundant or unlimited supply of O_2.

Calculation of these multireaction equilibria can be done by means of Smith's method mentioned above, but a moreconcrete method of solution is based on the principle that on reaching equlibrium the changes in free energy of the different reactions acquire the same value. For 4 elements Si, Mn, C and Nb, 6 equations can be written showing the mutual equilibrium between any two elements, but among them only 3 of them are independent equations. Examples for the consideration of equilibria between Si-Mn, Mn-C and C-Nb are set forth as 3 equations with iteration results from the computer and are given in Table 11.

Table 11 Equations for solving the multireaction equilibria

Si	Mn	C	Nb	a_{SiO_2}	a_{MnO}	$a_{Nb_2O_5}$
x	y	z	0.08	1	0.086	1×10^{-10}

$\lg x - 2\lg y + 0.1304x + 0.002y + 0.376z = 0.4777$ (Si − Mn)

$- 2\lg y + 2\lg z + 0.1904x - 0.028y + 0.506z = 2.6251$ (Mn − C)

$- 2\lg z + 0.5700x + 0.132y - 0.804z = - 3.4690$ (C − Nb)

Iteration number = 23

Solution results:

$x = 0.206178E + 00$

$y = 0.115303E + 01$

$z = 0.334552E + 01$

For the above calculations free energies of formation of oxides are quoted from Turkdogen[17]. e_i^j and the free energy of solution from Sigworth and Elliott[5], except those for Nb. the values for which are taken from our own previous research papers[18~20]. Values of e_i^j for 1573K are calculated from values of 1873K with the inverse proportion rule with respect to temperature. These thermodynamic data are listed in Table 12. As for the composition and activity of slags formed from blowing, which are given in Table 13, three categories are assumed: (1) soft blowing with the O_2-lance rather high above the melt and characterized by its high FeO-content; (2) hard blowing with the O_2-lance deep down into themelt. or bottom blowing with O_2 through the melt, in both cases the slag being characterized by its low FeO-content; and (3) moderate blowing characterized by the moderate amount of FeO-content in the slag. The corresponding activity values of the slag ingredients are estimated from the iso-activity curves of the FeO-MnO-SiO_2 system after Bell[21], the temperature effect being neglected.

Table 12 Thermodynamic data used for blowing of hot metal at 1573K

$[Si] + O_2 = SiO_{2(s)} \quad \Delta G^\circ = -474478J$

$2[Mn] + O_2 = 2MnO_{(s)} \quad \Delta G^\circ = -425011J$

$2[C] + O_2 = 2CO \quad \Delta G^\circ = -410860J$

$\frac{4}{5}[Nb] + O_2 = \frac{2}{5}Nb_2O_{5(l)} \quad \Delta G^\circ = -420450J$

$e_{Si}^{Si} = 0.13 \quad e_{Si}^{Mn} = 0.002 \quad e_{Si}^{Nb} = 0.29$

$e_{Si}^{C} = 0.21 \quad e_{Mn}^{Mn} = 0 \quad e_{Mn}^{Si} = -0.0002$

$e_{Mn}^{Nb} = 0.08 \quad e_{Mn}^{C} = -0.083 \quad e_{C}^{C} = 0.17$

$e_{C}^{Si} = 0.095 \quad e_{C}^{Nb} = -0.071 \quad e_{C}^{Mn} = -0.014$

$e_{Nb}^{Nb} = 0.26 \quad e_{Nb}^{Si} = 0.95 \quad e_{Nb}^{Mn} = 0.13 \quad e_{Nb}^{C} = -0.58$

Table 13 Composition and activity of slag

		SiO_2	MnO	FeO	Nb_2O_5
I	%	40.6	36.8	17.6	5.0
	N	0.46	0.36	0.17	0.01
	a	0.78	0.12	0.18	—
II	%	44.3	10.5	41.3	3.9
	N	0.50	0.10	0.39	0.01
	a	1	0.02	0.37	—
III	%	47.4	42.2	6.4	4.0
	N	0.53	0.40	0.06	0.01
	a	1	0.086	0.05	—

Table 14 shows the results of calculations of free energy change at different stages during blowing of Si, Mn, C and Nb in addition to the equilibrium compositions of the respective elements if present. Starting with the initial composition of the melt, it can be shown clearly that Si is to be oxidized first in preference to Mn, C, and Nb. As oxidation continues, equilibrium between Si and Mn is reached, from then on they both are oxidized simultaneously in preference to C and Nb, until equilibrium between Si, Mn and C is reached. The oxidation continues further up to the attainment of equilibrium of four elements Si, Mn, C and Nb. With a view to reduce the Nb-content, all elements should be oxidized side by side until a new equilibrium with the desired Nb-content is reached. The sequence of oxidation for these four elements is better illustrated in a schematic representation in Fig. 10. Obviously, this is also a case of selective oxidation of a certain element followed by its simultaneous oxidation with other elements, a characteristic feature of oxidation of multireactions as already discussed under the deoxidation of iron by Al and Si. Attention should be paid to the fact that reducing Nb from 0.08% to 0.03% has made C in the melt drop from 3.4% to 2.16%. The preservation of C with elimination of Nb has not been well fulfilled in this blowing operation. The final equilibrium composition of the melt depends not only upon the temperature and the initial composition of the melt. But also depends upon the activity of the products a_{SiO_2}, and a_{MnO} as well as $a_{Nb_2O_5}$. From the calculation results shown in Tables 15~17, it can be shown that:

(1) to retain a high C-content in the final melt. a low activity of Nb_2O_5 in the order of 10^{-8} to 10^{-10} would be desirable.

(2) to keep the residual Mn-content high, a slag high in a_{MnO} accompanied by a small value of a_{FeO} in the slag is desired.

(3) the same tendency also holds true for Si. Although the effect of its acitivity is not so manifest as that for Mn. It is suggested that with a view to extracting Nb from the hot metal as effectively as possible, hard blowing or bottom blowing would be more desirable to suppress the oxidation of Mn and at the same time to prevent the P from being oxidized into the slag. Whether the activity of Nb_2O_5 should be able to reach a value of the order of $10^{-8} \sim 10^{-10}$ needs further experimental evidence, but it is in conformity with the estimated values from our previous research work based on industrial data.

Table 14   Free energy change during blowing of hot metal

Composition/%				ΔG/kJ				Equilibrium
Si	Mn	C	Nb	Si	Mn	C	Nb	
0.70	2.00	3.40	0.08	−498.1	−482.0	−479.6	−465.7	
0.22	2.00	3.40	0.08	−481.1	−482.0	−479.6	−454.7	Si, Mn
0.16	1.65	3.40	0.08	−476.7	−476.9	−479.6	−452.2	Si, Mn, C

Continued Table 14

Composition/%				ΔG/kJ				Equilibrium
Si	Mn	C	Nb	Si	Mn	C	Nb	
0.069	0.71	2.48	0.08	-459.5	-459.5	-459.5	-460.0	Si,Mn,C,Nb
0.050	0.52	2.16	0.03	-452.7	-452.7	-452.8	-452.9	Si,Mn,C,Nb

$\frac{2x}{y}[M] + O_2 \rightleftharpoons \frac{2}{y}(M_xO_y), [M] = Si, Mn, C, Nb$

$a_{SiO_2} = 0.78, a_{MnO} = 0.12, a_{Nb_2O_5} = 1 \times 10^{-8}, p_{CO} = 1, T = 1573K$

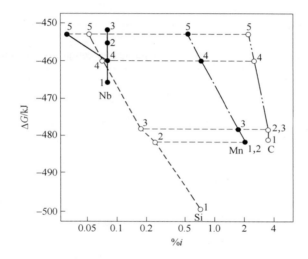

Fig. 10 Schematic representation of oxidation of Si, Mn, C and Nb, in hot metal

Table 15 Effect of a_{SiO_2}, a_{MnO} upon the equilibrium composition of the melt

Composition/%				Activity			ΔG/kJ			
Si	Mn	C	Nb	a_{SiO_2}	a_{MnO}	$a_{Nb_2O_5}$	Si	Mn	C	Nb
0.16	1.59	3.36	0.08	0.78	0.12	10^{-10}	-476.4	-476.2	-476.2	-476.7
0.12	1.13	2.96	0.03	0.78	0.12	10^{-10}	-469.5	-469.0	-469.1	-469.3
0.19	0.24	3.21	0.08	1	0.02	10^{-10}	-474.5	-474.3	-474.8	-475.2
0.15	0.18	2.86	0.03	1	0.02	10^{-10}	-468.6	-468.3	-468.2	-468.4
0.21	1.15	3.35	0.08	1	0.086	10^{-10}	-476.8	-476.5	-476.7	-476.6
0.15	0.81	2.94	0.03	1	0.086	10^{-10}	-469.2	-469.1	-469.2	-469.3

$a_{Nb_2O_5} = 1 \times 10^{-10}, p_{CO} = 1, T = 1573K$

Table 16 Effect of a_{SiO_2}, a_{MnO} upon the equilibrium composition of the melt

Composition/%				Activity			ΔG/kJ			
Si	Mn	C	Nb	a_{SiO_2}	a_{MnO}	$a_{Nb_2O_5}$	Si	Mn	C	Nb
0.069	0.17	2.48	0.08	0.78	0.12	10^{-8}	-459.5	-459.5	-459.5	-460.0
0.050	0.52	2.16	0.03	0.78	0.12	10^{-8}	-452.7	-452.7	-452.8	-452.9
0.087	0.115	2.42	0.08	1	0.02	10^{-8}	-458.9	-459.0	-458.8	-459.4
0.063	0.085	2.11	0.03	1	0.02	10^{-8}	-452.2	-452.4	-452.1	-452.5
0.090	0.52	2.49	0.08	1	0.086	10^{-8}	-459.8	-460.0	-459.9	-459.8
0.065	0.37	2.15	0.03	1	0.086	10^{-8}	-452.9	-452.6	-452.8	-452.9

$a_{Nb_2O_5} = 1 \times 10^{-8}$, $p_{CO} = 1$, $T = 1573K$

Table 17 Effect of $a_{Nb_2O_5}$ upon the equilibrium composition of the melt

Composition/%				$a_{Nb_2O_5}$	ΔG/kJ			
Si	Mn	C	Nb		Si	Mn	C	Nb
0.19	0.24	3.21	0.08	10^{-10}	-474.5	-474.3	-474.8	-475.2
0.15	0.18	2.86	0.03	10^{-10}	-468.6	-468.3	-468.2	-468.4
0.087	0.11	2.42	0.08	10^{-8}	-458.9	-457.9	-458.8	-459.4
0.063	0.085	2.11	0.03	10^{-8}	-452.2	-452.2	-452.1	-452.5
0.036	0.055	1.74	0.08	10^{-6}	-442.9	-443.2	-443.0	-443.5
0.025	0.040	1.48	0.03	10^{-6}	-436.0	-435.9	-436.2	-436.2
0.014	0.026	1.16	0.08	10^{-4}	-426.7	-426.5	-426.3	-426.9
0.0089	0.019	0.95	0.03	10^{-4}	-419.0	-419.1	-419.2	-419.1

$a_{SiO_2} = 1$, $a_{MnO} = 0.02$, $p_{CO} = 1$, $T = 1573K$

6 Role played by Fe during oxidation

Owing to the abundant amount of iron atoms as a solvent in the melt, on meeting with O_2, it is quite natural that Fe will be first oxidized. The FeO once formed dissolves in the melt and dissociates into Fe and [O]. But the latter when met with any deoxidizing agent stronger than Fe, say Si, would be at once scavenged to form stable SiO_2. This is equivalent, expressed thermodynamically, that Si would be oxidized in preference to Fe. Thermodynamic calculation, see Table 18, shows that the equilibrium composition of Si in contact with the solvent Fe at 1873K ranges from 6 to 126 ppm, depending on the values of a_{SiO_2} and a_{FeO}. That means, at any concentration of Si higher than equilibrium-composition, Si would be oxidized preferably to Fe, and as soon as the equilibrium composition has been reached, in case of further supply of O_2, Si and Fe would be oxidized simultaneously. But, owing to the kinetic aspects, in case that the FeO formed joins the slag and floats up to the surface of the melt as a separate phase, the reduction by Si would be much handicapped and retarded.

Table 18 Equilibrium of Si with Fe at 1873K

[Si] + 2(FeO) = [Fe] + (SiO$_2$)		
a_{SiO_2}	a_{FeO}	Si/×10^{-6}(f_{Si}=1)
0.18	0.80	6
1	0.40	126
1	1	20

It is due to this reason that a greater or smaller amount of Fe loss does occur in accordance with the different ways of blowing.

A few words might be added concerning the problem of direct and indirect oxidation. Direct oxidation usually means the oxidation with O_2-gas, while indirect oxidation means this with [O] or FeO. But in reality. as shown by Robertson and Jekins[22] with the aid of high-velocity camera, that when O_2 meets liquid iron at 1600℃, the period of direct oxidation is extremely short, an instant of 0.002~0.003s. Later the surface of the melt is covered with a thin film of FeO. So it is most likely that most of the oxidation is to be carried on indirectly. So far as the topic of selective oxidation is concerned, it does not matter whether oxidati on is carried on directly or indirectly, because transition temperature as well as simultaneous oxidation are independent of the forms of oxygen present. be it O_2, [O] or(FeO), so long as the comparison basis of the same amount of oxygen persists.

7 Concluding remarks

The study of the oxidation of elements in metal melts reveals that two categories should be considered: first, the selective oxidation of one element in preference to the other based on the transition temperature, and secondly, the selective oxidation of one element followed by its simultaneous oxidation with other elements. The sequence of oxidation is governed by the principle of minimum free energy of the system. The concept of transition temperature as well as the equilibria study of multireactions could be applied beneficially in metallurgical industrial practices. For decarburization of stainless steel and high-C ferromanganese and for desulfurization of nickel matte, the blowing operation should be conducted at a bath temperature always higher than the transition temperature of the specific composition of the melt, while for winning of V or Nb and for dechromizing of hot metal, a bath temperature always lower than the transition temperature for the specific composition of the melt should be strictly followed. The final equilibrium composition of the melt depends on temperature and initial composition of the bath, as well as the activity of the ingredients of slag and the partial pressure of the gaseous product. Correct choice of blowing conditions could be ensured in order to attain the expected optimum results successfully.

References

[1] Bodsworth C, Bell H B. Physical Chemistry of Iron and Steel Manufacture, 2nd Ed. , Longman, London, 1972:198.
[2] Bodsworth C, Bell H B. ibid:203.
[3] Wei Shoukun. Physical Chemistry of Metallurgical Processes, Shanghai Science and Technology Press, Shanghai, 1980:21.
[4] Elliott J F, et al. Thermochemistry for steelmaking, Vol. II, Addison Wesley, Reading, 1963:620.
[5] Sigworth G K, Elliot J F. Metal Science, 1974(8):298.
[6] Wei Shoukun. In First China-USA Bilateral Metallurgical Conference Nov. 1981, Beijing, China: 45; ActaMetallurgica Sinica, 1982(18):115.
[7] Wei Shoukun, Hung Yianro. Nonferrous Metals(CSM), 1981, 33(3):50.
[8] see reference[3]:64.
[9] see reference[3]:73.
[10] Wei Shoukun. to be published.
[11] Shao Xianghua Sino-Japanese Symposium of Iron and Seel, Sept. 1981, Beijing, China:183.
[12] Wei Shoukun. Acta Metallurgica Sinica, 1964(7):240.
[13] see reference[3]:276.
[14] Rellermeyer H, Knüppel H, Sittard J. Stahl u. Eisen, 1957(77):1296.
[15] Smith J M, Van Hess H C. Introduction to Chemical Eng'g Thermodynamics, 3rd Edition, Mc Graw-Hill, 1975:424.
[16] Turkdogan E T. In Chemical Metallurgy of Iron and Steel, The Iron Steel Inst. , London, 1973:153.
[17] Turkdogan E T. Phy. Chem. of High-Temperature Technology, Academic Press, New York, 1980:5.
[18] Wei Shoukun, et al. Rare Metals(CSM), 1983, 2(1):10; Chin. J. Metal Sci, Technol. , 1985(1):14.
[19] Zhang Shengbi, Tong Ting, Wang Jifang, Wei Shoukun. Acta Met. Sinica 1984(20):A 348.
[20] Wei Shoukun, et al. Solid State Ionics, 1986(18 & 19):1232.
[21] Bell H B. J. Iron Steel Inst, 1963(201):116.
[22] Robertson D G C, Jekins A E. In: Heterogeneous Kinetics at Elevated Temperatures, Plenum Press, New York, 1970:393.

Thermodynamic Study on Process in Copper Converters (The Slag-making Stage)[*]

Chen Chunlin Zhou Tuping Zhang Jiayun Wei Shoukun
Liu Xingxiang Bai Meng Jiang Jinhong

Abstract: The so-called Goto's model was modified by introducing a parameter of the oxygen efficiency from industrial trials, as well as the selected and newly re-assessed thermodynamic data. The application of the model to copper converters in Guixi Smelter has been carried out by the combination of thermodynamic calculations with the mass and heat balance using the plant data obtained in industrial trials for many heats. For the slag-making stage, good agreements have been reached between the calculated and measured temperature, blowing time as well as the contents for main elements in the matte and the slag. Relatively large deviations for contents of Zn and Pb in the slag may be caused by the complex chemical composition of the real molten slag, which may result in a large difference of γ_{Zn}, and γ_{Pb} adopted with their real values. It is noted that the model can simulate the slag-making stage of copper converting process in industrial Pierce-Smith converters well.

Key words: copper converter, thermodynamic study, slag-making stage

It is known that the chemical composition in a multi-phase and multi-component system in equilibrium can be calculated using two approaches: Gibbs free energy minimization as well as the so-called the equilibrium constant method by Blinkley. Combining the composition evaluation in a multi-phase and multi-component system with the method by Blinkley with the mass and heat balance, Goto proposed a thermodynamic model for the process of copper flash smelting process[1]. Assuming that the oxygen. blown into the converter was entirely consumed in the reactions of oxidation, Goto's model was extended to simulate the process in copper converters lately[2]. In the application, the calculated results were employed to analyze the converting process, however only the calculated blowing time, copper and Fe_3O_4. contents of the slag in slag-making stage were compared with the plant data for one heat of the operation in Naoshima[3]. It was shown

---

[*] 原刊于《J. Univ. Science & Technology Beijing》,1999,6:187~192.

that with very close blowing time, the differences between the plant data and calculated results were quite large. It is questionable that such a simple comparison could provide a proper measure indicating in what a level the model can simulate the processes in copper converters.

In this study, Goto's model was modified by introducing a parameter of the oxygen efficiency determined statistically from the plant data of a series of heats with stable converting operations[4] and using re-assessed thermodynamic data. The modified model has been applied to the interpret the converting process in the converters in Guixi Smelter. A systematic comparison between the calculated and the plant data in slag-making stage is illustrated in this paper.

1 Model modification

It is known that the converting process is strongly affected by oxygen efficiency. In the present study, the above mentioned model was modified by introducing a parameter of the oxygen efficiency. The definition of oxygen efficiency, η_O, is presented as

$$\eta_O = (Q_{O,rea}/Q_{O,all}) \times 100\%$$

Where $Q_{O,rea}$ is the oxygen amount consumed in the reactions for producing the matte or blister copper with chemically analyzed grades of matte or copper, and $Q_{O,all}$ is that supplied by air-blowing into the molten bath. Based on the industrial trials of stable converting operation practice, in Guixi Smelter, the oxygen efficiency of each converting period in each heat was obtained from the mass balance calculations, and its arithmetic mean values of all the testing heats are listed in Table 1[4].

Table 1 Oxygen efficiency data obtained from industrial trials in Guixi Smelter

Converting period	Oxygen efficiency / %
Slag-making stage, first slag	93
blowing	90
Slag-making stage, 2nd slag	—
blowing	—
Copper-making stage	81

The newly re-assessed standard Gibbs free energies of formation[5~8] were employed in this study rather than those in Goto's work[3] which are listed in Table 2. The following correlation for activity coefficient of FeS in mole[9,10] with its mole fraction has been used in this study

$$\gamma_{FeS} = 0.882(1 - x_{FeS})^{0.5}$$

The calculated α_{FeS} values for the matte agree well with those published by Bale and

Toguri[11] and differ from Goto's early publication[3]. These alternations could enable the present thermodynamic calculations more precisely.

Table 2 The newly reassessed Standard Gibbs free energy of formation using in this work different from those used by Goto et al. [3]

Component	$\Delta_f G^\circ / J \cdot mol^{-1}$	Reference
FeO	$-232714 + 45.31T$	[5]
ZnS	$-375382 + 191.57T$	[6]
PbS	$-111838 + 51.04T$	[6]
Cu_2O	$-195200 + 92.58T$	[7]
SO_2	$-361665 + 72.68T$	[5]
SO	$-57780 - 4.98T$	[5]
CO_2	$-395350 - 0.54T$	[8]

Regarding the calculation procedure, please refer to reference [2].

2 Industrial trials

The industrial trials of copper converting operations for the converters No. 1 and No. 3 were simultaneously carried out in Guixi Smelter in Aug. of 1996. The time schedule for the trials was registered. All the charged materials, products as well as by products such as molten matte, blister copper were sampled, quenched and then chemically analyzed. The mineralo graphical analysis was employed to give the composition for all the cold charge materials, quenched slag, matte as well as blister copper samples for 2 representative heats[12]. Temperatures of the melts were measured every time when the molten matte or blister was tapped using Pt-Rh thermocouple. The initial temperatures of the matte at pouring into the converter at a start of the 1st and 2nd slag blowing period have been estimated (1100±15)℃. The temperature drop of the melts which remained in the converter during the rest between the 1st and 2nd slag blowing was estimated about 100℃.

As an example, the chemical composition of materials charged in heat 173 is presented in Table 3, and the corresponding time schedule of the operation in Table 4.

Table 3 Chemical composition of materials charged for heat 173 of converter No. 3 (mass fraction in %)

Element	Matte in 1st slag blowing	Matte in 2nd slag blowing	O_2 1st slag blowing	O_2 2nd slag blowing	Residue	Silica flux	Skull	Recycled white matte
Cu	59.52	57.43	0.000	0.000	19.52	0.00	58.94	54.25
S	21.34	21.86	0.000	0.000	0.29	0.00	12.19	13.97
Pb	0.28	0.32	0.000	0.000	2.19	0.00	0.52	0.26

Continued Table 3

Element	Matte in 1st slag blowing	Matte in 2nd slag blowing	O_2 1st slag blowing	O_2 2nd slag blowing	Residue	Silica flux	Skull	Recycled white matte
Fe	14.68	15.13	0.000	0.000	39.81	1.40	12.75	15.70
Zn	0.42	0.63	0.000	0.000	2.29	0.00	0.41	0.32
O	1.30	1.00	26.560	24.760	16.61	0.00	6.08	5.12
N	0.00	0.00	73.320	75.120	—	0.00	0.00	0.00
H	0.00	0.00	0.070	0.070	0.00	0.00	0.00	0.00
C	0.00	0.00	0.001	0.001	0.00	0.00	0.00	0.00
SiO_2	0.00	0.00	0.000	0.000	11.67	93.97	5.52	6.60
T/K	1373	1373	298	298	1373	398	298	1373

Table 4 Time schedule of the operation in heat 173 of converter No. 3

Period	Time(min) Operation steps
1st slag-blowing	0 Charge:matte 118.5t, silica flux 3t, skull 11t, recycled white-matte 8t, residue 5t and air 30300m³(STP) 2 Charge:silica flux 5.25t 60(end):The 1st slag blowing is stopped at about 77.83%(mass fraction)Cu in matte, 52.5t slag is tapped. The temperature of slag is 1555K Rest time:90min
2nd salg-blowing	0 Charge:matte 61t, skull 5.5t and air 23290m³(STP) 2 Charge:silica flux 4t 45(end):The 1st slag blowing is stopped at about 78.26%(mass fraction)Cu in matte. 50t slag is tapped. The temperature of slag is 1548K Rest time:30min

3　Results of simulation and discussion

3.1　Result of heat 173 in converter No. 3

The results of the calculation using the model described above with the input data of heat 173 of converter No. 3 are illustrated in Figs. 1~7. For a comparison, the measured temperature and composition values in matte and slag are also marked in Figs. 1~5.

Fig. 1 shows that the temperature of the matte continuously increases with time due to the oxidation of Fe in slag-making stage. The initial temperature drops at the start two minutes in the 1st and 2nd slag blowing period were caused by the cold charge-silica flux. Fig. 2 and Fig. 3 respectively show the continuous decrease of the content of Fe and the increase of the content of Cu. The increase of the content of Cu in the matte is due to oxidation of Fe to FeO and Fe_3O_4 to enter the slag during the slag-making stage. It can be seen in Fig. 2, 3 that the predicted contents of Fe and Cu in matte are in a good agreement with those of measured. As shown in Fig. 4 and Fig. 5, the increase of Fe in slag results in the decrease of SiO_2 during the slag-making stage. The calculated contents of SiO_2 and Fe in slag agree well with the plant data.

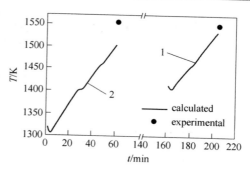

Fig. 1 Variation of furnace temperature
1—1st slag blowing; 2—2nd slag blowing

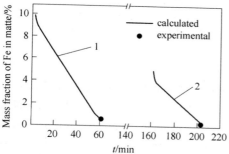

Fig. 2 Variation of copper content in matte
1—1st slag blowing; 2—2nd slag blowing

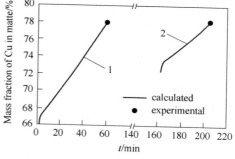

Fig. 3 Variations of iron content in matte
1—1st slag blowing; 2—2nd slag blowing

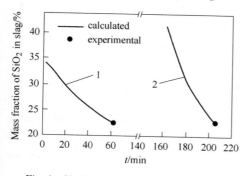

Fig. 4 Variations of SiO_2 content in slag
1—1st slag blowing; 2—2nd slag blowing

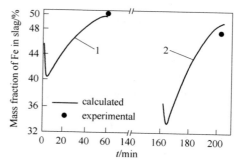

Fig. 5 Variations of iron content in slag
1—1st slag blowing; 2—2nd slag blowing

Fig. 6 and Fig. 7 respectively show the constant partial pressure of SO_2 and the gradual increase of the partial pressure of O_2 in gas phase during the slag-making stage. These may imply the decrease of the oxygen efficiency in the same period. Due to the limitation of the equipment conditions in the workshop, the composition of gas phase could not be measured, therefore a similar comparison for gas phase could not be provided.

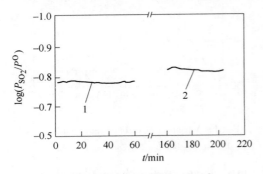

Fig. 6 Variations of SiO_2 partial pressure in gas phase
1—1st slag blowing; 2—2nd slag blowing

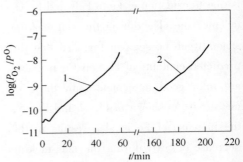

Fig. 7 Variations of O_2 partial pressure in gas phase
1—1st slag blowing; 2—2nd slag blowing

3.2 Other heats

In addition to heat 173 the model has been applied to calculate the other heats. For comparison, the calculated results with the relevant plant data are listed in Appendix 1 and 2.

The comparisons show that the calculated contents of main elements such as Cu, Fe and S in the matte are in a good agreement with those of the plant data. The calculated contents of SiO_2 and Fe in slag also agree with the measured data. The relative deviations between calculated and measured contents for Cu, Fe, S in matte respectively take the values of 0.6%, 0.3% and 0.1%, while for Fe, SiO_2 in slag the reasonably higher values of 1.7% and 1.6% were obtained.

As shown in appendix 1 and 2, the calculated mass fraction values of Cu in the slag are about 3% lower than the measured data. This difference may be caused by the ignorance of the mechanical entertainment of the matte droplets in the slag.

The differences between the predicted and measured contents for Zn, Pb in the slag are relatively high. These differences may be attributed to that the activity coefficients of $\gamma_{ZnS}, \gamma_{PbS}, \gamma_{ZnO}, \gamma_{PbO}$ used in the model differ from those in the real systems. Clearly, the accuracy improvements of the thermodynamic data for the molten slag and matte systems with a complex composition are desirable.

There are two forms of iron, ferrous and ferric ions, existing in the molten slag. Only the total iron amount in slag could be chemically analyzed in the industrial tests. For this reason, the amounts of the Fe_3O_4, calculated could not be compared with the relevant plant data. The range of the predicted results of Fe_3O_4 in the slag is from 28.56% to 36.54%, which agree with the data using the slag samples taken from the representative heats by means of mineral ographic analysis[12].

4 Conclusions

The so-called Goto's model was modified by introducing oxygen efficiency of copper converter and newly re-assessed thermodynamic data. The modified model has been applied to the converter operations in Guixi Smelter. For the slag-making stage, the comparison of the model predicted results with the plant data shows a good agreement in blowing time, bath temperature as well as the main elements Cu, Fe, S in the matte, Fe and SiO_2 in the slag.

The calculated mass contents of Cu in slag were about 3% lower than the measured values, which may be caused by the ignorance of the mechanical mixing of matte droplets on the slag. Rather large deviations between the calculated and measured contents of Zn and Pb in the slag were attained. The deviations may be reduced by the improvements of the accuracy for the relevant activity coefficients of components, such as ZnS, PbS in the matte and ZnO, PbO in the slag.

The modified model may provide a tool to analyze the slag-making stage of the process in copper converters and could be applied to monitor and optimize the operation practice of this period in industrial copper converters.

Acknowledgment

The financial support from the National Natural Science Foundation of China is gratefully acknowledged.

The authors would like to thank Professor S. See-tharaman, Department of Metallurgy Royal Institute of Technology, Sweden, for his valuable suggestion and the firm supports to the authors to start this study. Also, the authors would like to thank Professor Du Sichen, the same institution, for all his help in this work.

References

[1] Goto S. Inst. Min. Met., London, 1974:23.
[2] Goto S. [In:] Copper and Nickel Converters, TMS-AIME, Jonhnson R E. [eds.]. Warrendale, PA, 1979:33.
[3] Shimpo R, Watanabe Y, Goto S, et al. Advances in Sulfide Smelting, TMS-AIME, 1983(1):295.
[4] Zhou T P, Shi D P, Chen C L, et al. Nonferrous Metals, 1998(3):7.
[5] Kubaschewski O, Alcock C B Metallurgical. Thermo-chemistry, 5th ed., Pergamon Press, NewYork, 1979.

[6] Rao Y K. Stoichiometry and Thermodynamics of Metallurgical Processes, Cambridge University Press, 1985.
[7] Nagamori M. Met. Trans, 1994, 25B:839.
[8] Turkdogan E T. Physical Chemistry of High Temperature Technology, New York. Academic Press, 1980.
[9] Bjorkman B, Eriksson G. Can. Metall. Q. , 1982, 21:329.
[10] Krivsky W A, Schumann R. J. Metals. , 1957(9):839.
[11] Bale C W, Toguri J M. Can. Metall. Q. , 1976(15):305.
[12] Beijing General Institute of Mining and Metallurgy: Report of mineralographical analysis of the law material, inter-mediate products, products for the converters in Guixi Smelter, Aug. , 1996.

Appendix 1 Comparison between model predicted and plant data for Slag-making stage(I)

Blowing period	Molten phase	Element or component	Mass fraction of $i/\%$ (Heat 173, converter No. 3)		Mass fraction of $i/\%$ (Heat 50, converter No. 1)		Mass fraction of $i/\%$ (Heat 51, converter No. 1)	
			Measured	Predicted	Measured	Predicted	Measured	Predicted
the 1st	Matte	Cu	77.83	78.14	76.86	77.59	76.14	77.99
		S	19.02	18.92	18.64	19.18	18.90	18.77
		Fe	0.72	0.63	0.74	1.17	0.77	0.44
		Zn	0.052	0.20	0.021	0.18	0.014	0.05
		Pb	0.33	0.34	0.15	0.35	0.14	0.14
	Slag	Cu	3.34	0.36	3.57	0.30	3.11	0.46
		S	0.32	0.07	0.27	0.07	0.27	0.07
		Fe	50.42	49.96	49.64	51.01	51.33	48.37
		Fe_3O_4	—	31.60	—	31.18	—	30.79
		Zn	1.77	1.18	0.98	0.40	0.97	0.68
		Pb	0.49	0.49	0.13	0.19	0.28	0.48
		SiO_2	20.63	22.38	22.57	24.68	21.40	22.56
	T/K①		1555	1505	1532	1484	1550	525
the 2nd	Matte	Cu	78.26	78.14	79.25	78.89	76.83	78.01
		S	18.60	19.02	18.36	18.88	18.90	18.93
		Fe	0.46	0.62	0.58	0.51	1.16	0.50
		Zn	1.18	0.029	0.014	0.006	0.042	0.09
		Pb	0.20	0.30	0.14	0.04	0.21	0.15
	Slag	Cu	3.92	0.40	3.74	1.73	2.63	0.47
		S	0.30	0.08	0.27	0.09	0.29	0.10
		Fe	47.50	49.07	52.28	51.43	51.99	48.51
		Fe_3O_4	—	28.60	—	35.52	—	28.56
		Zn	1.53	1.40	0.90	1.78	0.93	0.005
		Pb	0.44	0.50	0.28	1.86	0.22	0.22
		SiO_2	23.15	22.87	21.05	17.44	21.51	22.35
	T/K①		1548	1536	1573	1585	1550	1562

① the temperature refers to that measured when start the tapping at the end of blowing, i represent element.

Appendix 2 Comparison between model predicted and plant data for Slag-making stage(II)

Blowing period	Molten phase	Element or component	Mass fraction of $i/\%$ (Heat 58, converter No.1)		Mass fraction of $i/\%$ (Heat 60, converter No.1)		Mass fraction of $i/\%$ (Heat 168, converter No.3)	
			Measured	Predicted	Measured	Predicted	Measured	Predicted
the 1st	Matte	Cu	78.08	77.73	77.24	77.69	79.05	79.07
		S	18.80	18.89	19.31	18.92	18.59	18.92
		Fe	1.12	0.68	1.00	0.63	0.60	0.43
		Zn	0.33	0.24	0.079	0.24	0.025	0.004
		Pb	0.33	0.39	0.37	0.28	0.21	0.02
	Slag	Cu	2.37	0.37	2.94	0.37	4.26	2.0
		S	0.27	0.07	0.39	0.07	0.24	0.08
		Fe	50.90	48.41	50.09	49.23	50.60	49.04
		Fe_3O_4	—	28.06	—	30.16	—	29.54
		Zn	1.18	0.28	1.61	1.40	1.15	0.014
		Pb	0.35	1.20	0.34	0.40	0.44	1.12
		SiO_2	21.27	23.88	21.41	22.77	20.23	20.74
	T/K[①]		1568	1528	1534	1515	1534	1552
the 2nd	Matte	Cu	77.77	78.79	77.06	77.21	78.11	79.03
		S	18.68	18.86	18.89	19.08	18.56	18.91
		Fe	1.43	0.49	1.00	1.16	0.77	0.50
		Zn	0.083	0.007	0.062	0.24	0.035	0.005
		Pb	0.36	0.06	0.32	0.28	0.25	0.20
	Slag	Cu	7.87	1.72	3.79	0.40	6.38	2.10
		S	1.00	0.09	0.69	0.10	0.60	0.07
		Fe	50.58	51.69	50.37	51.10	51.10	50.19
		Fe_3O_4	—	36.45	—	31.26	—	30.43
		Zn	1.28	2.19	1.36	0.80	1.12	1.63
		Pb	0.24	2.70	0.30	0.24	0.34	1.34
		SiO_2	18.48	17.47	20.84	21.96	19.45	17.89
	T/K[①]		1534	1594	1542	1575	1543	1531

① the temperature refers to that measured when start the tapping at the end of blowing, i represent element.

纯铁液中钡-氧、钡-硫平衡的研究*

王建生　韩其勇　魏寿昆

摘　要：采用钡蒸气平衡法研究了1600℃铁液中Ba-O和Ba-S的反应平衡，测得平衡常数和活度相互作用系数

$$K_{BaO} = 3.98 \times 10^{-8} \quad e_O^{Ba} = -46.5$$
$$K_{BaS} = 5.45 \times 10^{-8} \quad e_S^{Ba} = -47.8$$

1　引言

近年来，用于钢液终处理的合金越来越趋于多元化，钡就是其中的一种重要的合金元素。由于钡的加入，取得了较好的处理效果。但是，由于目前对终处理合金在钢中的作用机理认识尚不充分，致使在对终处理合金的研究中具有较大的经验性。众所周知，碱土金属中钡的沸点远高于钙、镁，更适合用于钢液中脱氧脱硫。但钡在钢中的热力学数据很缺乏，至今尚未见文献报道。因此，本文研究了纯铁液中$BaO_{(s)} = [Ba] + [O]$和$BaS_{(s)} = [Ba] + [S]$的反应平衡。

2　试验

试验装置见图1，铁料经去除氧化皮后放入TiN或BaS坩埚，装入钼反应室，在抽空器内抽真空密封。钼反应室用滑轮装置缓慢吊入钼丝炉中。钼丝炉控温电偶，上部为PtRh10-Pt，下部为PtPh30-PtRh6。由于实验中不能直接测定铁液温度，在模拟试验的条件下定期测温，确定钼丝炉下热偶控温毫伏值。钼丝炉下恒温区温差小于±1.5℃。反应平衡后迅速将钼反应室从炉内提出，放入水中，下部激冷，上部缓冷，以保证激冷过程中钼反应室内压力不至于下降得过于剧烈，防止钡蒸气从铁液中溢出。得到的样品剥去表皮，用盐酸、无水乙醇清洗，风干后取样分析成分。

实验用TiN及BaS坩埚系自制。加20%~30%氢化钛粉于氮化钛粉中作为黏结剂，混匀，等静压成型，脱模后在1750℃真空碳管炉中烧结1h，降温至1600℃，通氮气氮化2h，随炉冷却。坩埚断口呈金黄色，致密。除去表面氧

---

*　原发表于《1988年全国冶金物理化学年会论文集》，1988：206~211。

化层后，用于 Ba-O 平衡试验。硫钡研磨过 70 目筛，等静压成型，在碳管炉内 1700~1750℃ 烧结 1h，随炉冷却。在真空干燥器内避光保存，用于 Ba-S 平衡试验。

本试验中钼反应室的密封采用柔性石墨密封圈。常温模拟真空检漏，测得漏气率为 0.05~0.2torr·mL/s。高温试验表明此法密封效果很好。

平衡时间试验表明（表1、图2），3h 后反应已达平衡，但分析数据离差较大，5h 后趋于稳定。因此，平衡时间确定为 5h。

Ba-S 平衡试验用铁料经预脱氧。试验采用两种预脱氧方法。图 3（a）为 Ba 脱氧。将内装铁料和金属钡的 MgO 坩埚放入螺纹口密封的石墨坩埚内，在 1570~1630℃ 碳管炉内保温 3h，随炉冷却。脱氧后铁含氧小于 17ppm，成分均匀。图 3（b）为 Mg 脱氧，在碳管炉中保温 1h。脱氧后铁含氧小于 40ppm，成分不很均匀。

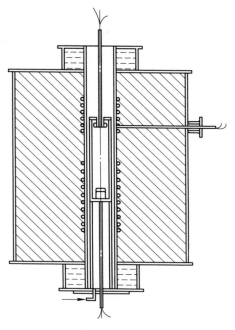

图 1 试验装置

表 1 钡、硫浓度随平衡时间的关系

elements \ time/h	3	5	7
[%Ba]	0.00019	0.00018	0.00019
	0.00015	0.00022	0.00018
	0.00027	0.00020	0.00024
	0.00024	0.00023	0.00029
	0.00027	0.00025	0.00022
		0.00028	
mean	0.00022	0.00023	0.00022
[%S]	0.00055	0.00063	0.00069
	0.00080	0.00060	0.00062
	0.00040	0.00058	0.00050
	0.00067	0.00068	0.00056
	0.00065	0.00052	0.00067
		0.00060	
mean	0.00061	0.0006	0.00061

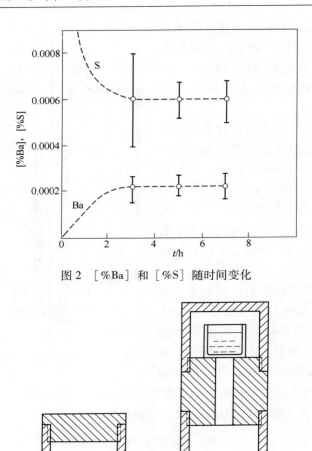

图 2 [%Ba] 和 [%S] 随时间变化

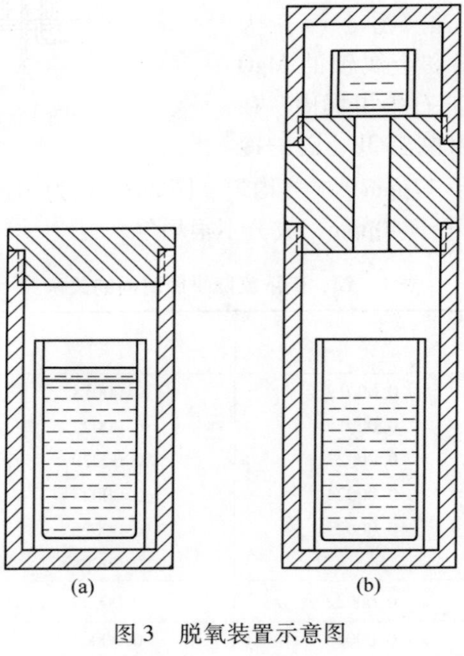

图 3 脱氧装置示意图

Ba-O 平衡试验中 TiN 坩埚内置少量 BaO 作为反应平衡固相。由于钡蒸气压大时，TiN 坩埚内铁液沿坩埚壁爬出，与钼反应室固溶，所以将 TiN 坩埚置于 MgO 坩埚内，以保证试验顺利进行。

试验用高纯铁料，杂质含量（%）：C 0.002, S 0.005, P 0.005, O 0.18, Mn<0.002, Si<0.005, N 0.038, Cu 0.007, Ni<0.02, Mo<0.008, Al 0.004, Bi

<0.001，Pb 0.0004。氮化钛粉粒度小于 10μm，杂质含量（%）：Fe<0.2，O<0.5，C<0.1。氢化铁粉-300 目，纯度大于 99.5%，杂质含量（%）：Fe<0.1，Si<0.02，Cl^-<0.08，C<0.02，N<0.04，O<0.4。BaO、BaS 为分析纯试剂，X 射线物相检测未发现有 $BaSO_4$、BaO 谱线。盐酸、无水乙醇系分析纯试剂。

铁中氧采用惰性载气熔融红外法分析，分析仪器为 TC-136 氧氮联合测定仪。硫采用高频熔融燃烧红外法分析，分析仪器为 CS-344 碳硫联合测定仪。钡采用高频等离子光谱-光电法（ICP-AES）分析，使用改进的旋流雾化进样品，灵敏度提高一个数量级[1]。铁中钡、氧、硫的分析检测限均为 0.1ppm。

3 实验结果及讨论

1600℃ 纯铁液中 Ba-O 及 Ba-S 反应的平衡数据见表 2、表 3。

表 2 1600℃ 纯铁液中 Ba-O 平衡数据

序号	[%Ba]	[%O]	[%S]	[%Ba] + 8.58[%O]	$-\lg[\%Ba][\%O] - e_{Ba}^{S}[\%S]$
1	0.00070	0.00310	0.00150	0.0273	5.971
2	0.00030	0.00044	0.00200	0.0041	7.289
3	0.00022	0.00065	0.00200	0.0058	7.254
4	0.00047	0.00085	0.00200	0.0078	6.808
5	0.00052	0.00155	0.00200	0.0138	6.503
6	0.00040	0.00349	0.00100	0.0303	6.065
7	0.00046	0.00014	0.00100	0.0017	7.399
8	0.00019	0.00100	0.00200	0.0088	7.143
9	0.00071	0.00411	0.00155	0.0360	5.852
10	0.00046	0.00152	0.00200	0.0135	6.567
11	0.00040	0.00212	0.00300	0.0186	6.686
12	0.00022	0.00025	0.00060	0.0024	7.383
13	0.00042	0.00022	0.00110	0.0023	7.260

表 3 1600℃ 纯铁液中 Ba-S 平衡数据

序号	[%Ba]	[%S]	[%O]	[%Ba] + 4.29[%S]	$-\lg[\%Ba][\%S] - e_{Ba}^{O}[\%O]$	deoxidiser
1	0.00106	0.00036	0.00109	0.0026	6.854	Ba
2	0.00086	0.00624	0.00179	0.0276	5.985	
3	0.00080	0.00239	0.00167	0.0111	6.384	
4	0.00011	0.00400	0.00071	0.0173	6.640	
5	0.00015	0.00200	0.00033	0.0087	6.655	
6	0.00037	0.00500	0.00088	0.0218	6.087	
7	0.00025	0.00044	0.00060	0.0021	7.198	
8	0.00008	0.00190	0.00050	0.0082	7.018	

续表3

序号	[%Ba]	[%S]	[%O]	[%Ba] + 4.29[%S]	$-\lg$[%Ba][%S] $-e_{Ba}^{O}$[%O]	deoxidiser
9	0.00175	0.00100	0.00340	0.0060	7.115	Mg
10	0.00059	0.00590	0.00093	0.0059	5.828	
11	0.00093	0.00173	0.00224	0.0084	6.686	
12	0.00110	0.00800	0.00186	0.0354	5.798	
13	0.00036	0.00039	0.00200	0.0020	7.651	
14	0.00022	0.00060	0.00080	0.0028	7.199	

在钡氧反应体系中，存在如下平衡

$$BaO_{(s)} = [Ba] + [O]$$

因为 $a_{BaO} = 1$，得到

$$\lg K_{BaO} = \lg[\%Ba][\%O] + e_{Ba}^{Ba}[\%Ba] + e_{Ba}^{O}[\%O] + e_{Ba}^{S}[\%S] + e_{O}^{O}[\%O] + e_{O}^{Ba}[\%Ba] + e_{O}^{S}[\%S]$$

由于 $e_{Ba}^{Ba}[\%Ba]$、$e_{O}^{O}[\%O]$、$e_{O}^{S}[\%S]$ 的数值相对较小，可忽略。代入 $e_{Ba}^{O} = e_{O}^{Ba} M_{Ba}/M_{O}$，整理上式得

$$-\lg[\%Ba][\%O] - e_{Ba}^{S}[\%S] = \lg K_{BaO} + e_{O}^{Ba}([\%Ba] + 8.58[\%O]) \quad (1)$$

用类似方法处理钡硫反应体系，得到

$$-\lg[\%Ba][\%S] - e_{Ba}^{O}[\%O] = -\lg K_{BaS} + e_{S}^{Ba}([\%Ba] + 4.29[\%S]) \quad (2)$$

由于不知 e_{Ba}^{S} 和 e_{Ba}^{O} 的数值，无法单独回归试验数据得到准确的热力学数据，因此，采用如下方法处理试验数据。

假设 $e_{Ba}^{S} = 0$，回归式（1），得到 e_{O}^{Ba}，将 e_{O}^{Ba} 值代入式（2）回归，得到 e_{S}^{Ba}，以此 e_{S}^{Ba} 值代回式（1）重新计算，如此迭代回归试验数据。计算表明，数值收敛很快。迭代回归得到

$$K_{BaO} = 3.98 \times 10^{-8} \quad e_{O}^{Ba} = -46.5 \quad |r| = 0.961$$
$$K_{BaS} = 5.45 \times 10^{-8} \quad e_{S}^{Ba} = -47.8 \quad |r| = 0.913$$

回归相关系数的统计分析表明，回归直线的置信度大于 $99.9\%^{[2]}$。以回归得到的 e_{O}^{Ba} 和 e_{S}^{Ba} 计算式（1）、式（2）两式左侧数值（列于表2、表3），并作图得到图4、图5。图中试验点分散的原因是由于本研究体系中元素含量很低，分析误差较大所致。图5表明，Ba 和 Mg 的脱氧铁均适用于钡硫平衡体系的研究，但 Mg 脱氧铁得到的试样含氧量较高。

由回归得到的 K_{BaO} 和 K_{BaS} 可求得 1600℃ 时

$$BaO_{(s)} = [Ba] + [O] \quad \Delta G_3^o = 63.5 \text{kcal/mol} \; ❶$$
$$BaS_{(s)} = [Ba] + [S] \quad \Delta G_4^o = 62.3 \text{kcal/mol}$$

---

❶ 1kcal = 4.184kJ。

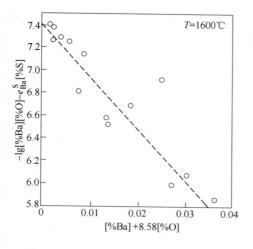
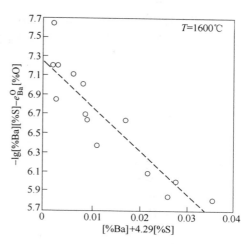

图 4 $-\lg[\%Ba][\%O] - e_{Ba}^{O}[\%S]$ vs $[\%Ba] + 8.58[\%O]$

图 5 $-\lg[\%Ba][\%S] - e_{Ba}^{O}[\%O]$ vs $[\%Ba] + 4.29[\%S]$

由文献 [3~5] 可查得 1600℃ 时

$$Ba_{(l)} + [O] = BaO_{(s)} \quad \Delta G_5^o = -58.3 \text{kcal/mol}$$

$$Ba_{(l)} + [S] = BaS_{(s)} \quad \Delta G_6^o = -56.1 \text{kcal/mol}$$

由钡氧反应体系可得到

$$Ba_{(l)} = [Ba] \quad \Delta G_7^o = \Delta G_3^o + \Delta G_5^o = 5.2 \text{kcal/mol}$$

由钡硫反应体系可得到

$$Ba_{(l)} = [Ba] \quad \Delta G_8^o = \Delta G_4^o + \Delta G_6^o = 6.2 \text{kcal/mol}$$

两个途径得到的金属钡在铁中的溶解自由能数值吻合较好。可以认为钡在铁中的溶解自由能

$$Ba_{(l)} = [Ba] \quad \Delta G^o = 5.7 \pm 0.7 \text{kcal/mol}$$

4 结论

（1）用钡蒸气平衡法研究了纯铁液中 Ba-O 及 Ba-S 的反应平衡，得到 1600℃ 时的平衡常数和活度相互作用系数为

$$K_{BaO} = 3.98 \times 10^{-8} \quad e_O^{Ba} = -46.5$$

$$K_{BaS} = 5.45 \times 10^{-8} \quad e_S^{Ba} = -47.8$$

（2）得到金属钡在纯铁中的溶解自由能

$$Ba_{(l)} = [Ba] \quad \Delta G^o = (5.7 \pm 0.7) \text{kcal/mol} \ (1600℃)$$

（3）柔性石墨是高温还原条件下理想的密封材料。

致谢：林源同志参加了本研究的部分试验工作。在试验过程中，李宝山同志曾给予大量的帮助，在此表示衷心的感谢。

参 考 文 献

[1] 何志壮, 吴廷照, 等. 分析化学, 1987, 15 (2)：136~141.
[2] Mravec J G. Electric Furnace Steelmaking, Ed. Sims C E V. Ⅱ, John Wiley & Sons, Inc. N. Y. , 1963：445.
[3] Barin Ⅰ, Knacke O. Thermochemical Properties of Inorganic Substances, Springer-Verlag, Berlin, 1973：77.
[4] Barin Ⅰ, Knacke O, Kubaschewski O. Thermochemical Properties of Inorganic Substances, Springer-Verlag, Berlin, 1977：69.
[5] Sigworth G K, Elliott J F. Metal Sci. , 1974 (8)：298~310.

Fe液中Ba-O、Ba-S、Ba-P平衡的研究*

王建生　韩其勇　魏寿昆　宋　波

摘　要：应用Ba蒸气平衡法，在密封Mo反应室内，用自制的TiN和BaS坩埚，进行了纯Fe液中Ba-O、Ba-S、Ba-P平衡的研究，得到反应：

$BaO_{(s)} = [Ba] + [O]$ 的标准自由能 $\Delta G^o_{wt\%} = 188500 + 38.3T(J/mol)$，$e^{Ba}_O = 624 - 1.25 \times 10^6/T$

$BaS_{(s)} = [Ba] + [S]$ 的标准自由能 $\Delta G^o_{wt\%} = 469100 + 111T(J/mol)$，$e^{Ba}_S = 826 - 1.64 \times 10^6/T$

1873K下，反应 $Ba_3P_2 = 3[Ba] + 2[P]$ 的 $K_{Ba_3P_2} = 7.76 \times 10^{-18}$，$e^P_{Ba} = -0.95$，$\Delta G^o_{wt\%} = 613200 J/mol$。

关键词：Ba，O，S，P，Fe液，热力学参数

钢中加入Ba，金属Ba不但是强脱氧剂和脱硫剂，还能有效地控制夹杂物的形状、尺寸及分布，改善浇铸和铸造的工艺性能，提高钢材质量[1~4]。金属Ba的沸点较高[5,6]，使其在钢液终处理合金应用上具有一定的优越性。由于缺乏Ba在铁液中的热力学数据，在钢液终处理过程中含Ba合金的物理化学行为还不大清楚。本文应用Ba蒸气平衡法，测定了铁液中Ba-O、Ba-S及Ba-P反应的热力学参数。

1　实验方法

实验用炉子为双恒温带钼丝炉，炉子下部为高温区，温度恒定在1873K，上部Ba液温度可在1473~1673K范围调节。钼反应室密封，盖上有一测温孔可准确测定Ba液温度。

炉子结构及实验方法同文献[7]。Ba-S平衡试验是在BaS坩埚、Ba-O及Ba-P平衡试验是在TiN坩埚中进行的。Ba-O、Ba-P平衡的Fe液中均分别放置有平衡反应产物相的BaO及Ba_3P_2，以保证达到平衡状态。Ba-P平衡所用的Ba_3P_2是将金属Ba与红磷在真空系统中1580℃合成的。Ba含量采用高频感应等离子光谱-光电法分析。

---

* 原刊于《金属学报》，1993，29（2）：B64~B69。

2 实验结果

2.1 纯铁液中 Ba-O、Ba-S 反应平衡常数及活度相互作用系数

图 1 为 1873K 铁液中 [%Ba] 与 [%S] 随时间的变化,可以看出,气液反应 3h 后反应即达到平衡,但数据波动较大;5h 后,波动较小。平衡时间定为 5h。

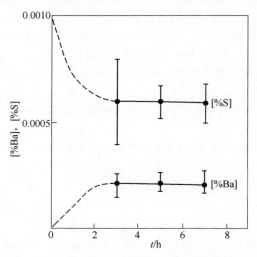

图 1 Fe 液中 [%Ba] 与 [%S] 随时间的变化

Fig. 1 Time dependence of [%Ba] and [%S] in molten Fe

不同温度铁液中 Ba-O、Ba-S 反应平衡数据分别列于表 1 和表 2 中。表 2 中最后一竖行为铁料脱氧所用的脱氧剂。

表 1 Fe 液中 Ba-O 反应平衡数据

Table 1 Experimental results of Ba-O equilibrium in molten Fe

T/℃	No.	%Ba	%O	%S
1550	6040102	0.00121	0.00300	0.00100
	6040103	0.00122	0.00275	0.00100
	6100701	0.00017	0.00244	0.00028
	6112103	0.00024	0.00085	0.00100
	7030402	0.00028	0.00213	0.00045
	7030504	0.00029	0.00107	0.00040
	7032402	0.00048	0.00137	0.00020
	7032502	0.00032	0.00029	0.00020
	7052501	0.00016	0.00061	0.00030
	7060206	0.00073	0.00045	0.00200
	7060307	0.00016	0.00034	0.00030

续表1

T/℃	No.	%Ba	%O	%S
1550	7060903	0.00039	0.00011	0.00020
	7071201	0.00018	0.00016	0.00100
	7071609	0.00066	0.00072	0.00160
	7072702	0.00015	0.00120	0.00110
	7072703	0.00008	0.00039	0.00060
	7072705	0.00007	0.00054	0.00030
1600	11-11	0.00070	0.00310	0.00150
	6092502-1	0.00030	0.00044	0.00200
	6092601	0.00022	0.00065	0.00200
	6101401	0.00047	0.00085	0.00200
	6101506	0.00052	0.00155	0.00200
	6112201	0.00040	0.00349	0.00100
	6112202	0.00046	0.00014	0.00100
	6112203	0.00019	0.00100	0.00200
	7012402	0.00046	0.00152	0.00200
	7030502	0.00071	0.00411	0.00155
	7060102	0.00040	0.00212	0.00300
	7071915	0.00022	0.00025	0.00060
	7071916	0.00042	0.00022	0.00110
1650	6042929	0.00014	0.00188	0.00280
	6101501	0.00033	0.00058	0.00200
	6101502	0.00034	0.00055	0.00200
	6101503	0.00033	0.00261	0.00300
	6120102	0.00033	0.00038	0.00100
	7012302	0.00014	0.00097	0.00100
	7030501	0.00021	0.00296	0.00330
	7030502	0.00012	0.00411	0.00045
	7032501	0.00032	0.00029	0.00020
	7060101	0.00062	0.00342	0.00300
	7060207	0.00044	0.00059	0.00200
	7060309	0.00016	0.00051	0.00030
	7060902	0.00058	0.00051	0.00030
	7071201	0.00020	0.00074	0.00120
	7071503	0.00019	0.00532	0.00160
	7071707	0.00045	0.00278	0.00300

表2 Fe 液中 Ba-S 反应平衡数据
Table 2 Experimental results of Ba-S equilibrium in molten Fe

$T/°C$	No.	%Ba	%S	%O	Predeoxidizer
1550	6092504	0.00011	0.00200	0.00043	Ba
	7030503	0.00024	0.00020	0.00107	
	7032504	0.00068	0.00050	0.00105	
	7051101	0.00032	0.00500	0.00051	
	7071203	0.00010	0.00050	0.00053	
	7071507	0.00010	0.00260	0.00073	
	6043019	0.00015	0.00300	0.00060	Mg
	7051102	0.00030	0.00400	0.00049	
	7060306	0.00017	0.00020	0.00046	
	7071508	0.00009	0.00080	0.00074	
	7071705	0.00056	0.00480	0.00134	
1600	6042802	0.00106	0.00036	0.00109	Ba
	6042902	0.00086	0.00624	0.00179	
	6043008	0.00080	0.00239	0.00167	
	6043020	0.00011	0.00400	0.00071	
	6101507	0.00015	0.00200	0.00033	
	6120101	0.00037	0.00500	0.00088	
	7060203	0.00025	0.00044	0.00060	
	7071704	0.00008	0.00190	0.00050	
	6040101	0.00175	0.00100	0.00340	Mg
	6042930	0.00059	0.00590	0.00093	
	6043002	0.00093	0.00173	0.00224	
	6101504	0.00110	0.00800	0.00186	
	7060204	0.00036	0.00039	0.00200	
	7071603	0.00022	0.00060	0.00080	
1650	6101508	0.00015	0.00200	0.00021	Ba
	7060111	0.00047	0.00030	0.00047	
	7060208	0.00027	0.00030	0.00064	
	7060301	0.00017	0.00800	0.00091	
	7060302	0.00014	0.00700	0.00095	
	7071204	0.00021	0.00100	0.00028	
	7071505	0.00033	0.00920	0.00044	
	7071506	0.00005	0.00430	0.00058	
	7071509	0.00024	0.00370	0.00077	
	7051001	0.00019	0.00700	0.00075	Mg
	7051102-1	0.00030	0.00080	0.00049	
	7060201	0.00045	0.00500	0.00048	
	7060205	0.00024	0.00042	0.00050	
	7060304	0.00023	0.01100	0.00061	
	7071504	0.00010	0.00760	0.00053	

Ba-O 平衡反应可表示为

$$BaO_{(s)} = [Ba] + [O] \quad (1)$$

由于 $a_{BaO} = 1$，得到

$\lg K_{BaO} = \lg[\%Ba][\%O] + e_{Ba}^{Ba}[\%Ba] + e_{Ba}^{O}[\%O] + e_{Ba}^{S}[\%S] +$
$e_{O}^{O}[\%O] + e_{O}^{Ba}[\%Ba] + e_{O}^{S}[\%S]$

由于 $e_{Ba}^{Ba}[\%Ba]$、$e_{O}^{O}[\%O]$、$e_{O}^{S}[\%S]$ 项的数值相对较小，可忽略。代入 $e_{Ba}^{O} = e_{O}^{Ba} M_{Ba}/M_{O}$，整理上式得

$$-\lg[\%Ba][\%O] - e_{Ba}^{S}[\%S] = -\lg K_{BaO} + e_{O}^{Ba}([\%Ba] + 8.58[\%O]) \quad (2)$$

用类似方法处理 Ba-S 反应

$$BaS_{(s)} = [Ba] + [S] \quad (3)$$

得 $-\lg[\%Ba][\%S] - e_{Ba}^{O}[\%O] = -\lg K_{BaS} + e_{S}^{Ba}([\%Ba] + 4.29[\%S]) \quad (4)$

由于不知 e_{Ba}^{S}、e_{Ba}^{O} 数值，无法单独回归式（2）和式（4），故采用如下方法处理试验数据：

假设 $e_{Ba}^{S} = 0$，回归式（2），求得 e_{O}^{Ba}。将 $e_{O}^{Ba} = 8.58 e_{O}^{Ba}$ 代入式（4），回归得到 e_{S}^{Ba}。再将 $e_{Ba}^{S} = 4.29 e_{S}^{Ba}$ 代回式（2）重新计算。如此迭代回归试验数据，计算结果见表 3。

表 3　Ba-O、Ba-S 体系计算结果
Table 3　Calculated results for Ba-O, Ba-S system

T_{Fe}/K	1823	1873	1923
$K_{BaO} \times 10^{-8}$	4.42	4.09	8.74
e_{O}^{Ba}	-57.8	-46.1	-21.4
$K_{BaS} \times 10^{-8}$	2.13	5.50	10.9
e_{S}^{Ba}	-75.8	-47.8	-28.9

由表 3 数据，可得反应式（1）和式（3）的平衡常数及活度相互作用系数与温度关系为

$\lg K_{BaO} = -2.00 - 9849/T,\ e_{O}^{Ba} = 624 - 1.25 \times 10^{6}/T,\ \Delta G_{1}^{o} = 188500 + 38.3T \quad (J/mol)$
$\lg K_{BaS} = 5.80 - 24516/T,\ e_{S}^{Ba} = 826 - 1.64 \times 10^{6}/T,\ \Delta G_{3}^{o} = 469100 - 111T \quad (J/mol)$

由文献 [8] 可得反应

$$BaO_{(s)} = Ba_{(l)} + \frac{1}{2}O_{2},\ \Delta G_{5}^{o} = 557180 - 102.68T \quad (J/mol) \quad (5)$$

$$BaS_{(s)} = Ba_{(l)} + \frac{1}{2}S_{2},\ \Delta G_{6}^{o} = 543920 - 123.43T \quad (J/mol) \quad (6)$$

由文献 [9] 可得反应：

$$\frac{1}{2}O_{2} = [O],\ \Delta G_{7}^{o} = -117150 - 2.89T \quad (J/mol) \quad (7)$$

$$\frac{1}{2}S_{2} = [S],\ \Delta G_{8}^{o} = -135060 - 23.43T \quad (J/mol) \quad (8)$$

由式（1）、式（5）、式（7）可得反应：
$$Ba_{(1)} = [Ba], \quad \Delta G_9^o = -251530 - 143.87T \quad (J/mol) \quad (9)$$
$$\ln\gamma^o = \Delta G_9^o/RT - \ln N^o = 22.81 - 30270/T$$

1873K 时，$\gamma^o = 772$。

由反应式（3）、式（6）、式（8）可得反应式（9）的
$$\Delta G_9^o = 60240 - 11T \quad (J/mol)$$
$$\ln\gamma^o = 7250/T + 4.18$$

1873K 时，$\gamma^o = 3136$。

2.2 纯铁液中 Ba-P 反应平衡常数及活度相互作用系数

对 Ba-P 反应
$$Ba_3P_2 = 3[Ba] + 2[P] \quad (10)$$

用上节类似方法可得：
$$-\lg[\%Ba]^3[\%P]^2 + 1187[\%O] + 615[\%S]$$
$$= -\lg K_{Ba_3P_2} + e_{Ba}^P(3[\%P] + 0.4511[\%Ba]) \quad (11)$$

以式（11）左边诸项为纵坐标，以 $3[\%P] + 0.4511[\%Ba]$ 为横坐标，作图2。

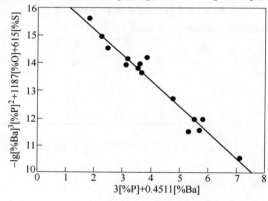

图 2 Ba-P 体系试验结果

Fig. 2 Experimental results for Ba-P system at 1873K

回归图2，得
$$-\lg[\%Ba]^3[\%P]^2 - 1187[\%O] + 615[\%S]$$
$$= 17.11 - 0.95(3[\%P] + 0.4511[\%Ba])$$

1873K 下，$\lg K_{Ba_3P_2} = -17.11$，$K_{Ba_3P_2} = 7.76\times10^{-18}$，$e_{Ba}^P = -0.95$。反应式（10）的 $\Delta G_{10}^o = 613200 J/mol$。

1873K 下铁液中碱土金属 Ca、Ba、Mg 与 O、S、P 体系的平衡常数及活度相互作用系数见表4。

表4　1873K 下 Fe 液中 Ca、Ba、Mg 与 O、S、P 的平衡常数 K 值和 e_i^j 值
Table 4　Values of equilibrium constants, K and e_i^j of Ca, Ba and Mg on O, S and P in molten Fe at 1873K

	Ca		Ba		Mg	
	K	e_i^j	K	e_i^j	K	e_i^j
O	5.5×10^{-9}	-457[10]	4.09×10^{-8}	-46	9.33×10^{-7}	-115[12]
	4.3×10^{-8}	-178[11]				
S	5.9×10^{-8}	-106[10]	5.50×10^{-8}	-48	[%Mg][%S]	$=2.0\times10^{-4}$[12]
P	7.0×10^{-14}	-3.1[10]	7.76×10^{-18}	-0.95		

Note: Values of K and e_i^j of Ba on O, S and P given by this study.

由表可见，按脱氧、脱硫能力，顺序为 Ca>Ba>Mg。

参 考 文 献

[1] 严铄，徐思盛，王毓麟. 钢铁，1981，16 (4)：28.
[2] Дерябин А А, Зайко В Ⅱ, Байрамов Б М, Семенков В Е, Сърейщкова В Е, Колосова З Л. СССР Патент，996501，1983.
[3] Чубиниазе Т А, Арсенищвии А Ю, Гоажилащвили Н А, Биркая Г Г, Косарев Л Н, Микеров Ю К. СССР Патент，981424，1983.
[4] 知水，杨印东，李万象. 特殊钢，1984（增刊）：10.
[5] Hultgren R, Orr R L, Anderson P D, Kelley K K. Selected Values of Thermodynamic Properties of Metals and Alloys. New York：Wiley，1963：43.
[6] Schins H E J, Van Wijk R W M, Dorpema B. Z Metallk，1971，62：330.
[7] 张晓东，韩其勇，陈冬. 金属学报，1990，26：B457.
[8] Turkdogan E T. Physical Chemistry of High Temperature Technology. New York：Academic，1980：6.
[9] Sigworth G K, Elliott J F. Met Sci，1974，8：298.
[10] Han Qiyong, Zhang Xiaodong, Chen Dong, Wang Pengfei. Metall Trans，1988，19B：617.
[11] 王鹏飞，韩其勇. 金属学报，1988，24：B7.
[12] 韩其勇，周德壁. 第6届冶金过程物理化学学术会议论文集，1986，重庆，1986：166.

Separation of Nb from Nb-bearing Iron Ore by Selective Reduction*

Chen Hong   Han Qiyong   Wei Shoukun   Hu Zhigao

Abstract: A new approach for extraction of Nb from Nb-bearing complex iron ore, found in North-West China, is proposed. Selective reduction of the complex iron oxides by CO-CO_2 gas mixture was conducted, and the reduced metallic Fe was magnetically separated from the Nb-bearing gangue residue. The Nb-rich product was leached with HCl to remove P. A highly concentrated Nb-bearing product with Nb/Fe = 6, Nb/P = 12 might be possible to make commercial grade of 60%~65% Nb ferroniobium.

Extensive deposit of complex Nb-bearing iron ore exists in the Baotou District, Northwest China. Ferroniobium is produced from this complex ore conventionally by the four-furnace process[1], namely: (1) reduction of the complex ore in the blast furnace; (2) air-blowing of the hot metal in the side-blown converter; (3) remelting of the Nb-rich slag for removal of the contaminated iron which contained much P in the electric furnace; and (4) reduction of the Nb-rich slag with ferrosilicon in the electric furnace to obtain ferroniobium with 10%~15% Nb and much Mn. Research on the demanganification of the Nb-rich slag with $CaCl_2$ with recovery of $MnCl_2$, and reduction of the demanganified Nb-rich slag with ferro-silicon with a yield of 60% Nb has been reported[2,3]. But the whole process is too long to warrant an economical production.

In this paper selective reduction of the ore with gas mixture of CO and CO_2 for separation of Fe from Nb-bearing iron ore was studied, while after magnetic extraction the Nb-rich product was dephosphorised by acid treatment.

1 Selective reduction of the iron ore with CO-CO_2

1.1 Theoretical basis

Using the thermodynamic data of reactions (1) to (5), the values of ΔG° for reduction of

---

* Steel Research 73 (2002) No. 5.

oxides of Nb with various valences can be evaluated as follows (with ΔG in J/mol):

$$C_{(s)} + 0.5O_{2(g)} = CO_{(g)} \tag{1}$$

$$\Delta G_1^\circ = -114400 - 85.77T^{[4]}$$

$$C_{(s)} + O_{2(g)} = CO_{2(g)} \tag{2}$$

$$\Delta G_2^\circ = -395350 - 0.54T^{[4]}$$

$$Nb_{(s)} + 0.5O_{2(g)} = NbO_{(s)} \tag{3}$$

$$\Delta G_3^\circ = -414200 + 86.6T^{[4]}$$

$$Nb_{(s)} + O_{2(g)} = NbO_{2(s)} \tag{4}$$

$$\Delta G_4^\circ = -783700 + 166.9T^{[4]}$$

$$2Nb_{(s)} + 2.5O_{2(g)} = Nb_2O_{5(s)} \tag{5}$$

$$\Delta G_5^\circ = -1888200 + 419.7T^{[4]}$$

Combination of equations (1), (2), (4) and (5) gives equation (6):

$$Nb_2O_{5(s)} + CO_{(g)} = 2NbO_{2(s)} + CO_{2(g)} \tag{6}$$

$$\Delta G_6^\circ = 39850 - 0.67T$$

$$\log K_6 = -2082.3/T + 0.035$$

In the pure CO-CO$_2$ mixture system and on assumption of pure solid substances and $(p_{CO} + p_{CO_2}) = 1$ atm:

$$K_6 = p_{CO_2}/p_{CO} = p_{CO_2}/(1 - p_{CO_2})$$

$$p_{CO_2} = \frac{K_6}{1 + K_6}, \quad p_{CO} = \frac{1}{1 + K_6}$$

Based on the values of ΔG_6° for reaction (6), the calculated values of K_6, p_{CO_2}, p_{CO}, V_{CO_2} and V_{CO} of the system in the equilibrium at different temperatures are listed in Table 1.

Similar calculations give the following results for reactions (7) to (9), Tables 2 to 4.

Table 1 Calculated values of K, p_{CO_2}, p_{CO}, V_{CO_2} and V_{CO} of reaction (6) at different temperatures

T/K	K	p_{CO_2}	p_{CO}	$V_{CO_2}/\%$	$V_{CO}/\%$
1573	0.051	0.049	0.951	4.9	95.1
1473	0.042	0.040	0.960	4.0	96.0
1373	0.033	0.032	0.968	3.3	96.8
1273	0.025	0.024	0.976	2.4	97.6

Continued Table 1

T/K	K	p_{CO_2}	p_{CO}	$V_{CO_2}/\%$	$V_{CO}/\%$
1173	0.018	0.018	0.982	1.8	98.2
1073	0.012	0.012	0.988	1.2	98.8
973	0.008	0.008	0.992	0.8	99.2
873	0.005	0.005	0.995	0.5	99.5
773	0.002	0.002	0.998	0.2	99.8
673	0.001	0.001	0.999	0.1	99.9

$$NbO_{2(s)} + CO_{(g)} = NbO_{(s)} + CO_{2(g)} \tag{7}$$

$$\Delta G_7^\circ = 88550 + 4.93T$$

$$\log K_7 = -4624.93/T - 0.258$$

$$NbO_{(s)} + CO_{(g)} = Nb_{(s)} + CO_{2(g)} \tag{8}$$

$$\Delta G_8^\circ = 133250 - 1.37T$$

$$\log K_8 = -6958.23/T + 0.072$$

$$\frac{1}{5}Nb_2O_{5(s)} + CO_{(g)} = \frac{2}{5}Nb_{(s)} + CO_{2(g)} \tag{9}$$

$$\Delta G_9^\circ = 96700 + 1.30T$$

$$\log K_9 = -5051.4/T - 0.068$$

The equilibrium curves and the phase stability area of niobium oxides during reduction with CO-CO$_2$ gas based on the data listed in Tables 1 to 4 are shown in Fig. 1.

Table 2 Calculated values of K, p_{CO_2}, p_{CO}, V_{CO_2} and V_{CO} of reaction (7) at different temperatures

T/K	K	p_{CO_2}	p_{CO}	$V_{CO_2}/\%$	$V_{CO}/\%$
1573	6.3×10^{-4}	6.3×10^{-4}	0.99937	6.3×10^{-2}	99.94
1473	4.0×10^{-4}	4.0×10^{-4}	0.99960	4.0×10^{-2}	99.96
1373	2.4×10^{-4}	2.4×10^{-4}	0.99976	2.4×10^{-2}	99.98
1273	1.3×10^{-4}	1.3×10^{-4}	0.99987	1.3×10^{-2}	99.99
1173	6.3×10^{-5}	6.3×10^{-5}	0.99994	6.3×10^{-3}	99.99
1073	4.9×10^{-5}	4.9×10^{-5}	0.99995	4.9×10^{-3}	
973	9.7×10^{-6}	9.7×10^{-6}		9.7×10^{-4}	
873	2.8×10^{-6}	2.8×10^{-6}	~1	2.8×10^{-4}	~100
773	5.7×10^{-7}	5.7×10^{-7}		5.7×10^{-5}	
673	7.4×10^{-8}	7.4×10^{-8}		7.4×10^{-6}	

Table 3 Calculated values of K, p_{CO_2}, p_{CO}, V_{CO_2} and V_{CO} of reaction (8) at different temperatures

T/K	K	p_{CO_2}	p_{CO}	$V_{CO_2}/\%$	$V_{CO}/\%$
1573	4.5×10^{-5}	4.5×10^{-5}	0.99996	4.5×10^{-3}	
1473	2.2×10^{-5}	2.2×10^{-5}	0.99998	2.2×10^{-3}	
1373	1.0×10^{-5}	1.0×10^{-5}		1.0×10^{-3}	
1273	4.0×10^{-6}	4.0×10^{-6}		4.0×10^{-4}	
1173	1.4×10^{-6}	1.4×10^{-6}		1.4×10^{-4}	~100
1073	3.9×10^{-7}	3.9×10^{-7}	~1	3.9×10^{-5}	
973	8.3×10^{-8}	8.3×10^{-8}		8.3×10^{-6}	
873	1.2×10^{-8}	1.2×10^{-8}		1.2×10^{-6}	
773	1.0×10^{-9}	1.0×10^{-9}		1.0×10^{-7}	
673	5.4×10^{-11}	5.4×10^{-11}		5.4×10^{-9}	

Table 4 Calculated values of K, p_{CO_2}, p_{CO}, V_{CO_2} and V_{CO} of reaction (9) at different temperatures

T/K	K	p_{CO_2}	p_{CO}	$V_{CO_2}/\%$	$V_{CO}/\%$
1573	4.1×10^{-17}	4.1×10^{-17}		4.1×10^{-15}	
1473	3.3×10^{-18}	3.3×10^{-18}		3.3×10^{-16}	
1373	1.9×10^{-19}	1.9×10^{-19}		1.9×10^{-17}	
1273	6.5×10^{-21}	6.5×10^{-21}		6.5×10^{-19}	
1173	1.4×10^{-22}	1.4×10^{-22}	~1	1.4×10^{-20}	~100
1073	1.4×10^{-24}	1.4×10^{-24}		1.4×10^{-22}	
973	5.2×10^{-27}	5.2×10^{-27}		5.2×10^{-25}	
873	5.6×10^{-30}	5.6×10^{-30}		5.6×10^{-28}	
773	1.0×10^{-33}	1.0×10^{-33}		1.0×10^{-31}	
673	1.4×10^{-38}	1.4×10^{-38}		1.4×10^{-36}	

The above calculations indicate that Nb_2O_5 can be reduced to NbO_2 at a temperature higher than 673K; NbO_2 can be reduced a little to NbO at 1573K; but NbO can never be reduced to metallic Nb.

Fig. 2 shows the classical equilibrium diagram of reduction of iron oxides by CO-CO_2[5] and the equilibrium curve of Bouduard's reaction $C + CO_2 = 2CO$. The equilibrium curve of reaction (9) at different temperatures coincides with the top line (CO% = 100) of the figure, showing that Nb_2O_5 can never be reduced to Nb under any conditions. So thermodynamically, in order to accomplish the selective reduction of iron oxides with keeping the Nb-oxide unreacted, the hatched area of Fig. 2 is the optimum area for choosing the best temperature and composition of $CO\text{-}CO_2$ for experimentation.

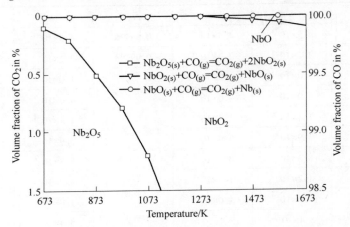

Fig. 1 Equilibrium curves and phase stability area of niobium oxides during reduction with $CO\text{-}CO_2$ gas

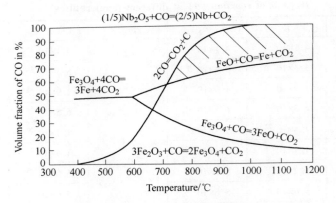

Fig. 2 Equilibrium diagram of reduction of iron oxides by $CO\text{-}CO_2$ gas with Bouduard's curve

1.2 Experimental

The experimental apparatus for the selective reduction is shown in Fig. 3. CO gas used in this experiment was generated by the Bouduard reaction of carbon dioxide with carbon, which was carried out in the carbon monoxide generating furnace charged with charcoal. At 1373K, the CO gas generated by this reaction was over 99%. To ensure more CO gas to be generated, the temperature in the high temperature zone should not be lower than 1373K.

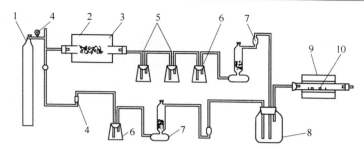

Fig. 3 Experimental apparatus for the selective reduction of
Nb-bearing iron ore by CO-CO$_2$ mixture gas

1—CO$_2$ gas cylinder; 2—CO generating furnace; 3—charcoal; 4—flow meter;
5—NaOH solution; 6—concentrated sulphuric acid; 7—gas sieve;
8—gas mixing bottle; 9—reduction furnace; 10—sample

The generated CO gas was passed to the gas mixing bottle, its CO$_2$ content being absorbed by NaOH solution and then dried with concentrated H$_2$SO$_4$. Another stream of CO$_2$ gas was also passed to the gas mixing bottle after purification. The gas flow rates of both CO and CO$_2$ were controlled by flow meters. The gas from the gas mixing bottle was passed to the reaction furnace to react with the sample.

The sample was made in the following way: The Nb-bearing iron ore powder (120 mesh in size) was stuck together with water glass, pressed into a ϕ9.7mm cylinder, and dried at 523K for 1hour.

The Nb-bearing iron ore used in this experiment was a concentrate from gravity concentration of No.2 ore in Baiyunebo, Baotou, China, its chemical composition is listed in Table 5.

Table 5 Chemical composition of the concentrate of No.2 iron ore (mass contents in %)

Nb$_2$O$_5$	TFe	FeO	SiO$_2$	TiO$_2$	CaO	MgO	MnO
1.77	53.7	16.7	2.42	1.13	4.95	2.37	0.40
Al$_2$O$_3$	P	S	F	RE$_x$O$_y$	K$_2$O	Na$_2$O	ThO$_2$
1.70	1.17	0.01	0.80	1.50	0.27	0.04	0.08

Based on Fig.2, a gas composition of 90%CO + 10%CO$_2$ was used in the experiments. The reduction temperature range was 1173 to 1423K; the gas flow rate was 200mL/min and 300mL/min, the reaction time was 10, 25, 45, 65, 85 and 105min.

1.3 Experimental results

From the results of preliminary experiment shown in Table 6, it can be seen that the flow rate of the CO-CO_2 gas being 200mL/min is good enough for reduction of the iron oxides, so that the flow rate of the gas used in the experiments therein after is 200mL/min. The experimental results for the relation between the reduction time, the reduction temperature and the metallisation percentage for the iron ore is shown in Fig. 4. The sample which had been reduced at 1373K for 1h was analysed by EDX. The SEM image is shown in Fig. 5. The dark part in the image is oxide and the bright part is metal. The results of analysis by EDX for spots A, B, C, D, E, F and G are listed in Table 7. After reduction at 1373K for 1h, the sample was ground into small particles (200 mesh in size), then the iron was separated from the ground sample by magnetic separation with laboratory magnet, the magnetic field intensity of which is around 0.05 Tesla, in water-free alcohol. The removal percentage of iron by magnetic separation was 96.5%. The composition of Nb-bearing product was analysed after drying, Table 8. Then the Nb-bearing product was leached with hydrochloric acid to remove iron, phosphorous and some other impurities. The composition of the final product is listed in Table 9.

Table 6 Results of preliminary experiment for the selective reduction by the CO-CO_2 mixture gas

Gas flowrate /mL · min^{-1}	Reducing time /h	Reducing temperature /K	Composition of reducing gas	Metallization percentage of iron ore/%
200	1	1273	90%CO + 10%CO_2	95.1
300	1	1273	90%CO + 10%CO_2	93.9

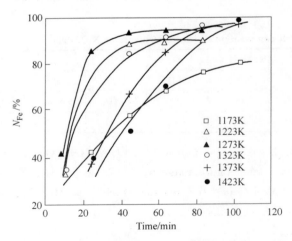

Fig. 4 Relation between the metallisation percentage of the iron ore, reduction temperature and reducing time

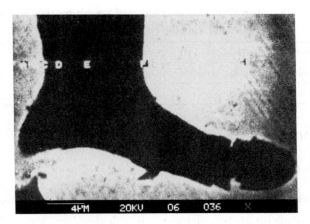

Fig. 5 SEM image of the sample reduced at 1373K for 1h by CO-CO$_2$ mixture gas and the spots analysed by EDX

Table 7 The results of the spot-analysis by EDX for the sample reduced at 1373K for 1hour

	A	B	C	D	E	F	G
Fe	98.18	98.483	98.664	40.155	22.259	64.44	98.988
Nb	0.028	0.174	0.275	6.322	7.977	3.167	0.035
P	0	0.085	0.159	3.674	4.907	1.967	0.19
Ca	0.385	0.48	0.241	16.957	23.810	8.671	0.049
Mg	0.542	0.283	0.038	7.745	8.653	7.943	0.275
Si	0.17	0.258	0.335	15.091	19.406	8.373	0.049
Ti	0	0	0.004	1.025	1.501	0.470	0.106
Mn	0.291	0.084	0.108	1.886	2.565	1.227	0.110
Ce	0.222	0	0	1.610	1.961	0.661	0
Al	0.183	0.166	0.168	4.188	5.308	2.283	0
K	0	0	0.019	1.397	1.634	0.805	0

Table 8 Total analysis for the Nb-rich product after removal of iron (mass contents in %)

TFe	FeO	SiO$_2$	CaO	Nb$_2$O$_5$	P
8.07	10.2	5.67	39.6	6.91	4.52
MgO	RE$_x$O$_y$	TiO$_2$	MnO	Al$_2$O$_3$	
4.85	1.03	<0.05	0.29	1.37	

Table 9 Main chemical composition of the Nb-bearing product after acid leaching (mass contents in %)

Composition	TFe	FeO	CaO	MgO	SiO$_2$	Nb$_2$O$_5$	P
Content in leached product/%	3.45	—	12.1	1.44	19.8	29.6	1.71
Removal rate/%	90.6	—	93.3	93.5	23.2	5.80	91.7

2 Discussion

2.1 Formation of Fe$_2$Nb

The phase stability diagram (also called the phase predominance area diagram), as well as the Ellingham-Richardson diagram, is calculated on the basis of pure substances, just for general discussion and for simplicity. Considering that the reacting substances are compounds, one might write equation (8) as follows:

$$2FeO \cdot NbO + 3CO = Fe_2Nb + 3CO_2 \tag{10}$$

Owing to a lack of thermodynamic data, it would be impossible to make any direct thermodynamic calculations for the above reaction. On the assumption that $2FeO \cdot NbO$ and Fe_2Nb might exist in solid solution and that pure substance standard would be chosen to express the activity of the components of the solid solution, the above reaction (10) might be written as follows:

$$(NbO) + CO = CO_2 + [Nb] \tag{11}$$

$$K = \frac{p_{CO_2}}{1 - p_{CO_2}} \cdot \frac{\gamma_{Nb} N_{Nb}}{\gamma_{NbO} N_{NbO}}$$

$$N_{Nb} = \frac{n_{Nb}}{2n_{Fe} + n_{Nb}} = \frac{1}{3}$$

$$N_{NbO} = \frac{n_{NbO}}{2n_{FeO} + n_{NbO}} = \frac{1}{3}$$

therefore: $N_{NbO} = N_{Nb}$

$$p_{CO_2} = \frac{K\left(\dfrac{\gamma_{NbO}}{\gamma_{Nb}}\right)}{1 + K\left(\dfrac{\gamma_{NbO}}{\gamma_{Nb}}\right)}$$

If we assume $\dfrac{\gamma_{NbO}}{\gamma_{Nb}} = 1$, equation (11) would have the same value of p_{CO_2} as equation (8) listed in table 3. If we assume $\dfrac{\gamma_{NbO}}{\gamma_{Nb}} = 100$ as an extreme case, at $T = 1373K$ (the highest temperature of reduction), $K = 1.0 \times 10^{-5}$, $p_{CO_2} = 1.0 \times 10^{-3}$, $V_{CO_2} = 0.1\%$. The predominance area of Fe_2Nb would be extremely small. From the above calculation it might be concluded that practically no formation of Fe_2Nb would be expected.

2.2 Formation of NbC

So far as the formation of NbC, left of the equilibrium curve of Bouduard's reaction in Fig. 2 is carbon stability area, the reaction being $2CO \rightarrow C + CO_2$; in the right area of the same curve, the reaction being $C + CO_2 \rightarrow 2CO$, there is no carbon deposition, so the formation of NbC is impossible and not to be considered.

2.3 Optimum conditions of selective reduction

From Fig. 4, we can see that under the same temperature, the metallisation percentage of the iron ore increases with the reaction time, and finally reaches a constant value. When the reactions are close to equilibrium at different temperatures, all the metallisation percentage for the iron ores is over 90% except at 1173K. When the reaction is close to equilibrium, the higher the reaction temperature, the higher the metallisation percentage of the iron ore. The speed for reaching equilibrium at 1173K and 1423K is the slowest, the speed at 1223K and 1273K is the largest, and that in the range between 1323K and 1373K gives intermediary values. During the reduction there are two factors which affect the sample, one is that Fe_2O_3 (hexagonal lattice) is reduced into Fe_3O_4 (cubic lattice), the lattice is changed and the sample expands. Another is that due to the sintering, the iron atom produced in reduction diffuses towards the centre of the sample, the sample contracts. High temperature speeds up the sintering rate[6]. So at higher temperatures, sintering dominates the whole process, the contraction plays a major role, making the sample dense. The sintered structure hinders the gas-solid reaction and decreases the reduction speed (at 1423K). At a slightly lower temperature (1223 to 1273K), because of the weak sintering effect and because the sample expands due to the change of the crystal lattice, the sample is loose. This case favours the gas reduction process, therefore in this temperature range, the reaction speed is the largest. The structures of the layers of reduced products are different. The dynamical mechanisms under different temperatures are different.

From the EDX analysis results at A, B, C, D, E, F and G spots in Fig. 5 (Table 7), it is shown that there is hardly any Nb in iron phase (high Nb content in spot C is because of the fact that spot C is in the boudary surface between metal and oxide). The content of Nb is rather high in the oxide phase, however, its distribution is not uniform. Although some of the iron still remained in the oxide containing high Ca, Mg impurities, the above results indicate that the selective separation of iron from Nb-bearing oxides could be achieved.

The composition of enriched niobium oxide after removing iron by magnetic separation is listed in Table 8. The total iron content in Nb-bearing oxide is about 10%. Nb_2O_5 is raised from 1.77% to 6.91%. The grade of Nb was enriched by a factor of 5. It is shown in Table 8 that the content of CaO in Nb-bearing oxide is high, nearly two-fifths of the whole. The iron is presented as FeO. Also there are P, MgO etc. in the Nb-enriched oxide product.

Though the Nb content is comparatively high after removing iron, Nb/Fe ratio is only 0.6. Compared with the requirement for making ferroniobium up to the industrial standard, the value of Nb/Fe ratio is too low. Therefore the remaining iron, phosphorous and other impurities in Nb-bearing oxide must be further removed.

2.4 Dephosphorisation of deironised product

Our experiments have shown that some oxides, especially P and Ca, in the Nb-rich deironised product can be removed by acid leaching for the enrichment of Nb[7]. When the Nb-rich product is leached respectively by nitric acid and hydrochloric acid, the Nb_2O_5 content in the leached oxides changed according to the leaching acid concentration. When nitric acid concentration is raised to a certain extent, the content of Nb_2O_5 in the leached product remains constant, but the content of Nb_2O_5 in the deirionized product increased with hydrochloric acid concentration. Obviously the leaching effect of hydrochloric acid is superior to that of nitric acid.

Therefore, based on these experiments, hydrochloric acid has been used as the leaching agent for the Nb-rich deironised product. The best conditions for leaching the Nb-bearing oxide using hydrochloric acid are as follows: The acid leaching concentration is 2.4M, the acid leaching time is 45min, the acid leaching temperature is 368K, solid liquid ratio is 1g : 15mL. Table 8 shows the composition of Nb-bearing oxide after removal of iron. The main chemical composition of Nb-bearing products after acid leaching is listed in Table 9. The removal percentages of P, CaO and MgO using hydrochloric acid are all over 90%. The content of Nb_2O_5 in the leached oxides is raised to 29.6%, Nb/Fe ratio is 6. This ratio is well beyond the requirement for making industrial standard ferroniobium (Nb>60%).

3 Conclusions

The Nb-bearing iron ore is selectively reduced by CO-CO_2 mixture gas, the metallisation percentage for iron ore is up to 95%, the reducing gas composition being 90% CO + 10% CO_2, the gas flow rate 200mL/min, the reducing temperature 1223 to 1423K.

The results of chemical analysis and SEM determination show that while the iron ore is reduced, the Nb oxides remain unchanged, thus making the separation of iron from niobium-bearing ore possible.

By magnetising separation of ground reduced sample in ethanol medium, the iron was removed. The content of Nb_2O_5 in the residue can reach as high as 6.91%, the Nb was enriched by a factor of 5.

The Nb-rich product can be leached with HCl to remove P. The final product might be possible to make commercial grade of 60%~65% Nb ferroniobium.

Acknowledgment

The authors are very grateful to Baotou Iron and Steel Corporation for offering the niobium-bearing iron ore.

(A 01 708; received: 30. July 2001; in revised form: 28. January 2002)

References

[1] Baotou Iron and Steel Corporation. Techn. Rep., 1985.
[2] Wei S K, Tan Z, Zhu Y. Proc. 4th Japan-China Symp. On Science and Technology of Iron and Steel, Kobe, Japan, 1987: 199~213.
[3] Tan Z, Wei S K, Zhu Y. Rare Metal (1989) No.2: 137~141.
[4] Turkdogan E T. Physical Chemistry of High Temperature Technology. New York, Academic Press, 1980: 5~26.
[5] Bogdandy L von, Ende H vom, Frohberg M G. Dic Physikalische Chemie der Eisen-und Stahlerzeugung, 1964: 177~123.
[6] Szekely J, Evans J W, Sohn H Y. Gas-Solid Reactions, Academic Press, London, 1976.
[7] Chen H, Han Q, Hu Z. Rare Metal (1995) No.2: 86~88.

附录 魏寿昆院士生平[1]

魏寿昆（1907~2014），天津人。冶金学家，教育学家。冶金物理化学学科奠基人之一。一级教授，中国科学院院士。1929年毕业于天津北洋大学矿冶工程系，获得学士学位，1935年获得德国德累斯顿工业大学化学系工学博士学位。1935~1936年在德国亚琛工业大学钢铁冶金研究所博士后进修。20世纪40年代，结合我国当时的矿产资源及技术改进等情况研究解决了生产问题，并获得五项发明专利；50年代应用活度理论对冶金反应进行了深入的热力学分析研究；60年代发展了炉渣离子理论，提出了高炉渣中计算硫离子活度系数公式；70~80年代在国内首先开展在冶金生产中应用固体电解质定氧技术，使之成为提高质量的重要手段，并用于热力学参数测定的理论研究；80~90年代针对我国富有的共生矿和贫矿研究了矿石中有害元素的分离去除和有益元素的提取富集。开展了大量理论研究并指导了生产。研究课题获得了国家自然科学奖三等奖一项，国家教委科技进步奖一等奖两项、二等奖一项，1997年获得何梁何利科学与技术进步奖，2006年获得金属学会颁发的冶金科技终身成就奖。发表学术论文140篇。在70余年的教学工作中，讲授过基础课、技术基础课和专业课共28门课程。在教学和科研工作中培养青年教师，指导培养20余名硕士生和博士生，出版了五部学术专著，曾获得学校、中国科学院、冶金工业部及国家教委部门颁发的"先进教师""教书育人先进工作者""老骥伏枥奖"等荣誉奖状和证书20项。曾任北京钢铁学院（今北京科技大学）教务处长、图书馆馆长、副院长。北京钢铁学院建校元老之一，中国金属学会创建人之一。曾任九三学社中央委员会常委。

一、成长经历

1. 弃商就学

1907年9月16日魏寿昆生于天津。魏家曾是殷实的大户人家，几经战乱后家道衰落，祖父辈的兄弟们迫于生计分家各谋出路。全家十几口人的生活全靠魏寿昆的祖父当雇员维持，日子过得很艰苦。

魏寿昆自幼敏而好学，让其祖父看到了复兴家业的希望，于是打算让魏寿昆读完私塾后，到商铺当学徒经商。但三年后，年仅10岁的魏寿昆却要弃商就学，

---

[1] 原发表于《20世纪中国知名科学家学术成就概览·化工、冶金与材料工程卷·冶金工程与技术分册（一）》，科学出版社，2015：99~112。撰写者：朱元凯。

他曾说:"读私塾、学买卖、当老板是我祖辈走过的从商道路。我抛弃从商之路,这是我家的一场革命。"这改变了魏寿昆的人生道路,他从私塾、小学、中学到大学,不仅刻苦好学,而且培育了高尚的品德。在私塾读《孟子》"孟子见梁惠王"篇时,虽然当时的他还不能完全理解其中的哲理,但随着生活阅历的增长,一直以不逐名利、仁和关爱、做正派人为座右铭以自勉。在北洋大学学习期间魏寿昆受到了专业方面的严格训练和实事求是的科学精神的影响,逐渐形成勤奋好学、求真务实、严谨创新的治学之本及为人之道,这也是他在几十年教书育人生涯中的准则。

2. 留德五年

在北洋大学(后改名为北洋工学院)学习期间,魏寿昆受到众多名师的教导,激发了他出国深造的愿望。当时考虑到如果将来报考化学学科,他在大学没有学过的有机化学课将是必考课程,因而在读大学三年级时,他便自学了有机化学。此外他在大学预科学过一年德语,随后数年中一直坚持自学。这两门课程在他后来考取留学生和读博士学位时起到了重要作用。"凡事预则立",这也是魏寿昆的成功之道。1930年秋魏寿昆成功考取了留德生。

1931年赴德后,他在德国柏林工科大学的德语学校预修德语,按规定需八个月毕业。但魏寿昆因有过去自学的基础,仅学习了两个月即获得进入大学的德语合格证书。他在柏林工业大学学习了两个学期,主攻有机化学、物理化学和实验课,1932年秋转入德国德累斯顿工业大学色染及纺织研究所,完成了染色原理及染料制备化学的学习,1935年6月获得工学博士学位。

由于在大学学的专业和志向是冶金学科,因此他又到科隆附近的好望钢铁公司的炼焦、烧结、高炉、转炉、平炉、电炉、铸锭、粗轧、型材、管材、板材、冷轧、材料检验及化学分析等车间及部门进行了全面实习,此后去亚琛工业大学钢铁冶金研究所博士后进修两学期,他除听专业课,参加专业实验及小课题研究外,还调查实验设备,搜集新技术、新设备及德国高等教育资料等。这一年的广泛学习和资料收集,对他后来的工作起了很大作用。1936年7月魏寿昆结束了在德国五年的学习,启程回国。

3. 辗转陕、黔、川

魏寿昆回国后重返母校,任矿冶系化学及冶金教授。1937年"七七事变"后,国民政府教育部明令天津的北洋工学院、北平的北平大学及北平师范大学迁至西安,组成西安临时大学(后改称西北联合大学)。魏寿昆随学校到达西安。次年西北联合大学解散改组。魏寿昆任教于由北洋大学工学院、东北大学工学院、北平大学工学院及焦作工学院组成的西北工学院。1937年8月西康技艺专科学校(大专及中专)在西康省(今四川省)西昌成立,魏寿昆被聘任为矿冶科及化工科主任,由陕西城固经12天长途跋涉到达西昌。1941年夏他又被聘为贵州工学院矿冶系主任兼教务主任,乃举家迁至贵阳;1942年夏受聘重庆经济部矿冶系冶金教授,全家又由贵阳迁居重庆。

魏寿昆在赴重庆工作的前五年中,辗转在五所学校任教,居无定所。到重庆后的四年中,曾在三个单位工作,搬了四次家,在当时的抗日战争环境下,工作条件差,生活艰苦,魏寿昆始终积极应对,克服来自工作和生活等各方面的困难,认真完成各项教务和教学工作。

4. 重展宏愿

经历了十几年的抗日战争和解放战争,新中国成立了,百废待兴。十年的"文化大革命"使中国的政治、经济、文化和教育各方面遭受了极大损失,魏寿昆坚信,历史车轮必定会前进。他利用一切可能坚守在教学与科研第一线。1976年"四人帮"倒台后,各项事业得到更好更快的发展,魏寿昆此时已年逾古稀,他迈开大步重登讲台,走进实验室,积极开展科研工作,为教书育人殚精竭虑。耄耋之年,他还与王之玺院士(原冶金工业部科技司司长,时年也年过八旬)率领调研组深入工矿企业,调查我国各大冶金企业的情况与规划,为我国钢铁工业的宏观和长远发展做出战略报告,由中国科学院上报中央,受到国家领导的高度重视。

二、主要研究领域和学术成就

(一) 教书育人

魏寿昆回国后任教于天津北洋工学院,不久随校西迁,在陕、黔、川三省各地奔波,先后在八所院校任教。新中国成立后,1952年院系调整,魏寿昆随北洋大学矿冶系教师和学生调至由五所院校(北洋大学、唐山交通大学、北京工业学院、西北工业学院和山西大学)组成的北京钢铁工业学院(1960年更名为北京钢铁学院,简称钢院),任教务长。他是钢院建校元老之一,1979年任副院长兼图书馆馆长。但无论担任何种行政职务,他从未卸下教师的职责。在70余年的工作中,魏寿昆针对不同教学对象——青年教师、研究生、大学生、大专生、工人班及"文化大革命"中的工农兵学员等,因材施教,讲授过基础课、技术基础课、专业课和专题讲座,共28门课程。

钢院建校之初,随五校调入的专业课教师和青年教师,其中多数是刚毕业的和53届提前一年毕业的学生。作为教务长,魏寿昆除自己讲课外,特别关注青年教师的培养。这是当时师资队伍建设的重要环节。他首先制订了全面培养计划:一是根据专业设置和课程要求,派出青年教师到外校进修,听苏联专家讲课;二是聘请外校有经验的教师来校授课,同时指导青年教师,使其在工作中提高;三是亲自指导青年教师制定教学大纲,讲解教学内容、重点和难点以及教学方法等。由于这些措施的稳步落实，师资队伍建设取得显著成绩，青年教师的讲课能力在教学实践过程中不断提高，逐渐成长为学校的主力，成为专业骨干教师。当年的青年教师和曾受业于魏寿昆的学生如今已是七八十岁的老人。回忆起自己的老师，他们一致认为其求真务实、严谨创新的学风及诲人不倦的精神让他们受益匪浅。举例来说，魏寿昆指导的一名博士生，因时间紧、原材料短缺供应不上，就用已有的实验数据及推导出的热力学公式，通过计算外推得到理想结果，认为可以达到任务要求了，但遭到了魏寿昆的否定。魏寿昆肯定了他的计算方法，但称若不经过实验验证，不能做定论。该生又干了两个月，得到的不仅是实验数据，而且更重要的是严格的训练和求真务实的作风。再举一例，一名学生口试后，满怀信心地认为"5分没跑了"（20世纪50年代学习苏联考试采用口试，记分为四级分制，即5、4、3、2），但魏寿昆又问了一个生产上的小问题，把学生难住了，结果"到手的5分"跑了，只得了4分。魏寿昆说在生产中不应放过任何小问题，看似一个"细节"，它可能影响整个生产。后来这位成为中国工程院院士的学生说，"一个'4分'使他养成了在工作中细致观察、认真研究的严谨作风"。

魏寿昆共培养了20余名硕士生和博士生，编著六部学术专著：《平炉炼钢厂设计》《专业炼钢学——平炉构造及其车间布置》《活度在冶金物理化学中的应用》《冶金过程热力学》《魏寿昆选集》《冶金过程物理化学导论》。

（二）主要科研成就

1. 五项专利

20世纪40年代，魏寿昆任重庆矿冶研究所室主任期间，针对生产的技术改进和矿产资源研发的具体情况，研究制备产品的新方法，并获得五项专利：①利用静置后处理法自白云石去钙提镁的新法；②利用碳酸钠或碳酸铵自白云石去钙提镁的新法；③人造镁氧制造镁砖的配料方法及加强黏性的风化法；④制造特纯钼酸铵或钼酸采用铝铁共沉淀新法；⑤提炼纯钼的二步还原新法。

2. 高温反应活度理论研究及应用

1907年G. N. Lewis提出活度概念，并将其应用于常温下非理想溶液的化学热力学计算。高温下的冶金熔体反应过程十分复杂。到20世纪30年代中期美国J. Chipman教授将活度概念成功地应用于冶金熔体反应，用实验测定了熔体中元素的活度系数，开创了高温反应的活度理论。中国学者李公达、陈新民和邹元爔等对高温反应热力学及活度理论研究都做出了贡献。

魏寿昆在留德期间即关注高温冶金过程的物理化学研究方面的进展，回国后限于客观条件，未能潜心于这方面的研究工作。20世纪50年代他受到苏联专家讲课时引用欧美文献的启发，研读了国外大量有关资料，开始对活度理论及其实际运用进行研究。1956 年他发表了《活度的两种标准态与热力势》和《活度在钢铁冶金二相的气体——金属液化学平衡反应中的应用》两篇论文。

在第一篇论文中，魏寿昆应用"活度"取代常温理想溶液中的"浓度"，计算高温熔体反应热力势，并讨论了计算活度的两种不同的标准状态（纯物质和1%溶液）。以某元素 A 溶于铁液生成含 A1% 溶液为例，推导出热力势变化（$\Delta\Psi$）：

$$A_{(液)} \to [A]$$

$$\Delta\Psi = RT\ln\left(fX\gamma^\circ \frac{0.5585}{M_A}\right)$$

此式为高温下计算溶液生成反应 $\Delta\Psi$ 一般关系式。式中，$\Delta\Psi$ 为反应热力势变化值，R 为气体常数，T 为绝对温度，f 为亨利活度系数，X 为 A 的摩尔分数，γ° 为拉乌尔活度系数，M_A 为溶液中 A 的摩尔质量。

当以 1% 溶液为标准态时，即 $X=1$，则 $f=1$，并有：

$$\Delta\Psi = RT\ln\left(\gamma^\circ \frac{0.5585}{M_A}\right)$$

此式是用于计算在适合亨利定律时，溶液生成过程的标准热力势变化值。

当适合于拉乌尔定律时，$\gamma^\circ=1$，则溶液生成过程的标准热力势变化值为：

$$\Delta\Psi = RT\ln\left(\frac{0.5585}{M_A}\right)$$

在第二篇论文中他总结了 49 篇 1936~1955 年发表的文献资料有关三种二相的气体——金属液化学平衡反应的研究工作。文中标准化了二元系和三元系平衡常数、活度及活度系数等参变数的符号，分析了 f_i^j 的物理意义；并利用活度分别计算了各种元素（第三元素）对铁液中的氧及硫的反应热力学，讨论了活度对铁液的脱氧和脱硫的实际意义，最后还指出由于多元系内各元素在铁液内彼此相互作用，它们的关系更为复杂，所以为阐明合金钢热力学性质，活度的研究更有必要，更有实际意义。

1959 年魏寿昆应曲英等教师的要求和推动，根据教学需要开设"活度在冶金物理化学中的应用"专题讲座，此后经过两次补充修改，于 1964 年完成《活度在冶金物理化学中的应用》专著并出版，受到冶金同行的重视与好评。该书中说，活度是人为的概念，是从实际角度出发，研究物理化学理论问题的得力工具。它反映真实溶液和理想溶液客观存在的偏差，用它计算溶体反应热力学函数，对化学冶金的理论研究具有重大意义。它可以进一步了解并阐明化学反应的机理，可促进发现不同物理化学现象的内在联系。由于近年来活度实验数据的积累，已经发现一些活度参数与元素的内部结构有一定关系的规律，说明活度的研究可提供探索合金溶液本质的途径。但这些规律尚嫌不足，也不够系统化，尚须深入研究。活度反映实际溶液与理想溶液存在偏差，但未说明其原因，也有待对溶液结构作进一步探讨。该书阐明了活度理论研究及其应用的现状与发展，是对冶金问题研究有重要指导作用的专著。

3. 炉渣离子理论

在文献中关于炉渣中 FeO 含量对脱硫作用的结论学界存在分歧与矛盾，很多学者皆用分子理论来解释，M. G. Froberg 曾用离子理论研究高炉型炉渣脱硫作用，但皆未能给出适合于多类型炉渣硫分配比公式。魏寿昆应用 Темкин 于 1945 年提出的炉渣完全离子化模型，并引用活度系数，研究炉渣氧化铁含量对脱硫的作用，推导出硫在渣与铁中的分配比公式：

$$\frac{(\%S)}{[\%S]} = \frac{32 L_S f_S}{\gamma_{Fe^{2+}} \gamma_{S^{2-}}} \times \frac{\sum n_+ \sum n_-}{n_{Fe^{2+}}}$$

$$L_S = \frac{a_{Fe^{2+}} a_{S^{2-}}}{a_{Fe} a_S}$$

式中，f_S、$\gamma_{S^{2-}}$、$\gamma_{Fe^{2+}}$ 分别为铁液中 S 及炉渣中 S^{2-} 和 Fe^{2+} 的活度系数；L_S 为平衡常数；$\sum n_+$、$\sum n_-$ 分别为 100g 炉渣中阳离子克摩尔数之和及阴离子克摩尔数之和；$n_{Fe^{2+}}$ 为 100g 炉渣中 Fe^{2+} 离子克摩尔数。

此式适用于任何含碳量的铁液（包括饱和碳的铁液）和任何成分的熔渣（包括高碱度、低碱度、高 FeO 或低 FeO）。它是较综合的公式，解决了文献中脱硫计算结论的分歧与矛盾。魏寿昆又根据文献中有关脱硫的实验数据，按炉渣完全离子理论推导出下式：

$$\lg(\gamma_{Fe^{2+}} \gamma_{S^{2-}}) = -53.5 N_{O^{2-}} + 2.12$$

式中，$N_{O^{2-}}$ 为炉渣中 O^{2-} 离子摩尔分数。由此式进一步计算证明了硫分配比公式适用于一般的低碱度渣，基本上不受温度的影响。

离子理论应用的实践——攀枝花矿山公司与攀枝花钢铁公司两公司曾因矿石作用观点上的分歧而引发矿石价格的纠纷。攀枝花矿山公司认为矿石中 TiO_2 属碱性，该矿属半自熔性，品质高，应当提价。攀枝花钢铁公司认为 TiO_2 属酸性，非半自熔性，矿石中的高 TiO_2 含量并未给炼铁过程带来好处，因而反对提价。魏寿昆利用高炉生产数据，按离子理论计算硫在渣与铁中的分配比，如 TiO_2 作为酸性氧化物，硫的分配比为 9∶1；TiO_2 作为碱性氧化物，则硫的分配比高达 60.7。实际高炉生产中硫的分配比一般不大于 10，从而说明 TiO_2 在高炉中属酸性。同时他也计算说明了该矿为非半自熔性矿石。魏寿昆便这样以实践与计算解决了两家的分歧，也验证了离子理论的应用价值。

4. 固体电解质浓差电池的研究与应用

固体电解质浓差电池定氧技术是以电化学理论为基础。由一个已知氧分压的参比电极和另一个待测含氧量的金属液为回路电极的两个半电池，中间用固体电解质（如 ZrO_2-CaO）连接而组成电池。由于两个半电池的氧分压不同，固体电解质是氧离子导体，在一定温度下两电极产生电动势，由测出的电动势则可计算出金属液中的含氧量。

1957 年 K. Kiukkol 和 C. Wagner 提出利用固体电解质组装成电池进行电动势测定之后，许多国家开展了大量的研究工作。从 1971 年开始，欧美国家、苏联及日本等的固体电解质钢液定氧电池先后问世，逐渐成熟后被大量用于炼钢操作过程中，对提高钢质量和成材率、降低铁合金消耗等起到了显著作用。1974 年固体电解质浓差电池成为当时钢铁冶金领域三大重大科研成果之一。

魏寿昆跟踪国际上固体电解质浓差电池的发展情况，总结大量文献成果，于 1972 年和 1973 年编写了《钢液直接快速定氧的固体电解质电池》及《浓差电池快速直接定氧法》两份内部资料，同时组成科研小组开始试制固体电解质测氧传感器。这是国内此方面起步最早的研究工作。1974 年以后，魏寿昆陆续发表论文并向全国冶金界宣传介绍和推广这项新技术。《近四年来固体电解质电池的进展及今后的发展趋势》（1978 年）一文从四个方面进行了分析：①固体电解质材料的物理及物理化学性质；②固体电解质定氧电池准确度及钢水低含氧量的测定；③固体电解质定氧电池在炼钢生产中的应用；④测定其他元素的固体电解质电池。最后该文提出国内开展固体电解质电池研究工作的意见，如进行电子导电与抗热震性的矛盾、电池的组装与测定准确度等问题的研究。1981 年在《固体电解质定氧电池的近况应用及展望》一文中，魏寿昆对固体电解质定氧电池的现况及前景作了总结性评述，对固体电解质的理论研究及其在生产中的应用起着指导作用。其内容包括：①顶吹氧气转炉终点的控制；②半镇静钢及沸腾钢脱氧的控制；③连铸镇静钢余铝含量的估计；④易切削钢硫化类夹杂物形态的控制；⑤配合计算机定氧电池炼钢过程的在线应用。

北京科技大学冶金物理化学教研室的固体电解质研究小组逐步发展成为由刘庆国、李福燊领导的固体电解质测试国家重点专业实验室，成功地试制出我国第一个可直接快速测定钢液中的氧的产品，通过冶金工业部鉴定。此外，他们还研究解决了连续定氧、快速定碳、快速定氧定硫、快速定硅、定铝以及测定烟道中的 SO_2 含量等问题，曾获得冶金工业部科技进步奖二等奖。在从研究小组成长为国家重点实验室的过程中，魏寿昆发挥了重要作用，在研发与推广应用等方面他也起了关键性的指导作用。

5. 选择性氧化理论与转化温度

火法冶金是由矿石、燃料、熔剂以及耐火材料等物质参与的高温复杂的化学反应过程。燃料——碳除提供热能外，也是重要的参与反应的物质。矿石中含有多种元素，根据冶金产品的质量要求或希望将其提取出来或希望将其抑制去除掉，因此，在生产中应控制氧化与还原反应过程的顺序。1964年魏寿昆在《炼钢过程中铁液磷碳元素氧化的热力学》一文中提出铁、碳、磷元素的氧化顺序问题。1976年和1978年他的《含钒铁水炼钢的一些物理化学问题》和《稀土钢冶炼的物理化学问题》两份报告，再次提出氧化顺序问题，提出"最低还原温度"和"氧化转化温度"概念。其后他结合系列生产研究总结性地提出"选择性氧化理论"，根据 G. K. Sigworth 和 J. F. Elliott 的数据绘出铁液中诸元素氧化的 $\Delta F^\circ\text{-}T$ 关系图，即改型的 Ellingham-Richardson 图，其标准态采用1%重量溶液。从图中各元素的 $\Delta F^\circ\text{-}T$ 线的位置高低可以判断处于较低位置的元素能比处于较高位置的元素优先被氧化。另外，图中碳氧化的 $\Delta F^\circ\text{-}T$ 线与一些元素氧化的 $\Delta F^\circ\text{-}T$ 线相交，此相交点所示温度即为"氧化转化温度"。在高于或低于此温度时，该元素与碳之间具有选择性氧化性质，据此可以选择分离某种元素。

根据转化温度和选择性氧化理论，相关工业生产实践取得了显著成果，例如，解决了上钢三厂冶炼不锈钢工艺的脱碳保铬问题、上钢一厂的摇包中铁水脱铬保碳的最佳工艺条件问题、攀枝花钢铁公司的提钒和包头钢铁公司（包钢）的提铌的最佳工艺条件问题、金川有色金属公司卡尔多炉的脱硫保镍问题（使镍的回收率超过95%）。1982年在《选择性氧化——理论与实践》一文中，魏寿昆简明讨论了选择性氧化过程中转化温度概念的意义及计算方法，并总结了近九年来其在生产实践中应用的实例。

6. 矿产资源综合利用

（1）炼钢过程脱磷的研究

磷（P）在钢中是一个有害元素，它使钢发生"冷脆"现象，即使钢易脆裂，冲击韧性降低。我国规定：普通钢 $w(P) \leqslant 0.045\%$，优质钢 $w(P) \leqslant 0.040\%$，高级优质钢 $w(P) \leqslant 0.035\%$，随着现代工业对金属材质的要求不断提高，如一些钢种要求 $w(P) \leqslant 0.010\%$，不锈钢 $w(P) \leqslant 0.005\%$，因此脱磷成为世界各国冶金工作者共同关心的问题，开展了铁水预处理、新型复合顶底吹炼及钢包冶金的脱磷工艺技术研究。

传统方法是氧化脱磷。高炉炼铁是还原条件——脱硫，炼钢过程是氧化条件——脱磷。当钢中含其他合金元素或冶炼合金钢时，若为脱磷所采用的氧化法同时会造成合金元素的氧化损失，因此生产技术路线由氧化脱磷转变为还原脱磷。魏寿昆首先进行了热力学分析，结合实验室和生产实际及国外有关资料研究各因素对还原脱磷的影响，诸如金属液中各因素的影响——氧含量、碳含量、铬含量和硅含量等，温度影响及体系气氛和坩埚材质的影响，对还原脱磷渣系的种类、脱磷剂的加入量、助熔剂 CaF_2 的作用、加入方法以及脱磷渣的后处理等作了分析归纳，对生产起到了有力的指导作用。

在《低磷钢生产》一文中魏寿昆分析了世界各国在生产低磷和超低磷钢方

面的研究成果，对用 Na_2CO_3 基渣系进行了铁水预处理的物理化学原理、过程影响因素和实际应用进行了阐述和讨论，对用 CaO 基渣系处理钢液的最新研究以及喷吹法脱磷进行了分析和讨论。该文指出用铁水预处理的方法是其生产低磷钢的一个合理途径。LD 渣脱磷和 Na_2CO_3 基渣系脱磷符合综合利用、节能的要求。后者更有脱磷、脱硫的优点，钢液的精炼可以使磷含量进一步降低。除了选择更加合适的渣成分外，也可通过喷吹的方法来获得更高的脱磷率，喷吹法还可使处理时间大大缩短。新法处理可以将钢中磷含量降到 0.01% 以下。

1990 年魏寿昆负责的"锰基合金热力学行为及其脱磷的研究"课题获得国家教委科技进步奖一等奖。

(2) 包钢铌冶金研究

包头白云鄂博矿是含铁、铌、稀土等几十种元素的共生矿。1952 年包钢建设之初按照苏联专家意见以开采铁矿为主建设钢铁生产线，1958 年投产后出现生产达不到设计能力、资源流失及污染严重等问题。资源综合利用的研究乃提上日程。

20 世纪 80 年代初，魏寿昆对包头铁矿提铌进行了系列基础理论研究——铌在铁液及熔渣中的热力学行为及铌铁合金去锰、铌铁分离新方法、铌在碳饱和铁液中的溶解度等，利用固体电解质定氧电池测定与计算 Fe-Nb 系、Fe-Nb-Mn 系和 Fe-Nb-Si 系中元素的活度和活度相互作用系数。

Fe-Nb 系中，通过实验测定计算得出铌在铁液中自身活度相互作用系数 e_{Nb}^{Nb}，改变了前人假定 $e_{Nb}^{Nb}=0$ 的处理方法，得出：

$$e_{Nb}^{Nb} = \frac{2274}{T} - 1.44$$

当 $T=1873K$ 时，$e_{Nb}^{Nb}=-0.22$。利用实验数据，求出 Nb 在铁液中的标准溶解自由能 (J)：

$$Nb_{(s)} = [Nb]\%$$

$$\Delta G^\circ = -134260 + 33.05T$$

$$Nb_{(l)} = [Nb]\%$$

$$\Delta G^\circ = -161160 + 42.84T$$

Fe-Nb-Mn 三元系，铁液中含 Nb 大于 1%，采用 Nb_2O_3 为渣料时，利用固体电解质定氧电池的 $[Nb]+2[O]=NbO_{2(s)}$ 反应式，通过 a_0 的测定研究 Mn 对 Nb 活度系数的影响，利用同一浓度法和同一活度法可得出基本一致的 e_{Nb}^{Mn} 数值：$T=1853K$，$e_{Nb}^{Mn}=0.18$；$T=1873K$，$e_{Nb}^{Mn}=0.11$。其温度关系式估计为：

$$e_{Nb}^{Mn} = \frac{12100}{T} - 6.35$$

Fe-Nb-Si 三元系，利用设计的定氧电池测定 a_O 值，由 $[Si]+2[O]=SiO_{2(s)}$ 求出 a_{Si}；然后根据同一浓度法及同一活度法计算三个温度（1823K、1853K 及 1883K）的 e_{Si}^{Nb}，再用惯用方法计算出相应的 e_{Nb}^{Si}，回归分析求出其温度关系。

同一浓度法：

1873K 情况为 $\quad e_{Si}^{Nb} = \dfrac{11454}{T} - 5.87 = 0.25$

1873K 情况为 $\quad e_{Nb}^{Si} = \dfrac{37763}{T} - 19.36 = 0.80$

同一活度法：

1873K 情况为 $\quad e_{Si}^{Nb} = \dfrac{8559}{T} - 4.37 = 0.20$

1873K 情况为 $\quad e_{Nb}^{Si} = \dfrac{28535}{T} - 14.60 = 0.63$

碳饱和铁液中 Nb 的溶解度对于含 Nb 铁矿热还原流程的工艺分析具有重要意义。国外研究者的实验测定与热力学计算数据相差较大。魏寿昆等采用扩散平衡法测定，并进行碳化铌过饱和析出实验，证实此法测定结果之可靠性。在 1300～1500℃ 范围内测得碳饱和铁液中 Nb 的溶解度与温度的关系为：

$$\lg[\%Nb]_{max} = -\dfrac{5899}{T} + 3.28$$

他们还进行了铌铁合金去锰的研究。包头含铌铁矿资源虽然丰富，但铌品位很低，从炼钢副产品——高锰低铌炉渣中回收锰和铌成为重要的研究课题。铁水提铌的火法冶炼不能分离铌和锰，其炼制的产品为含铌锰铁合金。北京有色金属研究总院采用制团氯化法从炼钢底吹渣中分取铌和锰取得了较好效果。但由于采用氯气为氯化剂，严重腐蚀设备，还存在反应产生的 CO、PCl_3、$POCl_3$ 等有毒气体的处理问题，此法的实际应用尚有困难。魏寿昆在指导博士生的研究工作中，采用了无腐蚀无污染的固体氯化剂 $CaCl_2$，从含铌锰炉渣中分离出 MnO 取得了很好的结果，并讨论了 MnO 的氯化机理和动力学条件。

SiO_2-MnO-Nb_2O_3 三元渣用 $CaCl_2$ 氯化分离的研究。MnO 生成 $MnCl_2$ 逸出，Nb_2O_3、SiO_2 保留在渣中。Nb_2O_5 回收率大于 90%，MnO 回收率大于 94%。动力学分析：反应级数在 1.6～2.6 范围内，表观活化能为 7.2cal/mol（1cal = 4.18585J）。反应主要由扩散环节控制，因之炉渣粒越小，反应速度越快。

SiO_2-FeO-Nb_2O_3 三元渣用 $CaCl_2$ 氯化分离的研究。FeO 呈 $FeCl_2$ 气体逸出，Nb_2O_3 和 SiO_2 保留于渣中。若氯化实际生产铌渣，逸出渣相则是 $MnCl_2$ 和 $FeCl_2$。其氯化趋势 FeO 稍大于 MnO。FeO 的氯化速率为 $v_{FeO} = KN_{FeO}^n$，K 值随温度升高而增大，反应级数在 1.6～2.4 范围内，扩散为控制环节。

SiO_2-FeO-Nb_2O_5 铌铁分离新方法。用 CO-CO_2 混合气体选择性还原含 Nb 铁矿。90%以上的铁矿物被还原为金属铁,而铌矿物不被还原。经磁选分离出金属铁,得到含 Nb_2O_5 为 6.91%的氧化物。其品位为原矿的 4 倍。再用盐酸浸取,90%以上的铁矿物、磷矿物及 CaO、MgO 进入浸液,而铌氧化物 Nb_2O_5 留于浸渣中含量近 30%,Nb:Fe(质量比)= 6,Nb:P(质量比)= 12。此富集后的含铌渣干燥后可用于炼制铌铁。此流程工艺条件:还原气体成分为 90% CO 和 10% CO_2,气体流量为 200mL/min,还原温度为 950~1150℃,金属化率大于 90%。此法可熔炼符合工业标准的铌铁(60%~65%Nb)。

还有铌渣相图的研究,为包钢提铌工艺的改进开展了热力学数据的测定。他们利用 DTA、SEM、X 射线衍射及固体电解质定氧技术等研究 MnO-Nb_2O_5、MnO-Nb_2O_5-SiO_2 等体系相图,并由相图得到该体系的自由能、熵等,测定 MnO-Nb_2O_5-SiO_2 渣中的 Nb_2O_5 和 MnO 活度。

他们的共生矿分离基础理论研究——"铌在铁液及渣中的热力学行为"于 1988 年获得国家教委科技进步奖二等奖。

(3)华南铁矿冶炼脱砷的基础理论研究

我国南部地区蕴藏着丰富的铁品位较高的铁矿资源,如广东大宝山褐铁矿、湖南连平磁铁矿,铁含量均在 48%~59%,但砷含量较高,如大宝山褐铁矿砷含量平均为 0.383%。

大宝山铁矿石中砷以臭葱石($FeAsO_4 \cdot 2H_2O$)为主,它在高温下易分解并产生剧毒性的 As_2O_3 气体,严重污染环境,危及群众健康。在高炉炼铁过程中,全部被还原进入铁水中,而炼钢过程中则留在钢水内。在耐蚀钢中少量砷能提高钢的耐腐蚀性能,但砷会增加钢的脆性,降低钢的低温冲击韧性。因此,为炼出高质量普通钢或合金钢,必须考虑脱砷问题。

为探讨钢铁的脱砷问题,魏寿昆及其课题组进行了大量热力学分析,关于 Fe-As 熔体中 As 的活度,采用熔化自由能法从 Fe-As 相图计算 a_{Fe},将 $\lg(\gamma_{Fe})$ 对 N_{As} 进行多元回归处理,得出 $\lg(\gamma_{Fe})$ 和 N_{As} 的关系式,然后利用 Gibbs-Duhem 方程式直接积分得到 γ_{As},求出铁液中砷的自身活度相互作用系数 e_{As}^{As} 在 1600℃时为 0.11。砷在铁液中的溶解自由能,以重量 1%溶液为标准态,为 $1/2(As_2)_g$ = (As) $\Delta G° = -64600-40.75T(J)$,前人的工作,计算 a_{As},未采用 α 函数,其结果不准确,有的计算采用有 α 函数的 G-D 方程用图解积分法求解 a_{As} 也不如直接积分准确。

Fe-AS-X-C 系中(X 为 Mn、Si 或 P)As 活度和活度系数的研究:通过 CaC_2-CaF_2(50wt%)系和 Fe-As-X-C 系平衡实验,测定渣中以 Ca_3As_2 形式存在的砷和 Fe 中的砷含量,通过热力学计算,确定铁液中其他元素(Mn、Si 或 P)对铁液中砷活度的影响:Mn、P、S 对砷活度影响较小,P、S 在生铁中含量一般小于

0.1%，Mn 含量约为 0.3%，故在一般铁水处理过程中，可以忽略它们对砷活度的影响，Si 能增加砷的活度。当[%Si]=1.75 时，铁水中砷含量从 0.131% 降到 0.062%，当[%Si]降低 0.1 时，不能脱砷。C 也是增加砷活度，利于脱砷，但为获得最佳脱砷效果，铁水中碳含量尚须进一步实验研究工作证实。

钢液中钙-砷平衡研究。利用合成 Ca_3As_2-CaC_2 二元渣进行钢液中的钙-砷平衡实验研究得到 $3[Ca]+2[As] = Ca_3As_2$，$\lg K = \dfrac{23780}{T} - 3.339$。$Ca_3As_2$ 的生成自由能：$3Ca_{(g)} + \dfrac{1}{2}As_{2(g)} = Ca_3As_2$，$\Delta G^\circ = -702300 + 130.46T(J)$。

铁水中脱磷、脱硫、脱砷的热力学分析说明只能还原脱砷，不能氧化脱砷，在铁水中虽然能同时脱磷、脱砷或同时脱磷、脱硫，但是用 CaO-CaF_2、CaC_2-CaF_2 渣系对铁水不可能同时脱磷、脱硫、脱砷。在还原条件下铁水中诸元素容易被脱除的顺序为硫、砷、磷。

利用 CaC_2-CaF_2 渣系对铁水预处理脱砷的实验研究可知：在添加合适的 CaF_2 渣量为 15g/100g 铁的条件下，对[%Si]=0.40~2.00 的饱和碳铁水的脱砷率可达 65%~80%。温度降低，脱砷率稍有增加。初砷量对脱砷率影响不大。在大约 1350℃，渣系中 CaF_2 的最佳含量为 40wt%~50wt%。[%Si]和脱砷率（η_{As}）的关系为 $\eta_{As}(\%) = 61.77 + 9.565[\%Si]$，渣量（$Q$）与脱砷率的关系为：$\eta_{As} = mQ^n$。式中 m 是与 f_{As} 有关的系数，$n = 0.52 \sim 0.62$。理论计算与实验结果十分符合。渣量大幅减少仍有较高脱砷率。

钢水二次精炼还原用 Ca 脱砷，反应式为 $3[Ca]+2[As] = Ca_3As_2$。

根据文献记载 Ca 在铁液中 1600℃ 的溶解度为 0.01%~0.03%，在 Ar 气保护下，通过喂入 Ca 丝，钢水中 As 可脱到双零水平。

以上对铁液中砷的热力学行为的研究为国内一些学者的研究及含砷铁矿的开发应用提供了有用依据。此项研究于 1993 年获国家教委科技进步奖一等奖。

魏寿昆在学术、生产技术以及教学工作中取得了突出成就。他的科研工作理论联系实际，为中国冶金工业的生产发展提供了理论依据。他还十分重视解决生产中的实际问题。在数十年的教学岗位上他传道授业，诲人不倦，将求实敬业的精神传递给一代又一代学子，为国家培养了大量建设人才，特别是在冶金战线上他们做出了成就。

三、魏寿昆主要论著

Wei S K. 1944. The decalcification of Szechuan dolomita. J Chinese Chem. Soc. (中国化学会)，11（1）：34-49.

魏寿昆. 1954. 平炉炼钢厂设计. 上海：商务印书馆.

魏寿昆. 1958. 专业炼钢学——平炉构造及其车间布置. 北京：冶金工业出版社.

魏寿昆. 1964. 活度在冶金物理化学中的应用. 北京：中国工业出版社.

魏寿昆. 1964. 炉渣氧化铁含量对脱硫的作用. 金属学报, (2): 157-164.

魏寿昆. 1966. 高炉型渣脱硫的离子理论. 金属学报, (2): 127-141.

魏寿昆. 1980. 冶金过程热力学. 上海：上海科学工业出版社.

魏寿昆, 洪彦若. 1981. 镍锍选择性氧化的热力学及动力学. 有色金属, 33 (3): 50-60.

Wei S K. 1981. Thermodynamics on the selective oxidation of elements in molten matte and metals. Beijing: Preprint of First China-USA Bilaterl Metallurgical Conferenca.

魏寿昆, 张圣弼, 佟亭, 等. 1984. 电化学法测定 Fe-Nb 熔体中 Nb 的活度. 钢铁, 19 (7): 1-8.

张圣弼, 佟亭, 王济舫, 魏寿昆. 1984. 含 Nb 及 Mn 的铁液中 Mn 对 Nb 活度系数影响的研究. 金属学报, 20 (5): A348-A356.

魏寿昆, 倪瑞明, 方克明, 等. 1985. 还原脱磷. 铁合金, (2): 1-7.

朱元凯, 董元篪, 彭育强, 魏寿昆. 1985. 钢液中钙-砷平衡的研究. 钢铁, 20 (10): 38-44.

魏寿昆. 1986. 砷在铁中的作用及在铁熔体中的热力学行为. 庆祝周志宏教授九十寿辰暨从事冶金工作七十年论文集. 上海：上海交通大学: 74-81.

朱元凯, 董元篪, 彭育强, 魏寿昆. 1986. CaC_2-CaF_2 渣系对铁水脱 As 的研究. 北京钢铁学院学报, (2): 103-112.

董元篪, 彭育强, 魏寿昆, 等. 1986. Fe-As 的活度. 钢铁, 21 (10): 11-13.

佟亭. 魏寿昆, 张圣弼. 等. 1987. Nb、Si 在铁液中活度相互作用的研究. 金属学报. 23 (2): B47-B54.

魏寿昆. 1988. 从炉渣离子理论计算的硫分配比看攀钢钒钛磁铁矿中 TiO_2 的属性. 钢铁钒钛, (4): 1-3, 65.

Wei S K. 1988. Selective oxidation of elements in metal melt and their multireaction equilibria. Steel Research, 59: 381-393.

谭赞麟, 魏寿昆, 朱元凯. 1989. 包头铁水铌渣的氯化及脱锰机理. 稀有金属, 13 (2): 137-141.

Wei S K, Ni R M, MaZ T, et al. 1989. The solubility of Ca in Mn melt and the 3rd-element interaction effects. Steel Research, 60: 437-441.

许家仪. 1990. 魏寿昆选集. 北京：冶金工业出版社.

Wei S K. 1992. Interaction coefficients in multi-component metallic solutions at constant activity and constant concentration. Chiba, Japan: The 6th Japan—China Syinposium on Science and Technology of Iron and Steel, Nov. 17-18: 1-10.

Chou K C, Wei S K. 1997. A new generation solution model for prediction thermodynamic properties of a multicomponent system from binaries. Met. & Materials Trans B, 28B: 439-445.

Chen H, Han Q Y, Wei S K, et al. 2002. Separation of Nb from Nb-bearing iron Ore by selective reduction. Steel Research, 73: 169-174.

参 考 文 献

[1] 魏寿昆. 魏寿昆年谱、奋斗人生. 自撰, 2005.
[2] 吴石忠, 姜曦. 魏寿昆传. 北京: 科学出版社, 2009.

后 记

感谢"魏寿昆科技教育奖"大奖获得者徐匡迪、张寿荣、殷瑞钰、周国治院士，长期支持魏寿昆院士系列文集著作的出版编纂整理工作。特别感谢徐匡迪院士对晚辈后学的鼓励，是本书得以出版的最大动力。

本文集由北京科技大学教育发展基金会、魏寿昆科技教育基金资助出版。感谢北京科技大学校领导武贵龙、杨仁树、张欣欣、罗维东、徐金梧、王维才，对魏寿昆院士科技文献出版的重视和指导。感谢北京科技大学科技教育发展基金会吕朝伟老师对魏寿昆院士系列文集出版工作的高效组织协调和细致管理。感谢北京科技大学朱元凯、曲英、林勤高龄知名教授，以严谨务实的学术精神、精益求精的工作态度，校对文稿，家属代表魏文宁教授对老先生们给予文集编纂提供的长期无私帮助表示敬佩和感激。

感谢北京科技大学冶金与生态工程学院李晶教授、史成斌老师、焦树强教授、张国华教授、佘雪峰老师、吴胜利教授、寇明银老师、包燕平教授等各学术团队为主体完成基础工作，师生们克服了原稿不清晰、重制图难描画等困难进行了重新录入、整理和校对，他们是：李晶教授和史成斌老师团队李首慧、张梦德、侯玉婷、王昊、梁伟、肖龙鑫、卢嘉枫、郑顶立、徐昊驰同学；焦树强教授团队肖翔、郑朝亮、李鑫、郭丰、李晓琳、杜洋同学；张国华教授团队徐瑞、汪宇、朱金辉、刘军凯、宋成民、吉鑫鹏同学；佘雪峰老师团队易万里、由晓敏、安振龙、钟海同学；吴胜利教授团队寇明银老师及洪志斌同学。

感谢北京科技大学冶金与生态工程学院党委书记张建良教授，对魏寿昆院士有关工作的长期支持。感谢冶金与生态工程学院院长张立峰教授；张百年、王春义、耿小红、宋波等老师关心指导。感谢冶金工业出版社社长谭学余对本书指导帮助；感谢任静波总编辑对老院士资料整理工作的长期关切;感谢钢铁冶金材料编辑中心李培禄主任及同事们,竭尽全力,精心赶制文稿校对、版面和封面设计等工作,使得本文集如期出版发行。

本文集所选部分学术文献发表时间为1956~1999年,部分或未达到现有出版要求,编委会全部重新进行了再加工和文字整理,过去的单位制和现行法定单位制有出入,了解过去存在过的单位仍有意义,所以忠实原貌未进行转换,为避免原稿传递混乱和处理疏漏,本书稿旨在聚焦再现先生早年学术成就,并面向冶金学界展示学科发展的历史风貌,还请学科研究者按图索骥寻找原文进行深入研究。

魏寿昆院士是中国冶金学界的奠基人之一。魏先生受后学们广泛敬重,被尊为"捧着一颗心来,不带半根草去"的真正的大师,先生为人品格高尚,广为学界推崇;为师见解精辟,学术影响深远。魏先生2014年6月30日以107岁高龄仙逝,曷胜怀念,特志数句,敬表全体学界敬仰尊重。全书力图保留作者原意,不予以刻意修饰,囿于从第一卷(1929~1949)以及本卷本原稿多为笔者博士期间跟随魏寿昆院士收集整理,偶有斑驳不清,难免存在记录误植与疏误,尚待读者辨析,还请学界指正。最后,望以魏先生对中国钢铁工业和冶金教育奉献和热爱之精神,恳请与钢铁学人共勉之。

姜 曦

2019年5月15日